Complex Science for a Complex World
Exploring Human Ecosystems with Agents

Complex Science for a Complex World

Exploring Human Ecosystems with Agents

Pascal Perez and David Batten (Editors)

ANU
THE AUSTRALIAN NATIONAL UNIVERSITY

E PRESS

ANU
E PRESS

Published by ANU E Press
The Australian National University
Canberra ACT 0200, Australia
Email: anuepress@anu.edu.au
Web: http://epress.anu.edu.au

National Library of Australia
Cataloguing-in-Publication entry

Complex science for a complex world : exploring human
ecosystems with agents.

ISBN 1 920942 38 6
ISBN 1 920942 39 4 (online)

1. Human ecology. I. Perez, Pascal. II. Batten, David F.

304.2

Cover design by ANU E Press.
Cover image, Canberra Bushfires, copyright © Pascal Perez 2003.

Contents

List of figures

List of tables

Foreword

Mankind has now entered the Anthropocene, the era in which human activities play as significant a role in shaping the biosphere as do natural processes. We see the signs of this in many places, perhaps most pervasively in the climate change brought on by rapid human perturbations to the planetary carbon cycle. The reaction of thoughtful governments to these signals has been to apply principles of sustainability, resilience and triple-bottom-line accounting to the problem of managing and regulating the interaction of humans and their environment. The science to underpin these efforts must understand and ultimately predict the dynamic behaviour of coupled systems embodying human behaviour and biophysical responses. Unlike the natural systems that environmental and earth sciences have traditionally addressed, these human dominated systems display learning, adaptation and complex non-linear feedbacks. They are 'Complex Adaptive Systems'.

Traditional approaches to modelling and understanding such systems have treated the natural and human parts quite differently. Natural biophysical processes have been approached with confidence by modellers who understood that, however complex a system like the earth's climate might be, it could still be expected to obey physical laws and its behaviour was, at least in principle, predictable. The human component, in contrast, was generally treated as entirely contingent and not subject to regular laws (with the notable exception of economics, whose practitioners make draconian simplifying assumptions about human choices with limited predictive success). This situation has changed drastically in the last decade with the growth of complexity theory and its application to human behaviour and decision making. Many aspects of human behaviour at the levels of large groups or whole societies prove to be amenable to simulation with remarkable fidelity by these techniques.

Still in its infancy, complexity theory tends to employ an eclectic collection of theories and methodologies designed to deepen our limited understanding of the properties of complex adaptive systems. Among such dynamic techniques, agent-based modelling (ABM) is being used increasingly to simulate human ecosystems. Its major advantage is an ability to generate system-wide dynamics from the interaction of a set of autonomous agents interacting in the silicon world of the computer. ABM is particularly well suited for representing social interactions and autonomous behaviours, and for studying their environmental impact at different scales. It also helps us to study the emergence of and interactions within hierarchical social groups, as well as the emergence of adaptive collective responses to changing environments and environmental management policies.

Human ecosystems constitute a subset of complex adaptive systems. They are characterised by very strong, long-term interactions between human communities and their environment. They process flows of matter, energy and information. Nowadays, research on human ecosystems straddles the social, computer and environmental sciences. It has created a space where anthropologists and sociologists meet with programmers and physicists. Until recently, such a creative space could not be found in Australia. This is exactly why the Human Ecosystems Modelling with Agents (HEMA) network was created in early 2002. A growing number of scientists needed a place to breathe in, to debate and to share ideas—a forum in the true Greek tradition.

From a handful of Australian and French scientists at the outset, the HEMA network has become an internationally recognised, steadily growing entity, closely connected to complementary groups like the European Social Simulation Association (ESSA) and the Multi-Agent-Based Simulation (MABS) community in Europe. This success has been achieved with the help of dedicated sponsors such as the Australian Department of Education, Science and Training and the French Embassy in Canberra. Several other research institutions deserve a special mention for their committed efforts: Centre de Coopération Internationale en Recherche Agronomique pour le Développement (CIRAD) and Centre d'etude du Machinisme Agricole du Génie Rural des Eaux et Forêts (CEMAGREF) in France, and The Australian National University (ANU) and the Commonwealth Scientific and Industrial Research Assocation (CSIRO) in Australia. The latter provided the backbone for the HEMA network, by way of its Centre for Complex Systems Science (CCSS) and, more specifically, the CSIRO Agent Based Modelling (CABM) working group.

This book aims to synthesise the synergistic collection of ideas and applications that have emerged from the HEMA network and the CABM working group over the last three years. In particular, it draws upon the work presented at a series of workshops co-organised by CABM and HEMA scientists. The first CABM/HEMA workshop was held in Melbourne from 11-12 July 2003. Its theme was 'Agent-based modelling of social, economic and biophysical systems'. Then, in May 2004, the 2nd workshop was convened at The Australian National University. On this occasion, the theme was 'Exploring human ecosystems: the next scientific challenge'. Finally, the latest workshop was held from 21-25 March 2005 in Bourg-St. Maurice, a scenic location in the French Alps. Organised by the Modèles et systèmes multi-agent pour la gestion de l'environnement et des territoires group (SMAGET), its theme was 'Multi-agent modelling for environmental management'.

In essence, there is an unbroken line between these 3 meetings: a consistent focus on techniques for modelling and managing human ecosystems with the help of virtual agents. By nurturing interest in the application of tools and techniques from complexity theory to the sensitive issue of human intervention in various ecosystems, the HEMA network has championed this important new area of scientific enquiry among a growing community of Australian and European scientists.

John J. Finnigan
Director, Complex System Science Program
CSIRO, Canberra, Australia

Acknowledgments

This book is born from the fruitful collaboration between:

The Human Ecosystems Modelling with Agents Network (HEMA), convened by Pascal Perez; and
The CSIRO's Agent Based Modelling Working Group (CABM), convened by David Batten

Editorial Board

Nigel Gilbert, University of Surrey, UK
Nils Ferrand, CEMAGREF, France
David Batten, CSIRO, Australia
Pascal Perez, ANU and CIRAD, Australia

Sponsors

The following sponsors supported the editing and publishing of this book:

Centre de Cooperation Internationale en
Recherche Agronomique pour le Developpement
(CIRAD)
42 rue Scheffer, 75116 Paris, France

Commonwealth Scientific and Industrial Research Organisation
(CSIRO)
Limestone Avenue, Campbell ACT 2612, Australia

Forum for European-Australian Science and Technology Cooperation
(FEAST)
National Europe Centre, ANU, Liversidge Street, Canberra ACT 0200, Australia

Ambassade de France en Australie
6 Perth Avenue, Yarralumla ACT 2600, Australia

Part I. Building a New Science for a Complex World

1. Complex Science for a Complex World: An Introduction

Pascal Perez and David Batten

Introduction

It is well known that human activities are endangering the stability and sustainability of many fragile ecosystems to such an extent that their future is in doubt. At the same time, these ecosystems are inherently challenging to manage successfully because of the complexity and uncertainty associated with their ongoing evolution. Much of this complexity and uncertainty may be attributed to the human dimension. Thus it is imperative that we deepen our understanding of how and why people choose to interact with one another and how this interactive behaviour affects these ecosystems as time passes. This book is a small contribution in this direction. It examines ways in which the collective behaviour of human beings can vary, and how this behaviour may affect the natural ecosystems which humans enjoy.

Fortunately, a new kind of science is helping us deepen our understanding of how human ecosystems might grow and change over time. Beyond a mere collation of various reflections and applications, the chapters in this book aim to convince the reader that this new kind of science is worthy of our attention. It is a science that fully embraces the complexity of our surrounding world. It is also a science that addresses the frontiers of interactions between human behaviour and environmental responses. Furthermore, it is a science that challenges our limited understanding and treatment of uncertainty. And, finally, because it is socially embedded, it is a science that can generate partnerships with local communities in a constructive manner.

The computational science we describe and discuss in this book is sufficiently different to normative science that we may call it a new kind of science. Observation and experimentation are still alive and well, but they have been joined by an entirely new breed of computational science: simulation. However, it should be recognised that science—old or new—is not a process for finding absolute truths or universal laws. As Roger Bradbury argues persuasively in the next chapter, science is not just a process, but is really a system—a complex adaptive system.

The complex adaptive system known as science consists of the scientists or agents and the knowledge system that they accumulate by way of an interactive and ongoing exchange of ideas. As Bradbury notes, the former is essentially a social system, whereas the latter is a playground of memes—with the interaction strongly mediated by the recipe. Science is a dialogue between humankind and nature, the results of which have been—and will continue to be—unpredictable (Prigogine 1996: 153). Like science, a human ecosystem is a complex adaptive system involving people, other living entities, an environment, information exchange and the co-evolution of all of these things over time. Although a human ecosystem's future state has always been unpredictable, the critical players (agents) on this adaptive stage today are the human participants. Thus it is imperative that we begin to understand human reasoning and its idiosyncrasies.

The complex world of human ecosystems

Understanding human behaviour

Deductive versus inductive reasoning

How do people make decisions in complicated situations? Many psychology texts argue that human reasoning is deductive. Deduction is reasoning from the general to the particular. A perfectly logical deduction yields a conclusion that must be true provided that its premises are true. Thus it involves specifying a set of axioms and proving consequences that can be derived from those premises. In reality, deduction is handy for solving a host of theoretical problems and a handful of simple practical problems, but it is less helpful for tackling complicated practical problems like those associated with the management of human ecosystems. Moreover, it is definitely of no help when the managers involved are likely to behave and react differently over time.

The truth of the matter is that each of us is a unique product of our own brain and our uniquely individual experiences. Our personal knowledge is honed by the concepts, notions and models which we choose to use to represent it. All of this has to be created, put together and revised over time by us as well as by others in society as a whole. Learning has both an individual and a collective dimension, which builds adaptively on the inductive and intuitive skills of a heterogeneous collection of minds. Erwin Schrödinger summed it up well:

> The world is a construct of our sensations, perceptions, memories. It is convenient to regard it as existing objectively on its own. But it certainly does not become manifest by its mere existence. Its becoming manifest is conditional on very special goings-on in very special parts of this very world, namely on certain events that happen in a brain.
> (Schrödinger 1967: 94)

As a matter of fact, human agents do more than make rational choices by way of simple deduction. In most cases, no one agent knows what all the others are doing, being forced to rely on a limited amount of shared information and unique experiences of its own. Hence, they make decisions on the basis of whims, hunches, heuristics or mental models and are willing to modify their preferred mental models (to a greater or lesser extent) and come up with new ones where necessary. In other words, when facing open-ended situations, human agents reason inductively.

In chapter 3, Pascal Perez draws upon philosophy of mind and semiotics to suggest, along with an increasing number of colleagues, that these contextual cognitive processes have deceived attempts to develop a predictive modelling formalism so far. He argues that the Artificial Intelligence paradigm needs to be replaced with more modest, but more robust approaches. Anne Dray and colleagues (chapter 12) provide a meaningful example of such innovative approaches through their description of a companion modelling experience in the South Pacific.

Sheep versus explorers

Inductive reasoning places different demands on we thinking individuals than the deductive metaphor. It involves pattern formation and pattern recognition, aided by intuition and creativity. Clearly some people are more intuitive or creative than others. They are better at seeking and discovering novel solutions to problems, being willing to experiment, adapt and instigate change. Others merely follow existing patterns, often resisting change under almost any circumstances. Like the spectrum of light, cognitive equipment consists of a mixture of cognitive skills of varying intensities.

We may classify agents in terms of two extreme forms of behaviour: those who actively search for new possibilities are *explorers,* and those who prefer to remain with the status quo are *sheep.* The above mentioned spectrum of cognitive skills implies that we all possess sheep and explorer qualities, albeit in different doses. Pure explorers tend to be imaginative, creative, highly-strung individuals who constantly search for better solutions to the problems they face. They are more inclined to reason inductively, to learn quickly and to adapt willingly to changing circumstances. Sheep are more placid, patient and resigned than explorers. Preferring to reason deductively, they are prone to choosing a well established pattern. They mostly cling to particular beliefs because they have worked well in the past. Sheep are slow learners who must accumulate a record of failure before discarding their favourite beliefs.

In chapter 4, David Batten explores sheep and explorer strategies among fishing fleets searching for profitable fishing zones. Non-equilibrium systems scientists such as Peter Allen, having studied the behaviour of fishing fleets, call the sheep Cartesians and the explorers Stochasts. As Allen notes, the first group makes good use of information, but the second generates it.

Self-referential situations

Conventional economic wisdom claims that agents have only one reasoning skill: the ability to process the information available to them in a purely logical, deductive manner to arrive at the best decision in a given situation. But this is useless in a self-referential situation like a stock market, where the best thing to do depends on what everyone else is doing. Self-referential situations often arise in human ecosystems, but are rarely recognised as such. A self-referential situation is one in which the forecasts made by the agents involved serve to create the world they are trying to forecast (Batten 2000). In many human decision-making situations, there is no optimal predictor. The best thing each agent can do is to apply the predictor that has worked best so far, then to be willing to re-evaluate the effectiveness of his set of predictors, and to adopt better ones as new information becomes available.

In chapter 4, David Batten develops the mechanisms of such 'perpetual experiments' and their consequences in terms of the collective behaviour of what he calls self-defeating systems. Drawing upon the powerful metaphor proposed by Brian Arthur (1994), the El Farol bar problem, the author suggests that many human ecosystems fall into this category and offer agent-based modelling an unchallenged field of exploration. To a certain extent, Ryan McAllister and colleagues (chapter 14) illustrate the ill-fated consequences of such self-defeating systems in the case of privately-owned pastoral regimes in Australian rangelands.

Complexity of human ecosystems

Adapting to co-evolution

Human ecosystems constitute a subset of complex adaptive systems. They correspond to real life systems characterised by very strong and long-term interactions between human communities and their environment. According to John Holland (1995), these systems display the following characteristics:

- *Emergence*: a system-level phenomena is emergent if it requires new categories to describe it, which are not required to describe the behaviour of the underlying components. In other words, interactive individual components instantiate emerging patterns at the level of the system.
- *Path dependency*: due to the highly non-linear relationships between individual components or parts of the system, a given system-level phenomena can be achieved, in theory, through an infinite number of combinations at the micro-level.
- *Non state equilibrium*: the Complex Adaptive Systems display an ever-changing dynamic equilibrium, driving back and forth the system between chaotic to ordered states. On the edge of chaos, these systems are very sensitive to any perturbation from the individual components.

- *Adaptation*: the evolution of the system is driven by the co-evolution of its individual components. They adapt to their environment and modify it in a recursive way. If the components are cognitive beings, the adaptation relies mainly on the individual and collective learning processes.

According to Stepp et al. (2003), human ecosystems not only process matter and energy flows, but—more specifically—information flows as well. Therefore, they display very specific characteristics. As stated in the previous section, human ecosystems are inherently complex and adaptive, due to the ability of human beings to switch from rational deductive reasoning to inductive pattern recognition. Besides, our ability to communicate and learn from others creates the conditions for co-evolutionary processes in which positive feedback loops follow negative ones, punctuation dispels equilibrium, chaos threatens order, and chance gives a hand to necessity. In chapter 2, Bradbury argues that, until recently, human beings had been able to adapt to changes and to cope with co-evolution through rather simple heuristics. But human activities have gradually strengthened the links—let's call it globalisation—between loosely connected environments and societies. The author suggests that:

> the balance of [our] effective adaptive strategies has shifted decisively
> and forever from heuristics to what we might call symbioses—the sorts
> of strategies that evolution favors in closely connected systems, the ones
> we see today inside cells and organisms, and between symbiots.
> (chapter 2, p. 23)

It is indeed remarkable to observe the analogies between what is known as human body functional integrity and what is nowadays called environmental sustainability. More information, more interactions, and shorter communication paths tend to create what David Newth refers to as *small worlds* in his network-centric analysis of social interactions (chapter 5). The symbiotic dimension of our current understanding of sustainable development is illustrated in two contrasted chapters of this book. In chapter 7, Katherine Daniell and colleagues present a tentative framework to assess sustainable urban development. Taking Christie Walk housing development in Adelaide (Australia) as a case study, the authors insist on the necessity of taking into account water distribution, gas emission, ecosystem health, waste management, economic viability, and social aspiration in order to derive relevant and interrelated sustainability indicators. In chapter 12, Anne Dray and colleagues describe the same necessity of taking into account complex social and spatial interactions in order to help local stakeholders to self-design sustainable water management on the atoll of Tarawa (Republic of Kiribati).

Coping with uncertainty

By admitting that many human ecosystems are complex and adaptive, we accept their inherent uncertainty. Indeed, if the system is sufficiently complex, it may not be practical or perhaps even possible to know the details of each local interaction. Obviously, the understanding of system-level patterns is often purchased at a cost:

> the observer must usually give up the hope of understanding the
> workings of causation at the level of individual elements.
> (Lansing 2003: 185)

As a matter of fact, uncertainty in human ecosystems can result from two different causes: unpredictable non-linear interactions or ill-defined predicates. The latter—more frequently encountered than usually admitted—relies on our limited ability to infer robust causality links among given sets of elementary processes. For example, Durkheim (1979: 58), in his famous study of suicide, concluded that no matter how much a researcher knows about a collection of individuals:

> It is impossible to predict which of them are likely to kill themselves.
> Yet the number of Parisians who commit suicide each year is even more
> stable than the general mortality rate.

A process that seems to be governed by chance when viewed at the level of the individual turns out to be strikingly predictable at the level of society as a whole. One would argue that statistics prevail in this case, others would invoke Richard Dawkin's memes (Dawkins 1976), but we could also admit that we don't know enough yet about the intimate social dynamics that control such a deviant behaviour. In chapter 10, Pascal Perez and colleagues describe a first attempt to simulate illicit drug use and local markets in Australia. Authors admit that blending together law enforcement, harm reduction, and treatment strategies already represents a daunting challenge; but, trying to infer users' or dealers' behavioural patterns—elusive and secretive by nature—needs to be dealt with through trans-disciplinary and consensual approaches.

On the other hand, unpredictable non-linear interactions are the *raison d'etre* of complex adaptive systems. The self-referential problem proposed by Arthur (1994) can drive a system to an equivalent situation to the one described by Durkheim. But in the El Farol case, individual behaviours are perfectly deterministic while totally unpredictable for an external observer. Conversely, some perfectly predictable individual behaviours interacting together can lead to unpredictable global behaviour of the whole system. In chapter 11, David Batten and George Grozev provide a clear illustration of the impact of such non-linear interactions in the case of the Australian National Electricity Market (NEM). They describe the NEM as an evolving system of complex interactions between human behaviour in markets, technical infrastructures and the natural environment and propose to explore plausible sustainable futures through agent-based simulations.

In chapter 3, Pascal Perez insists on the inherent uncertainty attached to human ecosystems. He argues that scientists studying these complex adaptive systems fall into two broad categories based on their epistemological postures: Legalists believe in a positivist and experimental science, while Revolutionaries claim that human ecosystems can only be apprehended through a post-normal approach to science. Drawing upon recent theoretical and methodological advances in agent-based modelling, he argues that an uncertainty-compliant science needs to overcome this dichotomy.

Scales and hierarchies

In their 2002 book entitled *Panarchy: Understanding Transformations in Human and Natural Systems*, Gundarson and Holling proposed a holistic and history contingent view of human ecosystems. The panarchy concept describes how a healthy system can invent, experiment, and survive through hierarchies and adaptive cycles that represent ecosystems and socio-ecosystems across scales. Each level of the panarchy operates at its own pace, protected from above by slower, larger levels but invigorated from below by faster, smaller cycles of innovation. The whole panarchy is therefore both creative and conserving. The interactions between adaptive cycles in a panarchy combine learning with continuity (Holling 2001). An obvious strength of the approach is to allow us to view resource management in a more structured way.

Indisputably, this theoretical framework helps clarify the meaning of sustainable development. But critics of the approach point out the direct filiation of panarchy from ecological modelling, in particular dynamical system modelling (DSM). While biological cycles are easy to define, much remains to be said about cycles within human societies. Stepp et al. (2003) provide a compelling list of human characteristics hardly tractable at the level of societies or human ecosystems. The way forward lies in our ability to transcend boundaries between systemic and atomistic approaches. In chapter 13, David McDonald and colleagues demonstrate that it is technically possible to blend system dynamics and individual (or group) behaviours of complex marine environments. In an entirely different context, Ang Yang and colleagues (chapter 8) analyse interconnections between several communication, observation and command networks on a battlefield in order to infer relevant conclusions for modern warfare, including fallible human judgement.

Exploring human ecosystems with agents

Autonomous and adaptive agents

For some years, the fields of software engineering and artificial intelligence have been making use of the concept of interacting autonomous agents. Although the term *agent* has been defined and used in several ways by other scientific groups, for the purposes of software engineering it has been sufficient to start with a minimal definition of an agent, such as the following (Ferber 1999): an agent is a physical or virtual entity that:

- is capable of acting in an environment;
- can communicate directly with other agents;
- is driven by an autonomous set of individual goals or objectives;
- possesses resources of its own;
- is capable of perceiving its environment to some extent;
- has only a partial representation of this environment;
- possesses skills and can offer services;
- may be able to reproduce itself; or
- whose behaviour tends towards satisfying its objectives, taking account of the resources and skills available to it.

The definition corresponds closely to a kind of living organism whose behaviour is aimed at satisfying its own needs and fulfilling its own objectives. Adaptation among such an assembly of agents may occur in two ways: by altering individual characteristics (learning), or as a collective process that brings reproductive processes into play (evolution). We can think of this combination as adaptation that is simultaneously individual and collective. A broad range of agents that are simulated in several different contexts, as described in the second part of this book, are summarised in the Table 1.1.

Table 1.1. Agents simulated in Chapters 7-14

Chapter	Author(s)	Simulation	Types of agents
7	Daniell et al.	Housing market	House occupants or households
8	Yang et al.	Warfare model	Command and control groups
9	Elliston and Beare	Pest incursion	Pests, farmers, contractors
10	Perez et al.	Illicit drug market	Users, dealers, wholesalers, police and outreach workers
11	Batten and Grozev	Electricity market	Generator firms, retailers, network service providers, customers and traders
12	Dray et al.	Atoll water	Landowners management
13	McDonald et al.	Coastal marine	Fishing, shipping, ecosystems; petroleum, environment
14	McAllister et al.	Rangeland	Pastoral enterprises

Hierarchies of autonomous agents

In his quest to find a general biology, Stuart Kauffman suggests that the biosphere got itself constructed 'by the emergence and persistent co-evolution of autonomous agents' (Kauffman 2000: 3). He goes further by suggesting that there may be a fourth law of thermodynamics that roughly states that biospheres maximise the average secular construction of the diversity of autonomous agents and the ways those agents make a living to propagate further. Although this raises a labyrinth of issues about the core features of autonomous agents and their abilities to manipulate the world on their own behalf, it also has important ramifications for the study of human ecosystems. Complex webs of interacting life within ecosystems need to be recognised and addressed in any simulation that attempts to explore alternative co-evolutionary pathways for the future.

Kauffman defines an autonomous agent as a self-reproducing system able to perform at least one thermodynamic work cycle. If true, this means that all free-living cells and organisms are autonomous agents. It also suggests that we lack a concept of propagating organisation. Complexity in organisation rises from the subatomic through the atomic, molecular, cellular, organismal, societal, and ecological system levels. Coevolving autonomous agents may be co-constructing and propagating organisations of work simultaneously at all of these levels. This poses a major challenge for our new kind of science. Nevertheless, more ambitious simulation models such as the Management Strategy Evaluation (MSE) approach for the North West Shelf region (see chapter 14) are beginning to grapple with several levels of the agent hierarchy.

Interconnected and embedded agents

The time when systemic Dynamical System Modelling (DSM) and atomistic (Agent-Based Modelling (ABM)) approaches were described as inherently conflicting is gone. Pioneering work has proven that these approaches are largely complementary (Carpenter et al. 1999; Janssen et al. 2000). DSM provides an elegant analytical framework to study dynamic equilibrium of a system, to the extent that the global parameters can be made explicit. On the other hand, ABM demonstrates an ability to demonstrate emergent phenomena, but is not a predictive tool. Hence, ABM could help discover some emerging values of system-wide parameters and DSM might provide bounding behavioural domains for agent-based simulations.

In chapter 5, David Newth describes how, from social systems to computer networks, graphs can be used to describe the way in which components in large systems interact. In mathematical terms, a network is represented as a graph, in which nodes represent network elements and edges define relationships between elements. According to the author, this powerful paradigm now needs to be coupled to agent-based approaches in order to refine our understanding of human ecosystems.

A comprehensive series of research tends to illustrate how social network structure can be used to determine the characteristics of individual actors, and how social network structure and associated dynamics can modify the behaviour of these actors (Borgatti and Foster 2003). But much remains to be said about the influence of autonomous and partly rational actors on the structure and evolution of these social networks. Here lies the crucial question of how social coalitions and factions form and dissolve over time. Coming back to our previous analogy between human body integrity and environmental sustainability, it is a matter of knowing how system structure and fuctions co-evolve in order to maintain the resilience of a given human ecosystem.

In chapter 9, Lisa Elliston and Steve Beare illustrate the impact of such an organic link with their simulation of a pest invasion in Northern Queensland. Ways in which wheat farmers and contractors interact shape the network over which Karnal bunt, a disease of wheat, diffuses and contaminates various parts of the countryside. Conversely, the presence of contaminated areas influences the behaviour of local farmers and the strategies deployed by the Quarantine services. In a different context, Pascal Perez and colleagues (chapter 10) describe an equivalent co-evolutionary process of interactions between illicit drug markets and law enforcement strategies. So called *hot spots* move around the urban environment due to endless adaptive strategies deployed by police forces and crime syndicates.

Exploring national research priorities with agents

The second part of this book provides meaningful examples of agent-based modelling applications. Ranging from warfare strategies to natural resources management, the different chapters display two characteristics that, too often, are lacking in descriptions of computer simulation. First, the context of application is carefully presented, as it seems essential to understand the reasons why decision-makers decided to fund or to participate in such projects. Then, the corollary is to give the opportunity to decision-makers or practitioners to comment on the usefulness of the approach. Hence, each chapter is introduced by a specific foreword from a personality who was actively involved or interested in using the outcomes of the modelling process.

Effective applied research doesn't need only to be deeply rooted in a context and to be accessible to stakeholders, it also needs to participate in the global backing of national capacities. In Australia, the National Research Priorities (NRPs) provide a general framework guiding research institutions. Four NRPs are currently developed: an environmentally sustainable Australia; promoting and maintaining good health; frontier technologies for transforming Australian industries; and safeguarding Australia.

One could argue that agent-based modelling and computer simulation are breakthrough sciences that deserve to fall entirely into the third priority. On the contrary, the variety of contexts discussed in this book demonstrate the large range of applications of these methods by linking each chapter of the second part of this book to a relevant NRP.

An environmentally sustainable Australia

Water — a critical resource

In chapter 12, Anne Dray and colleagues describe how they used multi-agent simulations in conjunction with a role playing game to develop a Negotiation Support System for groundwater management in Tarawa (Republic of Kiribati). Their Companion Modelling approach relies on three successive stages. First, a Global Targeted Appraisal focuses on social group leaders in order to collect different standpoints and their articulated mental models. Then, these contrasted models are merged into a single conceptual one using UML (Unified Modelling Language) formalism. This conceptual model is further simplified in order to create a role-playing game. This computer-assisted game is played during iterative sessions, generating innovative rules and water management scenarios among stakeholders.

Transforming existing industries

Ryan McAllister and colleagues (chapter 14) present an interesting reflection on Australian pastoral land-use systems that have been characterised by private-property regimes, which, to varying degrees, have created fragmented and disconnected landscapes. The authors argue that there are both environmental and economic risks associated with productive land fragments being too small, and these risks necessitate an understanding of fragmentation's driving forces. Understanding these forces, however, is made difficult because the problem involves social, economic and environmental factors, interacting over a range of temporal and spatial scales. Hence, the authors developed an agent-based model to explore rangeland dynamics that involve such complexity. The current model contains pastoralists, livestock, key ecological processes, and governance.

Sustainable use of Australia's biodiversity

In chapter 13, David McDonald and colleagues propose a Management Strategy Evaluation (MSE) framework to demonstrate practical science-based methods that support integrated regional planning and management of coastal and marine ecosystems of the North West Shelf region (Western Australia). Their multiple-use MSE has, so far, focused on four sectors: oil and gas, conservation, fisheries and coastal development. For each sector a selection of development scenarios, provided by the relevant interest groups, is represented. These scenarios include prospective future sectoral activities and their impacts, and the sectoral response to management policy and strategies. According to the authors, the agent-based modelling software InVitro is well placed for analysing prospective social and ecological impacts of multiple-use management strategies in a risk-assessment framework such as MSE.

Promoting and maintaining good health

Strengthening Australia's social and economic fabric

Katherine Daniell and colleagues (chapter 7) present a model that was used to evaluate the plausible futures of Christie Walk housing development in inner-city Adelaide, Australia. The authors underline the fact that assessing the sustainability of development proposals has become of great importance to policy and decision makers. However, effective methods of assessing the overall sustainability of housing developments (proposed or existing) have yet to be established. This chapter presents a new methodology to assess the sustainability of housing development systems. The methodology uses a Sustainability Scale approach for livelihood indicators, coupled with multi-agent simulations to represent the complex housing patterns.

Safeguarding Australia

Critical infrastructure

In chapter 11, David Batten and George Grozev describe the development of an agent-based simulation model that represents Australia's National Electricity Market (NEM) as an evolving system of complex interactions between human behaviour in markets, technical infrastructures and the natural environment. This simulator, named NEMSIM, is the first of its kind in Australia. Users will be able to explore various evolutionary pathways of the NEM under different assumptions about trading and investment opportunities, institutional changes and technological futures, including alternative learning patterns as participants grow and change. Simulated outcomes can help the user to identify futures that are eco-efficient—for example, maximising profits in a carbon-constrained future.

Protecting Australia from invasive diseases and pests

Lisa Elliston and Steve Beare (chapter 9) present a model investigating the impact of a potential incursion of Karnal bunt in wheat in a valuable agricultural producing region of Australia. An incursion management model was developed to estimate the regional economy effects of the exotic pest incursion in the agricultural sector. By developing an agent-based spatial model that integrates the biophysical aspects of the disease incursion with the agricultural production system and the wider regional economy, the model can be used to analyse the effectiveness and economic implications of alternative management strategies for a range of different incursion scenarios.

Protecting Australia from terrorism and crime

In chapter 10, Pascal Perez and colleagues have explored the complexity of illicit drug markets in Melbourne. The intricacy of multiple interactions between individuals, the various time lines linked to different aspects of harm reduction, and contrasted social rationalities observed among field practitioners (prevention, law enforcement, harm reduction) contribute to the creation of complex and unpredictable systems. In order to explore this complexity, an agent-based model called SimDrug was designed. The prototype model includes users, dealers, wholesalers, outreach workers and police forces. In order to overcome the limited knowledge we have of underground illicit activities, a trans-disciplinary group of experts regularly informed (and validated) the model with consensual rules.

Transformational defence technologies

Ang Yang and colleagues (chapter 8) bring us into the complex world of warfare for which a number of Agent-Based Distillation Systems (ABDs) have been developed and adopted to study the dynamics of warfare and gain insight into military operations. According to the authors, these systems have facilitated the analysis and understanding of combat. However, these systems are unable to meet the new needs of defense, arising from current practice of warfare and the emergence of the theory of Network Centric Warfare (NCW). The authors propose a network centric model which provides a new approach to understand and analyse the dynamics of both platform centric and network centric warfare.

Building a new science for a complex world

Towards holistic visions and polymorphic tools

Holistic visions

To develop a truly holistic view of human ecosystems—to identify feasible pathways towards their sustainability—it is necessary to make use of tools and techniques that are able to grapple with the complex interactions between human activities and the stability or resilience of the non-human ecosystem habitat. Various chapters in this book show that human behaviour can be complex in itself, necessitating its representation in a more sophisticated manner in computational approaches that aim to improve our understanding of the collective outcomes under alternative management strategies. Agent-based modelling is one promising way of exploring the possible collective implications of more complicated behaviour of human agents in an ecosystem context. Role-playing games are another valuable method of engaging with stakeholders and eliciting their views in an ecosystem context.

Opportunities for further, impressive progress exist. Recent developments in object-oriented programming languages have allowed users to create autonomous modules which can interact with each other even when they have been designed by different people, different teams or different companies (Ferber 1999). Agent-based models and Multi-Agent Systems (MAS) have an important role to play in this endeavour by serving as possible successors to object-oriented systems and combining local behaviours with autonomy, best-practice agent modules and distributed decision making. Thus it seems very likely that the software engineering of tomorrow for addressing more complex societal problems will be agent-oriented, just as that of today is object-oriented.

Polymorphic tools

To make our new kind of science truly polymorphic, the mapping of relationships between agents needs to be continuously updated depending on circumstances, proficiencies, perceptions, tasks to accomplish, or relational rules based on the social contracts established. This means that the simulation models need to re-cognise different views, biases or expertise. Such simulation tools are more participatory in nature, and should lead to further exciting developments in the future.

In the field of companion modelling, for example, convergent views may emerge after allowing several recognised experts to take the lead one after the other. This is similar to a cooperative team sport like football; for instance, the right-winger becomes the leader when the ball comes into his space, but it can happen that he becomes the goalkeeper when the situation requires it. This flexible kind of companion modelling forms the basis of AtollGame (chapter 12), in which the agent-based model and the corresponding role-playing game were designed ac-cording to different viewpoints—converging or conflicting—that were recorded during a series of interviews.

Towards a new epistemology of science

Integration and implementation

The sixth chapter of this book presents an argument by Gabriele Bammer to think differently about the future of current science and the way it engages with decision-makers. According to the author, developing a new specialisation—In-tegration and Implementation Sciences—may be an effective way to draw togeth-er, and significantly strengthen, the theory and methods necessary to tackle complex societal issues and problems. It would place research on human ecosystems in broader context and link it with a range of complementary concepts and skills.

There is indeed a pressing need to integrate not only the participation, but, more importantly, the engagement of local stakeholders in projects that concern their future. Aslin and Brown (2004) argue that local communities need to be involved in the analysis of the results (consultation) and the choice of the possible scenarios (participation), but also in the knowledge creation itself (engagement). This is a post-normal posture adopted, for example, within companion modelling approaches (Bousquet et al. 2002), for which Anne Dray and colleagues (chapter 12) provide a good illustration.

Research on sustainable development too often relies on deductive scientific approaches to reach outcomes that require more inductive and flexible solutions. But flexibility means that one must assume some uncertainty during implementation and must get away from traditional reliance on deterministic and predictable solutions. What is true at the technical level becomes paramount at the political level. As stated by Bradshaw and Borchers (2000: 1):

> One of the most difficult aspects of translating science into policy is scientific uncertainty. Whereas scientists are familiar with uncertainty and complexity, the public and policy makers often seek certainty and deterministic solutions.

This call for a meta-integration of science takes us back to the beginning of this book. Roger Bradbury's vision of a complex and adaptive science is nothing less than a fantastic opportunity for scientists to think differently about the goals of their own research and their relations to others.

A few last words

We hope that you will enjoy reading such a diverse *ouvrage* whose purpose is to attract more early career scientists into our field of research and to convince decision-makers that a growing contingent of colleagues working on complexity theory can provide useful tools and methods to better understand complex and adaptive environments. It is time to reassure you, the reader, that the rise of a *Complex Science for a Complex World* doesn't mean more complicated relationships between science and society.

References

Arthur, W.B. (1994) Inductive behaviour and bounded rationality. *The American Economic Review* 84: 406–11.

Aslin, H.J. and V.A. Brown (2004) Towards whole of community engagement: A practical toolkit. Canberra, Australia: Murray-Darling Basin Commission.

Batten, D.F. (2000) *Discovering Artificial Economics. How Agents Learn and Economies Evolve.* Oxford: Westview Press.

Borgatti S.P. and P.C. Foster (2003) The network paradigm in organizational research: a review and typology. *Journal of Management* 29(6): 991–1013.

Bousquet F., O. Barreteau, P. d'Aquino, M. Etienne, S. Boisseau, S. Aubert, C. Le Page, D. Babin and J.C. Castella (2002) Multi-agent systems and role games: collective learning processes for ecosystem management. In M. A. Janssen (ed.) *Complexity and Ecosystem Management: The Theory and Practice of Multi-agent Systems,* pp. 248–85. Cheltenham: Edward Elgar.

Bradshaw, G.A. and J.G. Borchers (2000) Uncertainty as information: narrowing the science-policy gap. *Conservation Ecology* 4(1): 7. Available at http://www.consecol.org/vol4/iss1/art7/

Carpenter, S., W. Brock and P. Hanson (1999) Ecological and social dynamics in simple models of ecosystem management. *Conservation Ecology* 3(2): 4. Available at http://www.consecol.org/vol3/iss2/art4

Dawkins, R. (1976) *The Selfish Gene.* Oxford: Oxford University Press.

Downs, A. (1962) The law of peak-hour expressway congestion. *Traffic Quarterly* 16: 393–409.

Durkheim, E. (1979) *Suicide: A Study in Sociology.* Transl. J.A. Spaulding, G. Simpson. New York: Free Press.

Ferber, J. (1999) *Multi-Agent Systems: An Introduction to Distributed Artificial Intelligence.* New York: Addison Wesley.

Gunderson, L.H. and C.S. Holling (eds) (2002) *Panarchy: Understanding Transformations in Human and Natural Systems.* Washington, D.C.: Island Press.

Holling, C.S. (2001) Understanding the Complexity of Economic, Ecological, and Social Systems. *Ecosystems* 4: 390–405.

Holland, J.H. (1995) *Hidden Order: How Adaptation Builds Complexity.* New York: Helix Books (Addison Wesley).

Janssen M.A., B.H. Walker, J. Langridge and N. Abel (2000) An adaptive agent model for analysing co-evolution of management and policies in a complex rangeland system. *Ecological Modelling* 131: 249–68.

Kauffman, S.A. (2000) *Investigations.* New York: Oxford University Press.

Lansing, J.S. (2003) Complex adaptive systems. *Annual Review of Anthropology* 32: 183-204.

Schrodinger, E. (1967) *What is life? With Mind and Matter and Autobiographic Sketches.* Cambridge, UK: Cambridge University Press.

Stepp, J.R., E.C. Jones, M. Pavao-Zuckerman, D. Casagrande, and R.K. Zarger
(2003) Remarkable properties of human ecosystems. *Conservation Ecology*
7(3): 11. Available at http://www.consecol.org/vol7/iss3/art11

2. Towards a New Ontology of Complexity Science

Roger Bradbury

Introduction

I think that this is a Promethean moment. I think that when we look back on today, we'll say 'that was when things became different'. This may sound a bit like King Henry on St Crispin's day at Agincourt:

> This story shall the good man teach his son;
> And Crispin Crispian shall ne'er go by,
> From this day to the ending of the world,
> But we in it shall be remembered;
> We few, we happy few, we band of brothers

Now I find this a worry, because aren't we the inheritors of a tradition that has striven for the last two and a half millennia to demonstrate that most moments are not Promethean, that our time and place in the universe are not special? Isn't it the politicians and spin-masters, not the scientists, who exhort us to think that now is different and that we are different? But the evidence worries me, and, what is more, the implications worry me. So indulge me while I lay out the evidence that, contrary to the null hypothesis, we really are living in a singular moment in the history of this bit of the universe. And then indulge me further while I unfold some of the consequences.

I think there are two big historical processes intersecting at the moment: the coming into being of a fully connected world, and the coming into being of a new way of doing science. Both only make sense from a complexity point of view, and, at base, that is why 'we happy few' matter in all of this. These two processes are going to interact incredibly strongly, I believe, and that will take us into radically new territory. But it is the novelty of each of these two processes that define our moment as Promethean, and it is to these that I will first turn.

A fully connected world

I want to argue that, for all of life's lease on the planet up to today, we, we living things, have lived with frontiers. Our ecosystems behaved as if there was an *out there* and a *round here*, a system in an environment. We know this by looking at the sorts of adaptations living things have. They show, broadly, organisms reacting with heuristics to the immediacy of their environment. It is true that there are some adaptations that show a subtlety that betrays close coupling—many interactions on coral reefs and symbioses of all sorts show this. But for most of the time, most organisms behave as if the world is loosely connected, that there is fuzziness and slop in the world, and that, to survive, organisms need to focus on just getting through the moment.

We can see that this is a sensible, efficient response to a physically dominated world, where grand biogeochemical cycles and forcing factors are more *geo* than *bio*. But as life evolved and radiated and became more complex, one might expect that evolution would turn its hand to how to survive in a world made complex by other living things. And here we see something curious. It is true that complexity—the emergence of complex adaptive systems—does encourage some closer coupling. Every naturalist's notebook is full of such wonders, and every diver and birdwatcher has seen them. But it is also true that heuristics—rules of thumb—account for much if not most of the interactions between living things. It seems that evolution demands that living things adapt frugally to the emerging complexity of the world, using the sorts of heuristics and work-arounds they used in earlier, simpler times. Even though living things are faced with the emergence of complexity, they are constrained in their response by such prosaic things as history, mutation rates and generation times.

There seem to be two canonical but conflicting rules that govern the way in which living things respond to the world around them. And neither insists that the complexity of the world be acknowledged in any deep way. The first is that, for most practical purposes, what happens next is, to a first approximation, a simple extrapolation of what just happened. Even complex non-linear processes look just like simple linear ones when taken one little step at a time (this, after all, is the key discovery of the calculus). The second is, for living things in a world made complex by other living things, after a few steps, something else will happen. Thus there is no need to get too involved adapting to (that is, predicting) the impact of some complex long term process when it could be overtaken at any time by the *Next Big Thing*.

By and large, evolutionary success for living things has meant coping with the monotony of endless moments as well as the occasional surprise. This is the law of the frontier, and reflects the fact that adaptation through tight couplings is found mostly within organisms—the things we call physiology—or within cells—the things we call biochemistry. The couplings between organisms are generally much looser and those between ecosystems looser still. The world, as a system, is really a bunch of loosely connected subsystems, where the connections, for *adaptational* purposes, are more or less indistinguishable from the random buffetings of *the environment* coming through the frontier.

We see this in the dynamics of all ecosystems on earth. This includes, of course, the dynamics of human societies. Those systems that behave as if they have a frontier can adapt with simple *here and now* heuristics and succeed, whether they are the Huns invading Europe, the English colonising the new worlds, some weed invading a rangeland, or just a barnacle on the shore facing tides and storms. What makes today different is that man 'knits up the ravelled sleave' of the world through all his activities, and it is just about complete. Globalisation is just a fancy term for closely connecting the world's economies, and hence its societies, and this has more or less reached the point where we can speak of a world economy as a single functioning entity. In parallel with this, our economic exploitation of the world's natural ecosystems is now more or less complete. And economic exploitation means close connection. There are really no wild, unexploited places left. The last coral reefs, seamounts, deserts and taigas are getting locked in as I speak. There are no frontiers.

We see the evidence all around us, even if we focus on the epiphenomena and not the underlying structural change in the way the world works. We see SARS, blackouts, the Asian financial bust, ballast water, the ozone hole, global warming, and even bushfires connected to red tides as a list of phenomena instead of symptoms of the emergence of a singularity: the fully connected world. We, we living things, now live in a world system where the balance of effective adaptational strategies has shifted decisively and forever from heuristics to what we might call symbioses—the sorts of strategies that evolution favors in closely connected systems, the ones we see today inside cells and organisms, and between symbiots. No doubt evolution will sort this out over the next several hundred million years, but in the meantime: 'Houston, we have a problem'.

A new kind of science

Hold that thought, and allow me to return to my second singularity that helps define today as a Promethean moment: the coming into being of a new kind of science. The phrase is Stephen Wolfram's, but I want to stretch it further than he did, because I think he only got halfway there (Wolfram 2002). He was really describing an old *new kind of science*. Let me explain. We can make a strong argument that the sort of science we do when we simulate, particularly when we simulate complex systems, is sufficiently different to normative science that we may call it a new kind of science. The argument (and Wolfram is by no means the first to make it) goes roughly that science first went through an observational stage, then an experimental stage and now is going through a simulation stage. These stages parallel developments in the tools and technologies we use, in particular mathematics, and also roughly track the sorts of phenomena that are understandable by different stages. This argument is countered by the logical positivist idea that experimental science is the real dinkum science, preceded in a developmental sense by observational science, and extended, reluctantly and in special cases, to simulation. And the most typical special case is complex systems. But in this view, experimental science remains the core of science, and we complex systems scientists remain beyond the pale with our success judged by how well we can eventually bring our problems into the core.

Our program, by these lights, is to make the complex simple enough to allow it to be handled eventually by normative science. Our simulation is a means to an end. Now Wolfram and others challenge this, arguing that we have been seduced by the success of experimental science at picking the low hanging fruit on the tree of knowledge. The real problems and real knowledge hang much higher up, forever beyond its reach. And if we condemn science to be what the logical positivists call normative, we condemn ourselves to ignorance. To that point I am in full agreement with the Wolfram view, but I put a different interpretation on the history of science, one which is based on thinking of the science enterprise itself as a complex adaptive system, and not necessarily as some sort of progression through stages: this is now normative, later that is. In this view, which I have elaborated elsewhere, I use the same evidence as a Wolfram might, but I do not come to a view that we are changing what we mean by normative science, that is changing in earlier times from observational to experimental and now to simulation. Instead I am rather more interested in just what science is.

My argument is that science is not a process—a method—for finding the truth, and hence does not really have stages (Bradbury 1999). It is rather a body of knowledge about the world *and* a recipe for growing that body of knowledge *and* a bunch of people busying themselves with that knowledge and that recipe all mixed up together. But it is important to realise that science is neither just the algorithm nor the output, but the whole lot including the people. It is also important to realise that we can push this analogy too far, that science is a special mix. Each part can change the other. The knowledge can change the recipe; the recipe can change the people.

Thus science is not a process for finding the truth or anything else. It is not a process at all, but a system, a complex adaptive system. It is a complex adaptive system made up of two interacting complex adaptive systems—the scientists and the knowledge system, the former a social system, the latter a playground of memes—with the interaction strongly mediated by the recipe. Now I need to make it clear that this recipe for growing the knowledge is not what is often called *the scientific method* or what we called earlier *experimental* or *normative* science—the approach used by most scientists since the seventeenth century. It is instead a recipe for discovering the scientific method, and *all* the other methods that achieve the same ends, which are to increase the body of scientific knowledge, to improve the recipe and to change the people.

What sets it apart from other such systems, such as art, religion or politics, is that the idea of *belief* is deliberately and persistently winnowed out of science; despite the persistent human tendency to keep putting it back in. Each of those other human enterprises has at its base a kernel of some fundamental and fundamentally unprovable, and hence illogical belief, whether it be belief in a god, an aesthetic or a manifesto, on which it then constructs the logic of itself. Even mathematics has this character, and is thus not scientific.

The practice of complexity

Seeing science as a complex adaptive system allows us to be sympathetic to Wolfram's idea of a new kind of science. It also allows us to see why the practice of complexity is at a Promethean moment. At the present time, our science broadly accepts the rules of engagement set by the experimentalists: experimental science still controls the way we think about complex systems. This is a deeply cultural problem with a history at least Newtonian and probably Platonic. It has to do with a philosophical belief in elegance, that the universe will be ultimately explicable in simple elegant ways, that behind the messiness of the world there lies the music of the spheres. This has driven mathematics and physics, in particular, into a compact from which we struggle to emerge. I see some hope for change—in computer-aided theorem proving, for example—but I also see a great reluctance to abandon the belief of an underlying simplicity in the way the universe works.

We see the belief at work whenever we see a complex systems model that deliberately keeps itself simple under the guise of tractability, or insists on simplified outputs, or whenever we see modelling that dwells on complex dynamics emerging from simple interactions. But make no mistake, it is a belief, and it holds us back. It reinforces the idea of a normative science to which complex systems science is subordinate. Science needs to renegotiate its concord with mathematics, to the ultimate benefit of both, before we can move on to our new kind of science.

What we would move on to is what I call *deep complexity*. It promises to be a science that accepts the contingent complexity of the world, adapts its ideas about explanation, prediction and control as goals for science, and learns a new concept of understanding. However, I am not talking about some post-modern deconstructed science, where everything is relative, where there is no such thing as absolute truth. I am talking about a new kind of science that completely fulfils science's necessary and sufficient conditions as a complex adaptive system. But, be assured, it will be very different from the logical positivist view. It will be complex not simple, drawing its strength more from the coming ubiquity and transcendent power of computing rather than the analytical power of mathematics. It will be a cooperative, even symbiotic, endeavour with the net as it evolves into a richer complex adaptive system. And the understanding it creates will not always be global—the gold standard of experimental science—but local and embedded, but still connected—still a part of the fabric of science.

This new science will relate to observational and experimental science on its own terms, not those dictated by the logical positivists. It will be data and parameter rich, drawing on both observational and experimental science, but in much more fluid ways than at present. This is the new science waiting to be born. We happy few, we band of brothers, we see our science poised, at a moment of profound change, but held back by the culture of normative science. What will bring on the birth, what will provide the push, I think, is a change in the interactions: firstly, those between science and the world—the demand for a new sort of science for a newly complex fully connected world; and secondly, those within science—the emergence of a set of tools that allows complex systems science to renegotiate its relationship with the rest of science.

Thus the abject failure of normative science to deal with the complex problems of a connected world will provide a push from the world. And the beginning of grid computing and sensor webs will allow for the first time for big complex problems to be modeled in their full complexity, but will also involve a renegotiated relationship with observational and experimental science and encourage the liberation of complexity science. The agenda is huge—to change science, to change society, to understand wholly new classes of problems, and to do that with new non-human partners. And we are few—perhaps a few thousands worldwide. But once, the calculus was understood and used by even fewer. We might be few, we band of brothers, but the battle needs to be joined today, so that tomorrow we can say with Henry:

> And gentlemen in England now a-bed
> Shall think themselves accursed they were not here,
> And hold their manhoods cheap whiles any speaks
> That fought with us upon Saint Crispin's day.

References

Bradbury, R.H. (1999) Just what is science anyway? *Nature and Resources* 35(4): 9–11.

Wolfram, S. (2002) *A New Kind of Science*. Champaign, IL: Wolfram Media.

3. Agents, Icons and Idols

Pascal Perez

Abstract

Since the early 1960s, Artificial Intelligence has cherished the ambition to design an artificial cognitive machine able to reproduce intimate aspects of human behaviour. Distributed Artificial Intelligence and its most recent avatars—Multi-Agent Systems—have developed the concept towards social interactions and societal dynamics, attracting the attention of sociologists and ethnographers who found new ways to elaborate or validate their theories. But populations of cognitive agents aren't the real thing, despite the efforts of their designers. Furthermore, one must cautiously examine the rationale behind these often incredibly complex arrangements of algorithms, in order to assess the usefulness of such exercises. As a matter of fact, Artificial Intelligence relies on a very positivist, and sometimes reductionist, view of human behaviour. For centuries, from Bacon to Pierce, philosophy of mind has provided meaningful insights that challenge some of these views. More recently, post-normal approaches have even taken a more dramatic stand—some sort of paradigm shift—where direct knowledge elicitation and processing override the traditional hardwiring of formal logic-based algorithm within computer agents. *Keywords*: Agent-Based Modelling, Artificial Intelligence, Icon, Idol, Philosophy of Mind, Cognition.

Introduction

Scientists developing Multi-Agent Systems, as part of Distributed Artificial Intelligence (DAI), tend to focus on individual components interacting within a given system (Gilbert and Troitzsch 1999). This is a purely bottom-up approach where representations of the individual components, the agents, display a large autonomy of action. Hence, system-level behaviours and patterns emerge from a multitude of local interactions. Intentionality is deliberately placed at the level of the agents to the detriment of the system itself, greatly limiting its ability to control its own evolution. In the case of human ecosystems, agents can represent individual actors or relevant social groups and communities (Bousquet and LePage 2004). The following definition of a Multi-Agent System (MAS) is generally admitted. A MAS is a conceptual model of an observed system that includes:

- an environment (E), often possessing explicit metrics;
- a set of passive objects (O), eventually created, destroyed or modified by the agents;

- a set of active agents (A). Agents are autonomous and active objects of the system;
- a set of relationships (R), linking objects and/or agents together; and
- a set of operators (Op), allowing agents to perceive, create, use, or modify objects.

An agent is a physical or virtual entity that demonstrates the following abilities: autonomy, communication, limited perception, bounded rationality, and decision-making process based on satisfying goals and incoming information (Ferber 1999). A Multi-Agent Based Simulation (MABS) is the result of the implementation of an operational model (computer-based), designed from a MAS-based conceptual representation of an observed system. The strength of MAS approaches consists in their ability to represent socially and spatially distributed problems. Meaningful examples of application are to be found in ecology (Janssen 2002), sociology (Conte and Castelfranchi 1995), or economics (Tesfatsion 2001).

Cederman (2005) asserts that generative process theorists in social science, shifting from traditional nomothetic to generative explanations of social forms and from variable-based to configurative ontologies, may find in Multi-Agent Systems relevant tools to explore the emergence of social forms in the Simmelian tradition, thanks to common foundations in both epistemology and ontology.

In the following sections of this chapter, we try to evaluate Cederman's assertion against evidence. First, we describe general features of cognitive agents as stated and used in Artificial Intelligence (AI) research. Then, we argue that our understanding of mental processes, from Bacon's idols to Tversky's prospect theory, is inherently limited. In the third section, we question the supposed objective autonomy of agents, drawing from Peirce's icons and Varela's enactive cognitive theory. Finally, we propose a way forward that encapsulates the designer and the modelling process into the observed system itself.

Cognitive agents

Kenetics (Ferber 1999), as a theory, aims to establish principles for conception, design, and implementation of computational Multi-Agent Systems. These systems of interacting agents are described in terms of components (agents), structure (network of agents), and organisation (ways and reasons for agents to interact).

As intentionality is embedded into the agents, they need mental-like processes for decision and action. Drawing from traditional psychology, AI tends to describe and explain human behaviour through mental states representing beliefs, desires, and intentions (Brazier et al. 2002). The Belief-Desire-Intention (BDI) paradigm, largely used in AI, states that individual decisions arise from the recursive exchange of information between these three mental states (Figure 3.1).

Figure 3.1. Belief-Desire-Intention (BDI) structure of a conative system

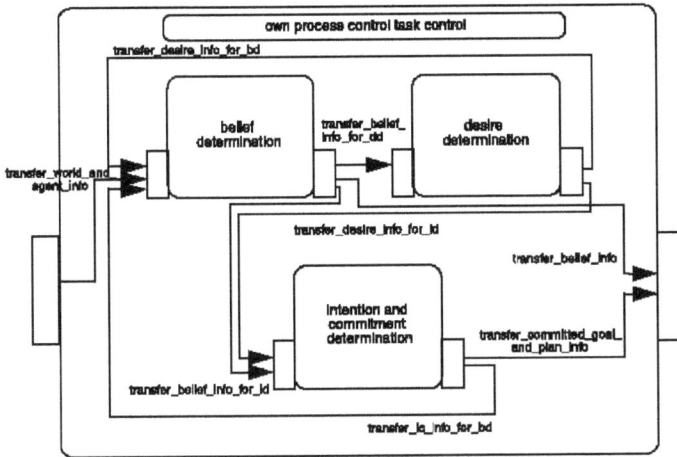

Source: Brazier et al. 2002

Jacques Ferber (1999: 242), in his design framework for MAS, proposes a more comprehensive classification of these mental states, he calls cognitons, for which Table 3.1 gives a partial list organised into categories. These different categories represent different sub-systems interacting during cognitive processes.

Table 3.1. Partial list of Cognitons proposed by Ferber

Category	Cogniton	Description
Interaction	Percept	Cogniton transmitted by external sensors
	Information	Cogniton transmitted by another agent
	Decision	Cogniton selecting action
	Request	Cogniton transmitted to another agent
	Norm	Cogniton imposed by the social organisation
Representation	Belief	Cogniton representing states of the world and self
	Assumption	Possible representation not yet believed
Conative	Tendency	Cogniton resulting from impulse or demand
	Impulse	Internal need coming from the conservative system
	Demand	External need resulting from request or percept
	Intention	Internal duty for decision
	Command	Cogniton selecting decision
	Engagement	External constraint on decision
Organisation	Method	Set of rules and techniques to implement action
	Task	Set of stages needed to implement action or method

Source: Ferber 1999

Interaction system

The interaction system enables the agent to perceive and acquire information from the surrounding environment. This individual perception contributes to the elaboration of a subjective, limited, and contextual representation of the world.

From a philosophical perspective, there are two conflicting theories on perception. On one hand, the Aristotelian view assumes that perceived objects actively 'impregnate' our senses. Thus, we passively receive this imprint and integrate it to our cognitive system. This causal conception of perception asserts that we can access the objective qualities of surrounding objects. This model is widely accepted in cognitive science (logic theory) and computer science (shape recognition). On the other hand, the Kantian view asserts that percepts are constructed by the observer and depend upon previous experiences of perception. This active conception of perception constitutes an axiomatic principle in semiotics and it is consistent with major experimental results in neurobiology. Unfortunately, its application to computer science raises several technical problems that have, so far, limited its use in AI despite valuable experiments such as the 'Talking Heads' (Kaplan 2001). Figure 3.2 presents two computational systems of perception for artificial agents based on active or passive perception.

Figure 3.2. Passive and active perception systems. (Adapted from Ferber 1999)

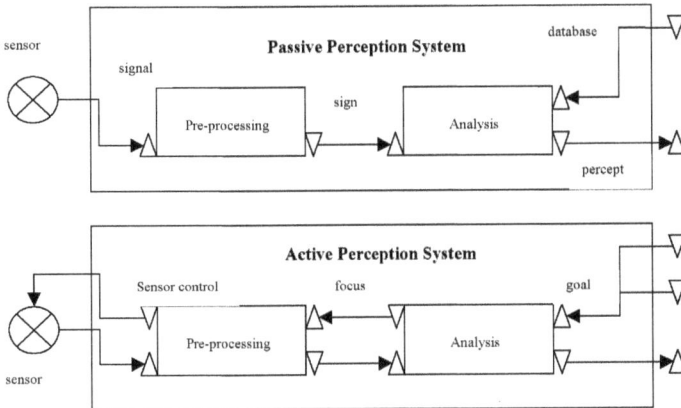

System of representation

The system of representation enables the cognitive agent to store and manipulate acquired knowledge and beliefs. AI tends to group knowledge, know-how, experience, facts, and memories into a single set of information called 'beliefs'. These beliefs help the agent to decide and to implement actions. As a matter of fact, much theoretical work has been concerned with ways of representing, classifying and manipulating these beliefs for action (pragmatic dimension) rather than focusing on the very essence of 'knowledge' (epistemological dimension).

The Physical Symbol System Hypothesis, enunciated by Simon and Newell, is the founding principle of Symbolic Artificial Intelligence (SAI). Borrowing concepts from philosophy of mind (Kantian schemata) and semiotics (Peircean symbols), the principle states that any belief can be represented through a set of symbols and rules of inference (ontology). Four axiomatic propositions are generally accepted:

- Representations are independent from any underlying physical structure.
- Mental states are intentional: they are linked to a referent external to the agent.
- Representations are made of symbols or groups of symbols.
- Reasoning consists in manipulating symbols with rules of logic inference.

Evidence coming from neuro-biology has supported criticism of the first proposition by Connection Artificial Intelligence (CAI). The use of neural networks for task-oriented reasoning has indeed provided powerful alternate solutions. But, a more general criticism towards SAI relates to its implicit assumption of perfectly autonomous agents. As a matter of fact, social agents are embedded into an environment that not only supports and feeds individual reasoning but, more essentially, 'permeates' individual experiences through permanent interactions. Social psychology asserts that intelligence is culturally grounded and that knowledge evolves only through interaction with others by means of proposition, confrontation, and refutation (Cole and Scribner 1974). Interestingly, this same criticism gives some credit to the unique concept of belief used by SAI to describe different types of knowledge: from a social psychology viewpoint, any type of knowledge results from a historically contingent and socially built consensus. Hence, scientific knowledge and theories are themselves meta-beliefs, consensual models of a given 'reality'.

From an SAI perspective, agents continually use beliefs and assumptions (cognitons) from their system of representation to build descriptive or predictive models of their environment. Ferber (1999) proposes the following list of belief categories:

- Environmental belief: current or predicted state of the physical environment.
- Social belief: social norms and rules applicable within a given social group.
- Relational belief: competences and intentions attributed to other known agents.
- Personal belief: representation of self.

Conative system

The conative system defines the set of activities to be undertaken by an agent, based on available information and beliefs. The ways agents take their decisions, and the reason why they discard some options to focus on others, are questions that stretch well beyond Artificial Intelligence and nurture endless debates in philosophy and psychology.

As SAI relies upon logic inferences to describe an agent's behaviour, such as predicate logic or modal logic, causality links are meant to be rational. Hence, agents tend to display goal-satisfying decisions and, therefore, their actions are first driven by their needs and tendencies. According to Ferber (1999), these tendencies are themselves motivated by percepts, impulses, norms, or engagements, and trigger a decisional process based on existing beliefs and assumptions (Figure 3.3). Like beliefs, motivations can be separated into four categories:

- Environmental motivation: reflex or reinforcement due to percepts.
- Social motivation: engagement due to social norms or deontic rules.
- Relational motivation: engagement or hedonism linked to other interacting agents.
- Personal motivation: self-engagement or hedonism due to impulses.

Figure 3.3. Conative system and its two sub-systems. (Adapted from: Ferber, 1999)

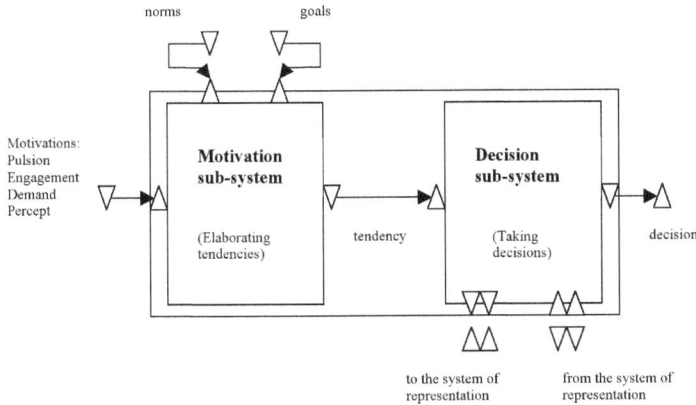

The way SAI agents take intentional decisions and eventually undertake subsequent actions is largely based on causal philosophy of action (Bratman 1987). Intention is altogether a choice and an engagement towards this choice. Hence, in AI, an agent X is said to have the *intention* to perform an *action* A if X wants a *proposal* P about the state of the world to be true, and:

- X believes that P is a consequence of A,
- X believes that P is not currently realised,
- X believes he is able to perform A,
- X believes that A is possible and, consequently, P will be satisfied.

Cohen and Levesque (1990) have proposed a formalism for rational action, based on modal logic that has been largely used in Distributed Artificial Intelligence (DAI). Their formalism gives way to necessary, possible, or contingent predicates. Likewise, it allows expressing the temporality of intentions as *planning to do something in the future,* and needs a different set of tasks compared with *deciding to do something now.* Applying such formalism to Multi-Agent Systems implies that each agent is able not only to predict the consequences of its intended action, but also to anticipate the results of the other agent's behaviour. Hence, the agent needs to carry in his social or relational beliefs some ideas about the other agent's commitments. This is where the concept of engagement becomes paramount: self and social engagement are needed to introduce some sort of regularities in the system that can be hopefully anticipated.

Organisation system

Finally, the organisation system, through its methods and tasks (cognitons), allows the agent to prioritise, halt, and resume the pending decisions provided by the conative system. External information channeled through the interaction system may alter the implementation of a decision into action. A suspended decision will eventually resume according to the persistence of its triggering intention(s) (Ferber 1999).

Exploring human ecosystems

It is one thing to understand how SAI is used to design rational and intentional cognitive agents. But we also have to question the reasons why, in the first place, we intend to create these artificial entities? I will leave aside DAI applications belonging to robotics or computer-oriented technologies where autonomous agents tend to 'mimic' intentional cognition in order to perform actions considered as rational by their designers. After all, these agents are not supposed to be, or even to represent, human beings.

Instead, I will concentrate on these Multi-Agent-Based Simulations (MABS) that are meant to represent actual human ecosystems. Only a small proportion of those applications are used as social virtual experiments to explore cognitive processes. Relying on robust cognitive architectures inherited from SAI, these models are designed to help theoretical breakthroughs:

> I believe that the contribution of [Multi-Agent-Based Social Simulation] to the theoretical development of the cognitive and social sciences could be really remarkable. SS can provide not only an experimental method, but good operational models of cognitive 'actors', of the individual social mind, of group activity, etc. Models that can be richer, more various, and more adequate than those provided by economics, without being less formal. In particular, my focus on the core relation between functions and cognition was aimed at pointing out how the coming 'agent-based' approaches to social theory, using learning but deliberative agents, could deal with very old and hard problems of the social sciences and could re-orient them.
> (Castelfranchi 2001: 35)

As a matter of fact, a large proportion of MABS applications are designed to explore and understand complex interactions between actual actors and their environment. Bousquet and LePage (2004) or Hare and Deadman (2004) provide comprehensive reviews of these models. Most of these applications depart from the SAI paradigm and implement over-simplistic, task-oriented, rule-based agents, focusing on spatial interactions, social communication and individual mobility. Often, the drift from internally consistent and Formal Logic Compliant (FLC) agents is justified by synthetic information coming from field surveys or expert knowledge.

I would argue that in both cases, formal logic compliance or not, MABS will fail to deliver if we cannot find innovative ways to link the model, its object, and its interpreter. For deceptive idols and socially constructed icons are conspiring against agents.

Deceptive idols

Cognitive agents are supposed to behave rationally, as they represent rational human beings. FLC agents abide by a positivist and scientific rationality, and need a consistent set of decisions to act. Non-compliant agents (FLN) are usually designed according to the phenomenological interpretation of behaviour given by experts (i.e. sociologists, anthropologists). In both cases, the question is not about the acceptance of a rational behaviour. The question is about the axiomatic predicates used by the agents, or interpreted by the experts.

Bacon's idols

In his *Novum Organum*, Francis Bacon classified the intellectual fallacies of his time under four headings which he called idols (Nova Organum 1620: 345:39):

> There are four classes of Idols which beset men's minds. To these for
> distinction's sake I have assigned names, calling the first class Idols of
> the Tribe; the second, Idols of the Den; the third, Idols of the Market
> Place; the fourth, Idols of the Theater.

An idol is an image, in this case held in the mind, which receives veneration but is without substance in itself. Bacon did not regard idols as symbols, but rather as fixations. In this respect he anticipated modern psychology.

'Idols of the Tribe' are deceptive beliefs inherent in the mind of man, and therefore belonging to the whole of the human race. They are abstractions in error arising from common tendencies to exaggeration, distortion, and dispro-portion. First of all, our tendency to let emotions rule reason can give us false impressions of the truth based on our feelings at the time. Another common idol lies in our tendency to seek out evidence of that which we already believe to be true. Bacon suggests that we become affectionate to ideas we have found and carried with us for some time; we become attached to them and collect evidence that supports them while throwing out that which contradicts them. Interest-ingly, Castelfranchi (2001) proposes a similar mechanism to explain the emergence of social functions among agents through 'learning without understanding' processes, superseding the intentional cognitive processes. But, so far, emotional agents remain out of reach of the current developments in SAI.

'Idols of the Den' (also called 'Idols of the Cave' in some editions) are those which arise within the mind of the individual. Like in Plato's allegory, thoughts of the individual roam about in this dark cave and are variously modified by temperament, education, habit, environment, and accident. Thus an individual who dedicates his mind to some particular branch of learning becomes possessed by his own peculiar interest, and interprets all other learning according to the colors of his own devotion. In our case, this idol may affect directly the designer rather than the artificial agent. Lissack and Richardson (2001: 101) give an illustration of this bias in their criticism of Wolfram's claim that most complex systems can be accurately represented by rule-based atomistic models:

> The act of interpreting differs from the act of observing, and both may differ significantly from the underlying phenomenon being observed. In their failure to respect this distinction, strong MSM proponents [Wolfram and colleagues] are implicitly suggesting that the interpretation is reality. However, while a good model of complex systems can be extremely useful, it does not allow us to escape the moment of interpretation and decision.

Obviously, deterministic and limited expert knowledge used to design FLN agents suffers the same type of criticism. Causal rules of behaviour inferred by an expert are merely subjective interpretations of a given reality. Putting it simply, there are as many realities as there are experts. Agar (2005) recently proposed a way forward by means of coupled emic/etic approaches to social simulations. The author advocates a constant feedback between what makes sense for the actual actors depicted in the model (emic) and what seems meaningful to the designer (etic).

'Idols of the Marketplace' are errors arising from the false significance bestowed upon words, and in this classification Bacon anticipated the modern science of semantics. The constant impact of words variously used without attention to their true meaning often betrays their purpose, obscuring the very thoughts they are designed to express. Acknowledging the volatility of the 'true meaning' of words, Bacon just caught a glimpse of the active perception, theorised by Kant and Peirce later on. Words, as elementary percepts, do not carry any specific and intrinsic meaning when they are perceived. They have to be re-interpreted internally by the receiver, according to previous knowledge and environmental hints. Maturana and Varela (1980: 32) propose to drop the denotative understanding of language altogether in favour of a connotative approach:

> So long as language is considered to be denotative it will be necessary to look at it as a means for the transmission of information, as if something were transmitted from organism to organism, in a manner such that the domain of uncertainties of the 'receiver' should be reduced according to the specifications of the 'sender'. However, when it is recognized that language is connotative and not denotative, and that its function is to orient the orientee within his cognitive domain without regard for the

cognitive domain of the orienter, it becomes apparent that there is no transmission of information through language. It behooves the orientee, as a result of an independent internal operation upon his own state, to choose where to orient his cognitive domain; the choice is caused by the 'message', but the orientation thus produced is independent of what the 'message' represents for the orienter. In a strict sense then, there is no transfer of thought from the speaker to his interlocutor; the listener creates information by reducing his uncertainty through his interactions in his cognitive domain.

As stated earlier, mainstream SAI satisfies itself with passive perception processes. Hence, most FLC agents can receive intelligible and meaningful information from other agents, without having to engage into deciphering and re-interpretation stages. Somehow, it makes agents' lives seem much easier than ours!

'Idols of the Theatre' occur due to sophistry and false learning. These idols are built up in the fields of theology, philosophy, and science, and, because they are defended by learned groups, are accepted without question by the masses. When theories have been cultivated and have reached a sufficient level of consensus they are no longer questioned. The long standing hegemony of the symbolico-cognitivist paradigm cannot hide the fact that relevant alternatives have challenged SAI's dominion: connectionist and evolutionary theories perform better on learning processes (Kaplan 2001); the autopoietic theory, by refusing the conveyance of information through linguistic interaction, provides a unified and unchallenged approach to signaling interactions (verbal, non-verbal, or extra-verbal) through structural coupling between individuals (Maturana and Varela 1980).

Prospect theory

Prospect theory focuses on cognitive and psychological factors that determine the value of risky prospects (Kahnemann and Tversky 2000). Its initial assumption is that subjective values attached to gambling are carried by expected changes of wealth (gains or losses) rather than ultimate states of wealth. More importantly, prospect theory replaces the traditional concept of risk aversion by a more intuitive one, called loss aversion, by which people tend to consider that a loss of $X is more averse than a gain of $X is attractive. This assumption explains why people might be risk seeking, and no longer risk averse, in the domain of losses.

Though trivial from an espistemological viewpoint, Prospect theory tends to reconcile theoretical development with empirical facts. According to Kahnmann and Tversky (2000:1):

> The study of decision addresses both normative and descriptive
> questions. The normative analysis is concerned with the nature of

rationality and the logic of the decision making. The descriptive analysis, in contrast, is concerned with people's beliefs and preferences as they are, not as they should be.

Although based on formal logic predicates, Prospect theory recognises the fact that part of the knowledge necessary to complete the theory is inherently elusive. It is possible to design FLC agents founding their decisions on this theory, but the axiomatic predicates that will tell us about discrepancies between gains and losses, or the way different items will be affected by these discrepancies remain, by far, out of reach.

Furthermore, prospect theory threatens directly two logical pillars of decision theory traditionally used by SAI: preference invariance and value coincidence. Invariance requires that the preference order between prospects should not depend on the manner in which they are described. Kahneman and Tversky (2000:5) have demonstrated that invariance cannot generally be satisfied: 'invariance is normatively essential, intuitively compelling, and psychologically unfeasible' due to the framing of outcomes through formulation effects. The framing effect also affects the relation between experience and decision values. But rational agents seldom make a difference between experience values (direct outcomes of actual actions) and decision values (expected outcomes of an anticipated choice). These two values tacitly coincide, despite Kahnemann's and Tversky's (2000: 16) warning:

> Some factors that affect experience are not easily anticipated, and some factors that affect decisions do not have a comparable impact on the experience of outcomes.

Cognitive dissonance

For Bacon, knowledge is intimately mixed with the idols, hence prefiguring our modern concept of belief. More importantly, Bacon draws visionary consequences from the presence of the idols, in terms of communication (Nova Organum 1620: 346:35):

> enter quietly into the minds that are fit and capable of receiving it; for confutations cannot be employed, when the difference is upon first principles and very notions and even upon forms of demonstration.

Individuals and groups exhibit varied responses when faced with new information. If such information is consistent with extant behaviours and beliefs, it can be readily accepted and integrated. However, if the new information conflicts with behaviour and belief, the resulting state is described as 'cognitive dissonance' (Bradshaw and Borchers 2000). According to the theory, the inconsistency and psychological discomfort of cognitive dissonance can be reduced by changing one's beliefs, values, or behaviour. Dissonance can also be avoided by rejecting or avoiding information that challenges belief systems or by interpreting dissonant information in a biased way. In this regard, most SAI structures force agents to discard new information conflicting with a given set of consistent predicates. It is only through reinforcement (punishment or reward) due to experience that contrasted set of predicates can be established. Elaborating on the conflicting views upon 'uncertainty' between scientists and policy-makers to explain the science-policy gap, Bradshaw and Borchers (2000: 3) outline the complexity of cognitive dissonance:

> Dissonance between existing beliefs and new information may be shaped by a host of factors, all of which inhibit the rate at which scientific findings are assimilated into policy. In what we have called the 'volition' phase of the science-policy gap, public debate around an emerging scientific consensus may derive from a combination of cultural, psychological, and economic interests threatened by the policy inferences of dissonant scientific findings.

The authors particularly point at the contrasted rhetorical figures used by scientists when they are in charge of policy-making compared with their usual handling of scientific uncertainties. Designers in DAI have tried to encapsulate these internal cognitive conflicts by implementing Agent-Group-Role structures in which one agent belongs simultaneously to several socio-cultural groups and plays different roles accordingly (Ferber, 1999). But the internal conflict resolution mechanisms—for example, between friendship engagement and professional commitment—rely on formal and individualistic logic once more.

Overall, despite the increasing interest in, and use of Multi Agent Systems to represent human ecosystems, the SAI paradigm remains mostly unchallenged as a theoretical framework used to develop cognitive agents. But looking at the real world through the lenses of the social sciences, we have to acknowledge the fact that a meta-theory of human behaviour doesn't exist yet. Rational decision theory, prospect theory, social learning theory, and others give us partial clues about human behaviour, and none can stand as an overarching and unified framework.

Hence, we must handle cautiously Cederman's assertion about the 'New Deal' offered by Multi Agent Systems to social scientists. These tools and their current states of application generally rely upon reductionist views of the world: symbolico-cognitive hypothesis, formal predicate logic, rational decision, and autonomy. As a matter of fact, the SAI paradigm stands out as a nomothetic meta-model for agent's behaviour while its foundation doesn't represent a meta-model of human behaviour but merely a partial interpretation of it.

Socially constructed icons

Beside the symbolico-cognitive paradigm, Multi Agent Systems explicitly emphasise the autonomy of agents through physical integrity, autonomous decision, and rational action. But social psychology (Cole and Scribner 1974), psychology of development (Papert 1980) and collective action theory (Oliver 1993) insist, for different reasons, on the fact that human beings are social beings before anything else: learning processes, cognitive inferences, or individual actions are shaped and dictated by our environment. Recent developments in DAI have tried to incorporate this 'sociality' dimension into agent's behaviour. We'll come back in the next section to Hogg's and Jennings' (2001) proposal to include social rationality into agents' expected utilities, or Jager's and Janssen's (2003) attempt to consider basic human needs and uncertainty as the driving factors behind decision making processes of their agents. But first, we need to understand the elements of social cognition that weaken the foundations of the SAI paradigm.

Peirce's icons

Charles Sanders Peirce, a founder of modern semiotics, has asserted that:

- we cannot think without signs,
- we have no ability for intuition, all knowledge flows from the former knowledge,
- we have no ability for introspection; all knowledge about the inner world is produced by hypothetical reasoning on the basis of observation of outer things. (Peirce Edition Project 1998)

On this solid foundation he builds up his entire theory of signs. Being a pure positivist himself, at least during his early career, Peirce is convinced to build a theory based on formal logic, independent from particular minds, and characterised by a triadic relation (Peirce Edition Project 1998: 411):

> [semiosis is an] action, or influence, which is, or involves, a cooperation
> of three subjects, such as a sign, its object, and its interpretant, this
> tri-relative influence not being in any way resolvable into actions
> between pairs.

Peirce suggests that people have not and cannot have direct access to reality. Signs are nothing else than the universal medium between human minds and the world. Semiotics defines different categories of signs according to their level of abstraction, from icons to symbols (Peirce Edition Project 1998). As a primary and elementary sign, an icon looks like it is signified. There is no real connection between an object and an icon of it other than the likeness, so the mind is required to see the similarity and associate the two. A characteristic of the icon is that by observing it, we can derive information about its significance. Hence, semiotics embrace the Kantian concept of active perception as stated previously.

Since signs are not private but socially shared, it is society that establishes their meaning. Therefore, the transcendental principle in semiotics is not (divine) intuition, but community, and the criterion of truth, social consensus. According to Peirce, any truth is provisional, and the truth of any proposition cannot be certain but only probable. Umberto Eco, in his exhilarating *Kant and the Platypus*, provides a vivid description of this fallibilism principle (Eco 1997: 98):

> Naturally at this point transcendentalism will also undergo its Copernican revolution. The guaranty that our hypotheses are 'right' will no longer be sought for in the a priori of the pure intellect but in the historic, progressive, and temporal consensus of the Community. Faced with the risk of fallibilism, the transcendental is also historicized; it becomes an accumulation of interpretations that are accepted after a process of discussion, selection, and repudiation. This foundation is unstable, based on the pseudo-transcendental of the Community.

In order to sort out conceptual confusions born from consensual uncertainties, Pierce uses a 'pragmatic' approach by linking the meaning of concepts to their operational or practical consequences. This pragmatism will act later on as a corner stone for SAI's intentionality of beliefs. But, when conventional SAI says: 'beliefs *are* intentional', semiotics tell us: 'we *have to consider* beliefs as intentional'. Somehow, SAI has developed an hyper-positivist approach of cognition, through passive perception, intentionality, and formal modal logic that has gone beyond the scope of its theoretical background. In this regard, FLN agents are more likely to 'mimic' social behaviour compared with their FLC counterparts as the rationale for their actions is inferred from an holistic, though subjective, view of the social system constructed by their designer. In other words, some behavioural rules have to be considered for the FLN agents to interact in a consistent way (system level), without prerequisite conditions on cognitive consistency (agent level). But we'll see in the next section that it comes at a cost.

Enactive cognitive theory

We human beings are living systems that exhibit cognition, and there is no way for us to address, much less explain, our cognitive abilities without employing those same cognitive abilities. To date, the primary response to this paradox has been to ignore it and proceed with respect to a presumably fixed fundament, external to our act(s) of cognition. Where the presumptive fundament is an 'objective reality', the mediation between situation and action is explained in terms of ordered inference with respect to a model of that reality.

Maturana and Varela (1980) questioned this conventional approach when it is confronted with the tacit, extralinguistic, or emotive character of human behaviour. They further disputed the 'objective reality' concept by considering:

- Ourselves operating in multiple 'worlds', particularly socio-cultural ones.
- A 'world' being molded by contextual factors intertwined with the very act of engaging it.

Their autopoietic theory considers living beings as living systems embedded into larger systems constituted by themselves and the environment they interact with. The theory addresses the basic configural and operational circularities of these living systems (Maturana and Varela 1980). Unlike other more positivist approaches to complex systems, the autopoietic theory focuses on the observer himself. It accomplishes this by shifting explanatory focus from atomic units in an objective world to essential relations among processes operating in circular ways to constitute the organism as a living system and the observer as a cognitive organism. As clearly stated by Maturana (Maturana and Varela 1980: 7):

> Cognition is a biological phenomenon and can only be understood as such; any epistemological insight into the domain of knowledge requires this understanding.

Later on, Varela and colleagues brought phenomenological concerns into the world of cognitive science (Varela et al. 1991). Their goal was to incorporate everyday experience into the scope of studies which had heretofore addressed cognition in terms of disembodied rational processes, circumscribed by abstract beliefs purported to mirror an objective milieu. All concepts fully accepted by symbolico-cognitivist approaches. Varela and colleagues proceed from the assumption that experience necessarily predates and underpins enquiry. Maintaining a focus on experience as action allows inspection and reflection on the manner in which 'mind' and 'body' reciprocally engage to consummate experience. Midway between cognitivist and connectionist paradigms, their Enactive Cognitive theory considers that (Varela et al. 1991: 148):

> context-dependent know-how [shouldn't be treated] as a residual artifact that can be progressively eliminated by the discovery of more sophisticated rules but as, in fact, the very essence of creative cognition...Knowledge depends on being in a world that is inseparable from our bodies, our language, and our social history—in short, from our embodiment.

As such, enactive cognitive theory does not address cognition in the currently conventional sense as an internal manipulation of extrinsic information or signals, as semiotics or symbolico-cognitivism would have us believe (we have seen above how Maturana and Varela treat linguistic interactions). Instead, it grounds cognitive activity in the embodiment of the actor and the specific context of activity. The theory fits very well with current trends toward emphasising contextualised studies of humans, their interactions, and their social systems. Unlike symbolico-cognitivist approaches, the enactive cognitive theory is consistently 'relativistic' in the sense that any given observation is observer-bounded, history-contingent, and socially embedded.

The theory implies an epistemology analogous to that of constructivism (Funtowicz and Ravetz 1993). For this reason, Multi Agent Systems built from FLN agents and based on expert knowledge never describe *the reality as it is* but rather *a probable reality* interpreted by a given observer in a given context. It means that such models cannot be evaluated through traditional scientific methods but need to be assessed, in context, against a set of criteria intelligible to the observed subjects and to the observer.

Social constructions

As mentioned above, very influential studies in social sciences have focused on the embodiment of cognitive activities into the actor's experiences. In the psychology of development, Piaget asserted that children only develop through progressive interactions with their environment (Papert 1980). In the early stages of development, Piaget's theory suggests that learning comes *only* from unexpected changes in the environment of the subject. Changes create a cognitive imbalance that must be adjusted through assimilation, the new experience is integrated into the current view of the world, or accommodation, the current view of the world must be modified to fit in the new experience. From this constructivist viewpoint, imbalance is no longer considered as a negative factor but rather as a driving force towards cognitive development. Hence, a perceived environmental instability seems to trigger some sort of phase transition, in a complexity theory sense, from one state of representation to another. It is interesting here to draw a link with the enactive cognitive theory, for which concrete experiences (know-how) are driving cognitive processes through connotative interactions (Maturana and Varela 1980).

Indeed, a new picture of cognitive processes takes shape; a mind submitted to perpetual changes and characterised by a dynamic equilibrium between past and current views of the world; a mind dominated by inductive inferences and desperately trying to make sense of incoming information through deductive and formal logic (Batten 2000). This description sharply contrasts with the well structured and perfectly organised mind proposed by SAI designers. As a matter of fact, FLC agents process information in a consistent way; their actions are driven by intentional cognition and the experienced outcomes are evaluated against expected utilities (Brazier et al. 2002).

Marvin Minsky, in his famous *Society of Mind* (1985), proposed an alternate solution to the symbolico-cognitivist theory. Drawing from an initial intuition that our mind is made of structural and independent units called *frames*, Minsky views the human mind itself as a vast society of individually simple processing *agents*. The agents are the fundamental thinking entities from which minds are built, and together produce the many abilities we attribute to minds. The advantage in viewing a mind as a society of agents, as opposed to the consequence of some formal logic-based system, is that different mental agents can be based on different types of processes with different purposes, ways of representing knowledge, and methods for producing results. The consistency of this atomistic mental system rises from the organisation of interactions between agents, not from a unified and coherent reasoning (Minsky 1985: 308):

> What magical trick makes us intelligent? The trick is that there is no trick. The power of intelligence stems from our vast diversity, not from any single, perfect principle.

In principle, these networks of mental agents display dynamic interactions where agents can be modified, created, or deleted in order to adapt or to respond to a given experience. This interactionist theory can accommodate—though not explicitly formulated—constructivist (Papert 1980) and connotative (Maturana and Varela 1980) standpoints. Unfortunately, to date, few Multi Agent Systems have been implemented according to Minsky's views, and a vast majority of these models concern robotics or computer-oriented technologies. Both fields have been excluded from the scope of this paper.

Another advantage of the interactionist theory is that it creates a continuum between individual and social actions. The same paradigm is used from mental processes to collective action, from personal engagement to societal norm enforcement. As a matter of fact, sociologists studying social movements often refer to Collective Action Frames to represent the process by which activists create and modify action-oriented sets of beliefs and meanings that inspire and legitimate the activities of a social movement (Benford and Snow 2000). A crucial aspect of collective action frames is that they are not merely aggregations of individual attitudes and perceptions but also the outcome of negotiating shared meaning. Collective framing necessitates complex interactions between activists, antagonists, bystanders, and observers (op. cit.: 614):

> Frames help to render events or occurrences meaningful and thereby function to organize experience and guide action. Collective action frames also perform this interpretive function by simplifying and condensing aspects of the 'world out there', but in ways that are intended to mobilize potential adherents and constituents, to garner bystander support, and to demobilize antagonists.

Cederman's quest for generative explanations of social forms based on configurative ontologies cannot be entirely satisfied by autonomous and rational agents, as formal modal logic constitutes a very specific and limited type of ontology. As we have seen, the interactionist paradigm offers a much richer context for cognition and *sociation*. Unfortunately, technical caveats have prevented so far to implement purely interactionist computer models beyond experimental and limited applications. Finally, Cederman's call to the Simmelian tradition questions the nature of these social patterns we are supposed to observe: if we accept Maturana's and Varela's views, these patterns are strongly observer-dependent and they need a constructivist epistemology for authentication.

Paradigm shift

From the previous sections, we can infer the following:

- The symbolico-cognitivist paradigm relies upon a limited, fragmented, and sometimes conflicting understanding of cognition and behaviour.
- Autonomy of cognition is inherently limited as beliefs are socially constructed and experience preempts cognition.
- A model of social behaviour – as a tentative representation of an objective reality – cannot be separated from its designer's experience and viewpoint.
- Multi Agent Systems are specific models that display autonomous and cognitive entities, hence, they are concerned with the three previous limitations altogether.

Systems built with FLC agents suffer mainly from the first and second limitations. They tend to limit the subjectivity of the representation by adhering to a consensual theoretical framework of reference. But the SAI paradigm itself could be considered, in a Baconian sense, as an idol of the theatre altogether. Systems built with FLN agents suffer mainly from the second and third limitations. They accept our limited understanding of cognition by way of pragmatism. But they are intimately related to their designer's experience.

In all cases, the power of explanation of a given model can hardly be assessed through traditional scientific positivism as soon as human cognitive processes are simulated. If it were the case, a set of explanatory hypotheses could be used to unambiguously validate the simulated processes against objective criteria. So far, in the absence of any meta-theory of human behaviour, only experimental economics and experimental psychology are used to inductively validate these models, with all the unrealistic constraints imposed to these experiments. Nevertheless, our next section provides meaningful examples of attempts to refine the symbolico-cognitivist approach in order to match the reality of socially embedded behaviours.

As mentioned above, Agar (2005), dealing with very elusive behaviours of illicit drug users, advocates a dual approach of modelling where actor-based components (emic) are used along with theory-based components (etic) in order to implement realistic behavioural and social models. Of course, this proposal assumes that ethnography or sociology can provide *in situ* replicable and explicit methods of knowledge elicitation. Dray et al. (2006) have recently tried to formalise such a process, using knowledge engineering techniques. If these techniques allow, in principle, a better traceability of elicited knowledge we have to consider Cole's and Scribner's (1974) findings from their cross-cultural studies: any given ethnographic experimental setting reflects its designer's own beliefs and intentions. Therefore, the authors advocate a constant cross-validation between the observer and the observed subjects in order to limit misinterpretation. Maturana and Varela (1980) would label these iterative and circular interactions between a designer, an observer, and observed subjects, a structural coupling process. It is the fundament of the post-normal approach to modelling proposed by Funtowicz and Ravetz (1994) and described below.

Normal views on social rationality

Kluver and colleagues (2003) point out the main difference in building theories between social sciences and natural sciences. The latter always proceed through increment, starting with rather simple models and enlarging them by successive steps, whenever the progress of research made it necessary. This *normal science* is guided by a certain paradigm that is transformed into an increasingly complex model as long as the paradigm makes it possible. By contrast, social sciences often adopt an all-at-once approach by which theorists try to capture from the beginning as much of social complexity as they can within their conceptual framework. Kluver and colleagues have no doubt about the fact that social sciences must embrace a *normal* posture in order to progress towards a better understanding of social complexity (2003: 3):

> That is not possible in computational or mathematical sociology, respectively: the basic models must be simple in order to understand their behaviour in principle. The enlargement of the basic models that is always necessary in advancing research can only be done if this basic understanding has been achieved. Therefore the social sciences have to adopt this methodical procedure from the natural sciences if a formally precise theory of social complexity is to be achieved.

The SAI paradigm has supported, through normal and positivist procedures, the implementation of always more complex models of individual decision and social behaviour. If emotional decisions or the 'embodiment' of cognition are still on the shelves, the social rationality of agents, as opposed to their goal-satisfying rationality, has recently attracted much attention. Hogg and Jennings

(2001: 382), recognising that individual and social concerns often conflict leading to the possibility of inefficient performances, proposed a framework for making socially acceptable decisions, based on social welfare functions, that combine social and individual perspectives:

> To be socially rational, an individual maximizes his social welfare
> function over the different alternatives. This function represents how
> an individual agent may judge states of the world from a moral or social
> perspective by taking into consideration the benefit to others of its course
> of action and weighing it against its own benefits.

The dynamic balance between individual and social utility functions is controlled by a metalevel controller that considers available resources (adaptation) and past experiences (learning) to tune the amount of cognitive efforts put into social rationality. An equivalent mechanism is used by Castelfranchi (2001) in order to formalise social functions (or roles) assumed by individuals as non-intentional mental processes. In this case, a metalevel controller supersedes the intentional and rational system of the agent. A learning-without-understanding process reinforces mechanically some individual beliefs, whether they are beneficial for the agent or not.

In both cases, the cognitive architecture becomes increasingly complex and the *normality paradigm* appears to act as a 'patching' process applied on a system of formal inferences that was not designed for such a purpose in the first place. Furthermore, these architectures and organisations seldom rely on direct evidence for validation. Instead, *virtual social experiments* are used to generate results that are evaluated against plausible utility values at the system and individual levels.

In order to overcome the increasing complexity of SAI architectures, and to facilitate, to some extend, the direct validation of the building assumptions, Jager and Janssen (2003) propose an alternate formalism. Their consumat theory is to be considered one of these conceptual jumps characteristic of the evolution of *normal* science: a new paradigm supports the creation of simpler models compared with the previous ageing generation. The consumat approach considers basic human needs and environmental uncertainties as the driving factors behind decision making processes. Agents engage in different cognitive processes, including social imitation and comparison, according to their perceptions of individual needs and environmental threat (Figure 3.4). Hence, the consumat paradigm overrides two structural limitations of the symbolico-cognitivist paradigm: experience preempts cognition and social rationality is directly built into intentional processes. Being at an early stage of development, the new theory relies on relatively simple rules that can be validated against experimental economics settings.

Figure 3.4. Structure of a Consumat agent

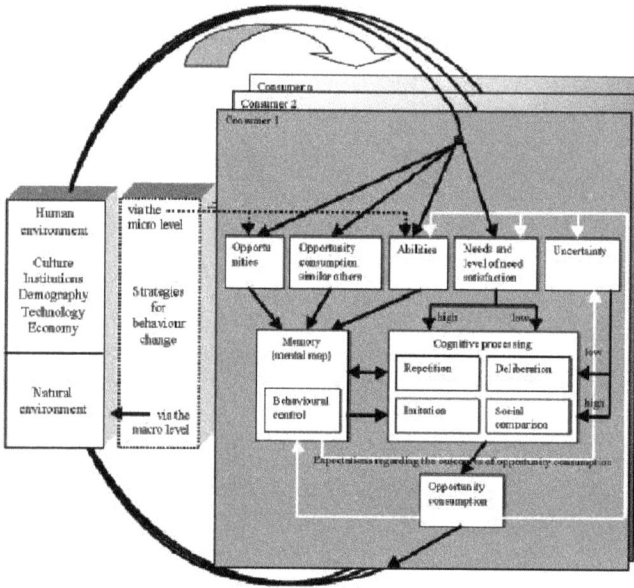

Source: Jager and Jannsen 2003

A post-normal temptation

Funtowicz and Ravetz (1993) also use environmental uncertainties, along with decision stakes, to analyse problem-solving strategies in the context of environmental and population risk policy issues. They argue that traditional scientific methodologies are ineffective when either attribute is high. Instead, they propose a new scientific posture they call *post-normal science* (op. cit.: 739):

> In those circumstances, the quality assurance of scientific inputs to the policy process requires an 'extended peer community', consisting of all those with a stake in the dialogue on the issue. Post-normal science can provide a path to the democratization of science, and also a response to the current tendencies to post-modernity.

The dynamic of resolution of policy issues in post-normal science involves the inclusion of an adequate set of legitimate participants in the process of quality assurance of the scientific inputs. For example, persons directly affected by an environmental problem will have a keener awareness of its symptoms. Thus, they perform a function analogous to that of peer-reviewers in traditional science, which otherwise might not occur in these specific contexts. Closer to the concern of this chapter, Funtowicz and Ravetz (1993: 745) challenge the commonly admitted rationality of decision and action:

Until now, with the dominance of applied science, the rationality of reductionist natural-scientific research has been taken as a model for the rationality of intellectual and social activity in general. However successful it has been in the past, the recognition of the policy issues of risk and the environment shows that this ideal of rationality is no longer universally appropriate. The activity of science now encompasses the management of irreducible uncertainties in knowledge and in ethics, and the recognition of different legitimate perspectives and ways of knowing.

Now, let's put this *post-normal* scientific posture into the context of our chapter. Among the majority of MABS used to explore human ecosystems, a significant number are meant to demonstrate how individuals and populations interact with their environment, as well as the environmental and social consequences of these interactions (Bousquet and LePage 2004). These situations are often characterised by:

- The presence of different groups of actors with contrasted, even conflicting, strategies.
- Irreducible uncertainties in representing and predicting responses from the environment.
- Individual and social rationalities based on multiple and competing utility functions.
- Self-referential conditions limiting goal-satisfying decisions to sub-optimal solutions.
- Emotional and cultural responses to policy incentives or penalties.
- Important framing effects and asymmetry of information.

We have to accept the fact that Multi Agent Models, even the more sophisticated ones, will always be pale copies of the original, subjective and partial represent-ations of a dynamic reality. But recognising this very peculiar fact doesn't mean that these models are useless, even Lissak and Richardson (2001: 105) in their criticism of computer-based social models admit that:

> There is no need for the models in question to have predictive power, despite the strong desire of both consultants and their clients that such models 'work'. The pedagogical value of exploring the interactions of complex relations through the manipulation of models is more than enough to justify the efforts that go into model development and proliferation. Clearly, it is easier to manipulate a computer model than a fully fledged 'in reality' laboratory experiment, but the limitations of such models must be remembered.

Chefs-d'Oeuvres are chiselled by talented craftsmen, not by their tools. Talent is all about sharing a vision and choosing the appropriate tools. In this regard, we must admit that the ultimate criteria of validation for this type of model is its actual appropriation by the final users (policy-makers, local communities, or else) and the consensual acceptance of the simulated outcomes. In this case, we can infer that:

Due to irreducible uncertainties and complex interactions within human ecosystems, social simulation designers should abide by a principle of subsidiary formalism. The principle acknowledges the fact that the most reliable source of knowledge about human decisions and behaviour are indeed the actors of the real drama themselves. Hence, the principle of subsidiarity stipulates that designer's cognitive efforts should be directed towards engaging with real stakeholders and eliciting their own mental models in the first place. Only when this option is unrealistic, the designers shall make it clear to everyone that decisional rules and algorithms are derived from theoretical or empirical predicates.

The subsidiarity principle offers a non-threatening opportunity for ethnography to accept the challenge of a new and dynamic formalism, beside narratives (Lansing 2003). The principle also mitigates the accusation of mere indexicality uttered by *System Thinkers* against individual-based simulations (Lissak and Richardson 2001). Finally, a post-normal posture might help solve the problem of validation that any complex system model is faced with: partial scientific validation associated with social authentication legitimate the modelling process and its outcomes (Funtowicz and Ravetz 1993).

Companion modelling

During the early 1990s, Granath (1991) introduced the concept of *collective design* in industry to define a process by which all actors involved in the production, diffusion, or consumption of a product are considered as equal experts and invited to participate to the design of the product. Each expert actively contributes to the collective process of transdisciplinary creation. Collective design usually faces two problems:

- socio-cultural barriers between different disciplines or social groups; and
- heterogeneous levels of knowledge and dissonant modes of communication.

Hence, in order to implement a collective design process, it is important to initially elicit specific knowledge and practices among experts. Then, the process itself must be grounded into successive *mediating objects*. These artifacts (ideas, models, products) are meant to channel creativity and to structure communication among experts. Collective design is to be considered as a social construct, no longer functional or rational. The final product emerges from conflicts, alliances, and negotiations.

During the late 90s, collective design and other approaches like system thinking or action learning were appropriated and adapted by scientists working on natural resource management (Hagmann et al. 2002; Barreteau 2003; D'Aquino et al. 2003). Interestingly, most of the cases were characterised by:

- direct engagement of science into field management issues (R&D projects);
- complex and adaptive socio-ecological systems (mid-scale with recursive interactions); and
- important cross-cultural contexts (rural Africa or Asia).

The co-construction of these models with local stakeholders didn't intend to provide normative models of reality, instead they were meant to enhance discussion and collective action through interactions around and about the mediating object (Lynam et al. 2002: 2):

> It is important to emphasize that, in the contexts in which these case studies are presented, the models were used more as part of a process of developing and exploring a common understanding of problems and possible solutions. They were not designed to be highly validated, predictive models in the sense in which systemic models are usually developed and used. We are not aware of other examples in which local people, who have no history of computer-based modeling, have been involved, not only in the use of computer models, but also in their development.

In these models, agents are designed according to the consensual information provided by their real counter-parts and the people they interact with. Decisional or behavioural rules are as complex and rational as the creators wish them to be. Likewise, agent's beliefs are tailored according to the phenomenological expression of the real stakeholder's mental models (Dray et al. 2006). Hence, this constructivist and post-normal modelling of cognitive processes doesn't intend to tell *How does it work?*, but rather *What is there that is so important?* But Becu and colleagues (2003) give evidence that the phenomenological expression of personal beliefs and intentions might not provide a consistent enough set of rules or assumptions to be directly encapsulated into the agent.

Recently, a group of scientists, *Collectif ComMod*, has decided to formalise their approach in order to establish deontic rules for developing companion models by direct interactions with local stakeholders (Collectif ComMod 2005). Companion modelling (ComMod) is an approach making use of simulation models in a participatory way to understand and facilitate the collective decision-making process of stakeholders sharing a common resource. The principle is to identify the various viewpoints and subjective referents used by the different stakeholders, and to integrate this knowledge into simulation models that serve as mediating objects. There is an iterative process of confrontation between factual evidence, model design, and scenario exploration (Figure 3.5). Other mediating objects or methods are usually used in conjunction with computer simulations, like role-playing games (Barreteau 2003; Dray et al. 2006). The different stakeholders, including researchers, aim at working out a common vision of the common resource management that highlights the diversity of interests. This approach has attracted growing attention from decision-makers and community-based organisations in order to engage more dynamically with stakeholders.

Figure 3.5. Traditional cycle of interactions during a companion modelling process

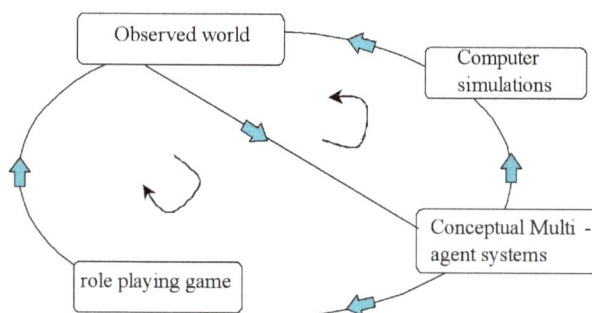

Conclusion

We have taken Cederman's assertion as a pretext for a journey through the world of artificial agents. From SAI to ComMod, we have drifted from positivism to constructivism, from a *normal* kind of science to a *post-normal* one. Despite the weaknesses and flaws pointed at in this chapter, the author, like Cederman, believes that Multi Agent Systems (MAS) offer a fantastic opportunity for the social, natural, and computer sciences to come together and engage in a 'new kind of science', much more challenging in epistemological terms than Wolfram's original proposal (see Lissak and Richardson 2001). Unlike Kluver and colleagues (2003), we think that there is a way for social sciences to appropriate computer formalisms without falling into a reductionist and positivist stand. The enactive cognitive theory (Maturana and Varela 1980) and the interactionist theory (Minsky 1985) need to be re-visited in a transdisciplinary manner. They can provide a theoretical substance to a vast majority of MAS applications built with non-formal logic compliant agents. The validation of these models needs to be embedded into a constructivist perspective where designers, users, and stakeholders not only evaluate the simulated outcomes, but also participate in the modelling process itself (Granath 1991; Collectif ComMod 2005).

Progressing on to the constructivist path doesn't mean that we have to discard the symbolico-cognitivist paradigm altogether. The strong, but limited, edifice proposed by SAI has shown undisputable capacities in producing replicable methods to test assumptions or benchmark findings on rational decision and action. For example, recent work from Castelfranchi (2001) or Hogg and Jennings (2001) are coming closer to designing realistic socially rational agents. But, paraphrasing Lissak and Richardson, 'the limitations of such models must be remembered'. Beyond the scope of this paper, the same type of warning must be addressed to social network theory and its extensive use of graph theory to explain social interactions (Borgatti and Foster 2003). Graphs are merely symbols, in a Peircean sense, displaying some sort of likeness with reality, they are not reality.

Finally, we have to agree with Cederman (2005) on the potential usefulness of Multi Agent Systems for social scientists. But we have to be aware of the fact that the symbolico-cognitive paradigm inherently limits the range of generative ontologies to be created. We have to take even more cautiously his reference to Simmelian social patterns. These patterns are subjectively construed by the observer, they are not the objective reality. Using MABS to replicate these patterns imposes conditions on the artificial agents that need to be socially validated as far as traditional scientific validation is no longer relevant in the case of complex human ecosystems.

Acknowledgement

The author wishes to thank David Batten, Francois Bousquet, and Roger Bradbury for their useful comments and advice.

References

Agar, M. (2005) Agents in living color: towards emic agent-based models. *Journal of Artificial Societies and Social Simulation* 8: 1. Available at http://jasss.soc.surrey.ac.uk/8/1/4.html

Bacon, F. (1620) *Novum Organum*. Trans. by Basil Montague. *The Works*, Vol. 3, pp. 343–71. Philadelphia: Parry and MacMillan (1854 edn). Available at http://history.hanover.edu/texts/Bacon/novorg.html

Barreteau, O. (2003) The joint use of role-playing games and models regarding negotiation processes: characterization of associations. *Journal of Artificial Societies and Social Simulation* 6(2). Available at http://jasss.soc.surrey.ac.uk/6/2/3.html

Batten, D.F. (2000) *Discovering Artificial Economics. How Agents Learn and Economies Evolve*. Oxford: Westview Press.

Becu, N., F. Bousquet, O. Barreteau, P. Perez and A. Walker (2003) A methodology for eliciting and modelling stakeholders' representations with Agent Based Modelling. *Lecture Notes in Artificial Intelligence* 2927: 131–49.

Benford, R.D. and D.A. Snow (2000) Framing processes and social movements: an overview and assessment. *Annual Review of Sociology* 26: 611–39.

Borgatti, S.P. and P.C. Foster (2003) The network paradigm in organizational research: a review and typology. *Journal of Management* 29(6): 991–1013.

Bousquet, F. and C. Le Page (2004) Multi-agent simulations and ecosystem management: a review. *Ecological Modelling* 176(3-4): 313–32.

Bradshaw, G.A. and J.G. Borchers (2000) Uncertainty as information: narrowing the science-policy gap. *Conservation Ecology* 4(1): 7. Available at http://www.consecol.org/vol4/iss1/art7/

Bratman, M. (1987) *Intentions, Plans, and Practical Reason.* Harvard University Press.

Brazier, F.M.T., C.M. Jonker and J. Treur (2002) Principles of component-based design of intelligent agents. *Data and Knowledge Engineering* 41: 1–27.

Castelfranchi, C. (2001) The theory of social functions: challenges for computational social science and multi-agent learning. *Journal of Cognitive Systems Research* 2: 5–38.

Cederman, L.E. (2005). Computational models of social forms: advancing generative process theory. *American Journal of Sociology* 110(4): 864–93.

Cohen, P.R. and H.J. Levesque (1990) Intention is choice with commitment. *Artificial Intelligence* 42: 213–61.

Cole, M. and S. Scribner (1974). *Culture and Thought: A Psychological Introduction.* New York: John Wiley and Sons.

Collectif Commod (2005) La modélisation comme outil d'accompagnement. *Natures Sciences Sociétés* 13: 165-8.

Conte, R., and C. Castelfranchi (1995) *Cognitive and Social Action.* London: UCL Press.

D'Aquino, P., C. Le Page, F. Bousquet and A. et Bah (2003) Using self-designed role-playing games and a multi-agent system to empower a local decision-making process for land use management: The SelfCormas experiment in Senegal. *Journal of Artificial Societies and Social Simulation* 6(3). Available at http://jasss.soc.surrey.ac.uk/6/3/5.html.

Dray A., P. Perez, N. Jones, C. Le Page, P. D'Aquino, I. White and T. et Auatabu (2006) The AtollGame experience: from Knowledge Engineering to a Computer-assisted Role Playing Game. *Journal of Artificial Societies and*

Social Simulation 9(1). Available at http://jasss.soc.surrey.ac.uk/9/1/6.html

Eco, H. (1997) *Kant and the Platypus*, edn 2000. London: Vintage, Random House.

Ferber, J. (1999) *Multi-Agent Systems: An Introduction to Distributed Artificial Intelligence*. New York: Addison-Wesley.

Funtowicz, S.O. and J.R. Ravetz (1993) Science for a Post-Normal Age. *Futures* 25(7): 739–55.

Gilbert, N., and K.G. Troitzsch (1999) *Simulation for the Social Scientist*. Buckingham, PA: Open University Press.

Granath, J.A. (1991) *Architecture, Technology and Human Factors: Design in a Socio-Technical Context*. Göteborg: Chalmers University of Technology, Architecture Workspace Design.

Hagmann, J. R., E. Chuma, K. Murwira, M. Connolly and P. Ficarelli. (2002) Success factors in integrated natural resource management R&D: lessons from practice. *Conservation Ecology* 5(2): 29. Available at http://www.consecol.org/vol5/iss2/art29

Hare, M. and P. Deadman (2004) Further towards a taxonomy of agent-based simulation models in environmental management. *Mathematics and Computers in Simulation* 64: 25–40.

Hogg, L.M.J. and N.R. Jennings (2001) Socially intelligent reasoning for autonomous agents. *IEEE Trans. on Systems, Man and Cybernetics Part A* 31(5): 381–99.

Jager, W. and M. Janssen (2003) The need for and development of behaviourally realistic agents. In J.M. Sichman, F. Bousquet and P. Davidson (eds), *Multi-Agent Simulation II* (Proceedings of the MABS 2002 Workshop, Bologna, Italy), pp. 36–50. New York: Springer Verlag.

Janssen, M.A. (ed.) (2002) *Complexity and Ecosystem Management: The Theory and Practice of Multi-Agent Systems*. Cheltenham: Edward Elgar Publishers.

Kahnemann, D. and A. Tversky (2000) *Choices, Values and Frames*. Cambridge: Cambridge University Press.

Kaplan, F. (2001) *La naissance d'une langue chez les robots* . Paris: Hermès Sciences Publications.

Kluver, J., C. Stoica and J. Schimdt (2003) Formal Models, Social Theory and Computer Simulations: Some Methodical Reflections. *Journal of Artificial Societies and Social Simulation* 6(2). Available at http://jasss.soc.surrey.ac.uk/6/2/8.html

Lansing, J.S. (2003) Complex adaptive systems. *Annual Review Anthropology* 32: 183–204.

Lissack, M.R. and K. Richardson (2001) When Modeling Social Systems, Models the Modeled: Reacting to Wolfram's 'A New Kind of Science'. *Emergence* 3(4).

Lynam, T., F. Bousquet, C. Le Page, P. d'Aquino, O. Barreteau, F. Chinembiri and B. Mombeshora (2002) Adapting science to adaptive managers: spidergrams, belief models, and multi-agent systems modeling. *Conservation Ecology* 5(2): 24. Available at http://www.consecol.org/vol5/iss2/art24

Maturana, H. and F. Varela (1980) *Autopoieisis and Cognition: the Realization of the Living*. Boston: D. Reidel.

Minsky, M. (1985) *The Society of Mind*. New York: Simon and Schuster.

Oliver, P. (1993) Formal Models of Collective Action. *Annual Review of Sociology* 19: 271–300.

Papert, S. (1980) *Mind-Storms: Children, Computers, and Powerful Ideas*. New York: Basic Books.

Peirce Edition Project (eds) (1998) *Peirce, C.S., The Essential Peirce, Selected Philosophical Writings, Vol. 2 (1893–1913)*. Bloomington and Indianapolis: Indiana University Press.

Tesfatsion, L. (2001) Introduction to the special issue on agent-based computational

Economics. *Journal of Economic Dynamics and Control* 25: 281–93.

Varela, F.J., E. Thompson and E. Rosch (1991) *The Embodied Mind: Cognitive Science and Human Experience*. Cambridge, MA: MIT Press.

4. The Uncertain Fate of Self-Defeating Systems

David Batten

Abstract

Complex patterns of human behaviour are difficult to capture in agent-based simulations of socio-ecological systems. Even knowing each individual agent's strategy at one point in time may not help when trying to predict the collective behaviour of certain systems—if it is in each agent's best interest to do the opposite of most other agents. In *self-defeating* situations, the collective population of agents may exhibit a panorama of simple or complex behaviour, depending on the extent to which information is shared. A well-known example is the El Farol bar problem, in which a simulated population of bar attendees oscillates in a seemingly random manner around a critical congestion level. In this paper, it is suggested that a family of resource management problems involving human interactions with ecosystems may possess a self-defeating character. This poses several new challenges for integrated resources management. A case in point is the potential over-fishing of fisheries, which is addressed in the paper and likened to a minority game.

Introduction

The use of agent-based simulation to assist in environmental management and decision-making has increased markedly in recent years. Simulation models representing various facets of human behaviour have emerged (Hare and Deadman 2004). Social scientists use this kind of simulation for several purposes, including the discovery of collective regularities. Some are built on strong behavioural foundations, where mental models of the silicon agents are based on empirical data or stakeholders' views. In other cases, agents' strategies are selected stochastically or even arbitrarily and the outcomes tested in computational experiments.

What is arbitrarily decided in such simulations is how to drive the artificial learning process that enables each agent to adapt to other agent strategies interactively and to decide whether and how to modify its own strategy over time. Imitation of other (e.g. nearby/visible) agents is a simple and common method of introducing adaptation over time. In other simulations, agents choose from a suite of heuristics or mental models made available to some or all of them. More sophisticated learning methods (such as neural networks and genetic algorithms) have also been adopted.

Largely overlooked in the simulation work to date is the fact that the intrinsic structure of the system being simulated can have a direct influence on ways in which the human agents are likely to behave, i.e. react and adapt over time. Some social systems that fall into this category are called *self-referential* problems—situations in which the forecasts made by agents serve to create the very world they are trying to forecast. In these complex systems, the *best* thing to do depends on what everyone else is doing.

The purpose of this paper is to look briefly at a special class of self-referential problems called *self-defeating* systems. A typical example is the bar problem, in which a simulated population of attendees oscillates in an apparently random manner around a critical congestion level. It is suggested that several resource management problems involving human interactions with ecosystems have a self-defeating character, posing new challenges for integrated resources management. An example is the potential over-fishing of fisheries, behavioural aspects of which are discussed in the paper and likened to a minority game. Such commons problems will require new institutional arrangements to overcome the self-defeating character of these situations.

Self-referential systems

Definition

A *self-referential* situation is one in which the forecasts made by the human agents involved serves to create the world they are trying to forecast. Such co-evolutionary systems have also been called *reflexive* and are generally *self-reinforcing*. They have attracted attention in several socio-economic contexts (Arthur 1994; Casti 1999; Batten 2000). A key feature of these systems is that the *best* thing to do (such as *to go* or *not to go*) depends on what everyone else is doing. As no individual agent knows that, the best thing that any agent can do is to apply the strategy or mental model that has worked best so far.

In self-referential systems, the mix of possible strategies is co-evolving incessantly over time, so much so that the choices made by individual agents seem to have little impact on collective outcomes. Sometimes, collective populations of agents exhibit emergent regularities. At other times, they display different types of complex behaviour. In the next section, we describe the El Farol bar problem, an early agent-based simulation of a complex adaptive system reported by Brian Arthur (1994).

The bar problem defined

Consider a system of N = 100 agents deciding independently each week whether or not to go to their favourite bar (called the El Farol) next Thursday. Space is limited, so the evening is enjoyable only if the bar is not too crowded (say $N_{max} = 60$). There is no collusion or prior communication among agents. Knowing the bar attendance over the past few weeks, each bar-loving agent simply decides independently to go if he expects less than N_{max} to attend or stay home if he expects more than N_{max} to go.

This problem is a metaphor for a broad class of self-referential situations: e.g. urban traffic congestion, canteen crowding or queue lengths at big sporting events. It has some interesting properties. First, if a decision model existed that agents could rely upon to forecast attendance, then a deductive solution would be possible. No such model exists. Irrespective of past attendance figures, many plausible hypotheses could be adopted to predict future attendance. Because agents' rationality is bounded, they are forced to reason *inductively*. Second, any shared expectations will be broken up. If all agents believe *most* will go, then *nobody* will go. By staying home, that common belief will be destroyed. If all agents believe *few* will go, then *all* will go, thus undermining that belief. Because any mental model that is shared by most of the agents will be *self-defeating*, agents' expectations must always differ.

Perplexed by the intractability of this problem, Arthur created a computer simulation in which his agents were given attendance figures over the past few months. He also created an *alphabetic soup* of several dozen predictors replicated many times. After randomly ladling out k of these to each agent, each kept track of his k different predictors and decided whether to go or not according to a preferred predictor in his set. This preferred predictor could be chosen in a variety of ways, although Arthur adopted the most accurate current predictor for each agent in his simulations.

The bar problem simulated

Once decisions have been made in Arthur's simulated bar, agents learn the new attendance figure, updating the accuracy of their own set of predictors. Then decisions are made for the following week. In this kind of problem, the set of predictors acted upon by agents—called the set of *preferred* predictors—determines the attendance. But the attendance history also determines the set of preferred predictors. We can think of this set as forming a kind of *ecology* (John Holland's term). Of interest is how this *ecology* evolves over time.

The simulations show that weekly attendance fluctuates unpredictably, but mean attendance always converges to 60 in the long run. The predictors self-organise into an equilibrium pattern or ecology in which (on average) 40 per cent of the preferred predictors forecast above 60, and 60 per cent below 60. This 40/60 split remains although the population of preferred predictors keeps changing in membership. The emergent ecology is rather like a forest whose contours do not change, but whose individual trees do. Similar results appeared throughout Arthur's experiments, robust to changes in the types of hypotheses.

There is another intriguing result. Although the computer-generated attendance results look more like the outcome of a random process than a deterministic one (see Figure 4.1), there is no inherently random factor governing how many people attend. Weekly attendance is a deterministic function of the individual predictions, themselves being deterministic functions of the past attendance figures.

Thus the bar problem is a relatively simple example of an emergent, self-defeating system. It is a situation in which a system of interacting agents can develop collective properties that are not at all obvious from our knowledge of the agents themselves. Even if we knew all agents' individual idiosyncrasies, we are no closer to anticipating the emergent outcome. Under the influence of a sufficiently strong *attractor*, individually subjective, boundedly rational expectations self-organise to produce a kind of *collectively rational* behaviour (Arthur 1994).

Figure 4.1. A simulated 100-week record of attendance at El Farol

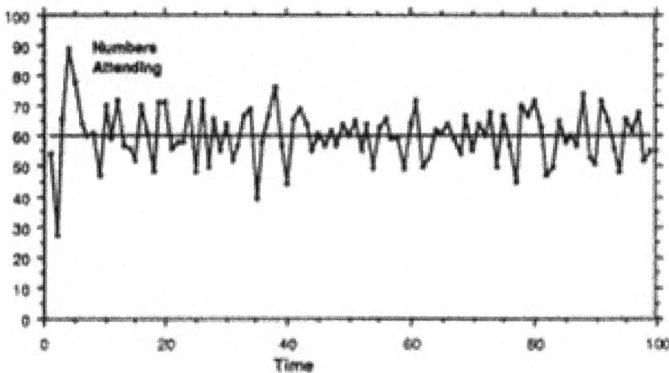

Source: Arthur, 1994

The bar problem extended

Given that agents do communicate with each other in the real world, we wonder what may happen if they are permitted to exchange information? Bruce Edmonds' work allows communication among agents before they make their final decisions whether or not to go to the bar (Edmonds 1999). Using a genetic programming algorithm to simulate adaptive learning, he allows each agent to competitively develop its models of what the other agents are going to do. Although the beliefs and goals of other agents are not known or represented by each agent, heterogeneity among the agents emerges in the form of non-uniform tactics and role-playing identities. These collective properties are features that emerge purely from the micro-dynamics.

Like the Prisoners' Dilemma, the bar problem is currently receiving more attention outside economics—as a metaphor for learning and bounded rationality. It has inspired a new literature in statistical physics on a closely related problem known as the minority game. A minority game is a repeated game in which N (odd) players have to choose one out of two alternatives (say A and B) at each time step. Those who happen to be in the minority win.

Seemingly simple at first glance, the game is subtle in the sense that if all players analyse the situation in the same way, all will choose the same alternative and therefore all will lose. Thus, players behave heterogeneously over time. Moreover, there is a frustration since not all players can win at the same time (this is an essential mechanism for modelling competition).

Since its introduction, this game has generated incredible interest. As there are only three parameters, the game is suitable for detailed numerical studies and analytical descriptions. Although each alternative is unspecified, for A or B one could read 'I am going to the bar', or, 'I am staying home'; 'I am choosing the motorway', or, 'I am going by the scenic route'; 'I am going to fish at the usual place', or, 'I am going to find a better spot'. Perhaps the most striking properties of the minority game are that:

- it is a model that addresses the interaction between agents and information;
- agents are able to cooperate and make use of the available information;
- there is a second order phase transition between a symmetric phase (in which no information is available to agents) and an asymmetric phase (in which information is available to agents); and
- if agents take their impact on the game into account, there is no phase transition and this case has an exact solution.

The minority game is an abstraction of the bar problem. Like simulations of the bar problem, numerical simulations of this game have displayed a remarkably rich set of emergent, collective behaviours (Challet and Zhang 1998).

Both the bar problem and the minority game contain the key elements of a *complex adaptive system*: firstly, a *medium* number of agents—a number too large for hand-calculation or intuition, but too small to use statistical methods applicable to very large populations; secondly, these agents are *intelligent* and *adaptive*, making decisions on the basis of rules of thumb or heuristics, like the bar predictors (needing to modify these rules or come up with new ones if necessary, they reason *intuitively*); and, finally, no single agent knows what all the others are (thinking of) doing, because each only has access to limited information.

Human ecosystems

In human ecosystems, human agents interact with one another and other life in a natural environment. Self-referential problems of the binary variety arise in these systems, but are seldom recognised as such. Since Hardin's influential article in Science (Hardin 1968), most are treated purely as commons dilemmas in which agents over-exploit a scarce resource in common. Examples include the over-fishing of fisheries, the degradation of national parks and the destruction of coral reefs. The question we address in the next section is to what extent the fisheries problem may be viewed as a complex adaptive system of the self-defeating variety.

Fisheries

In an insightful paper almost two decades ago, Allen and McGlade (1986) stressed that the key elements of fishing (like ancestral hunting) are *discovery* and *exploitation*, not simply the latter as is currently the case in western agriculture. As well as a desire to avoid congestion (like in the bar problem), fishing involves communication, adaptive learning and the emergence of non-uniform tactics and role-playing among fishing agents. Within such a socio-ecological system, key questions on our relationship with nature and the general problem of the management of a complex adaptive system can be explored.

Largely hidden from human view, the marine ecosystem is complex with respect to its species composition as well as to the processes occurring in it. As exploitation increases, Beddington and May (1977) showed that such an ecosystem moves towards instability and greater risk of collapse. Human responses tend to amplify random fluctuations, thereby further increasing the risk of collapse.

Ignoring the complex issue of ecosystem dynamics for the moment, let us concentrate on the strategies of the owners of the fishing vessels. From an owner's selfish viewpoint, if there are no institutional rules imposed on him, he aims to maximise his own vessel's catch. Obviously, the *best* thing for him to do will depend on what the other fishing vessels in the vicinity of his vessel are doing. Thus our fishing problem is a self-referential one.

Allen and McGlade (1986) developed dynamic, multi-species, multi-fleet models to explore the implications of different fishing strategies and information flows among fishing vessels in a Nova Scotia fishery. They identified two fishing strategies, calling them *Cartesians* and *Stochasts*. Stochasts search randomly for better sites or use their own intuition, without resorting to information that is shared between associates. They are risk-takers, seeking higher returns commensurate with the higher risks they take. At the other extreme we have the Cartesians, skippers who (for various reasons) are unwilling to take any risk and who only go to the zone promising the best *known* return. It does not matter how long it takes for a Cartesian to reach their chosen site, as long as the final catch is guaranteed.

Allen and McGlade found that less information exchange among the fishing vessels ensures a more random response on the part of boats (i.e. more Stochasts among the fleet). Because the Stochasts continue to explore less visited parts of the system, the fishery as a whole has a greater chance of survival. If boats refuse to take risks and go only to where they know there are fish, the end result can be disastrous for everyone. Discovery involves risk, but to abandon it invites disaster.

There is a powerful message here that goes beyond the bounds of fisheries alone. In any mobile population, for example, we find some risk-takers and some who are risk-averse. Like the number of fishing boats turning up at the same site, the number of vehicles turning up on a specific road each day is unpredictable. Of interest are the adaptive strategies of drivers exposed to regular traffic jams. Anthony Downs (1962) identified two behavioural classes of driver: those with a low propensity to change their mode or route strategy, called *sheep*, and those with a propensity to change, called *explorers*. Explorers search for alternative options to save time. They are quick to learn and hold several heuristics in mind simultaneously. Sheep are more conservative and prone to following the same option. Empirical work in North America has confirmed the presence of sheep and explorer behaviour in real traffic (Conquest et al. 1993).

The parallels between Cartesian or Stochast fishing strategies and sheep or explorer driving strategies may seem striking. Yet they are less surprising if thought of as symptomatic of a more general phenomenon. In the world of technology, risk-averters and risk-takers appear under different guises: *imitators* and *innovators*. If we allow for the coexistence of imitative and innovative mechanisms in a population, both must be treated as co-evolutionary variables dependent on the unfolding of events.

Evolutionarily stable strategies

Is the coexistence of Cartesian (imitative) and Stochast (innovative) strategies in a human population an *Evolutionarily Stable Strategy* (ESS)? An ESS is a strategy (or mix of strategies) which, if most members of a population adopt it, cannot be bettered by an alternative strategy. The reason is that the fitness of individuals adopting an ESS strategy is higher than the fitness of individuals adopting other strategies (Maynard-Smith 1982). Another way of putting it is that the best strategy for any individual depends on what the majority of the population is doing (Dawkins 1976). This last perspective helps us to grasp the idea that an ESS may be an important property of self-referential systems.

There are three classes of ESS: pure, mixed and conditional. A pure ESS is a strategy that is consistently exhibited by individuals throughout their lifetimes. A mixed ESS is a complex of two or more strategies varying either within or among individuals over time. In a conditional ESS, each individual's strategy varies under different conditions in the social or physical environment. If the evolutionary payoff for each strategy depends on what other individuals are doing and decreases as a greater proportion of the population adopts that strategy, other strategies may also be an ESS (frequency-dependent selection).

To apply the ESS idea to our fishery problem, consider the following variant of one of Maynard Smith's simplest hypothetical cases (hawks and doves). Here we draw extensively on the simulation results in Allen and McGlade (1986). Suppose that the only two types of fishing strategy in a fleet of vessels are Stochasts and Cartesians. We want to know whether Pure-Stochast or Pure-Cartesian is an evolutionarily stable strategy. As Stochasts and Cartesians compete for returns, we must estimate payoffs to the different fleet strategies in order to find an ESS.

Serving as the *eyes* of the fleet, Stochasts search for fishing zones of high return. A fleet with a high proportion of risk-taking Stochasts will discover successive zones of high return, eventually fishing out the high-return zones in a patchwork manner. But Stochasts ignore fish from nearby zones of intermediate returns. As it takes time for the high-return zones to recover, pure Stochast strategies tend to lead to boom-and-bust series of good and bad years. However, this patchwork approach guarantees that the fishery system survives rather than running the risk of collapsing.

Based on the results in Allen and McGlade (1986), we can assign a Pure-Stochast a cumulative payoff of 76 points (equivalent to 6 large circles plus 8 small ones) over a ten-year fishing period. This includes several bad years during which the stocks must recuperate.

Cartesians refuse to take risks and rely solely on the information passed on by the Stochasts. Thus they go only to zones where they know in advance that there are some fish. A fishing fleet dominated by risk-averse Cartesians will simply direct all the boats in the short term to a few of the *best* zones for fishing. Because the information about which are the best zones is generated by a small number of Stochasts, the tendency for Cartesians is to *lock* onto one particular location as the best. In the long run, this leads to the fleet exploiting a single location instead of the whole area, leading to a small catch and a small fleet. Based on figures found in Allen and McGlade, a Pure-Cartesian strategy yields a payoff of 45 points over the same ten-year period.

The above figures confirm our suspicions that Pure-Stochasts do better than Pure-Cartesians over time. They are rewarded for taking more risks. Pure-

Stochasts achieve better returns because they can maintain fishing activities over the whole area, obtaining a higher catch in the good years and maintaining a larger fishing industry overall. Pure-Cartesians become trapped in a few locations instead of spreading out across the whole area. They are limited each year to a smaller catch and a declining fishing fleet and industry over time.

Table 4.1. Strategies and payoffs for homogeneous fleets

Fleet strategy 10-year payoff	
Pure-Stochasts	76 points
Pure-Cartesians	45 points

But is a Pure-Stochast strategy evolutionarily stable on its own? To answer this question, let us suppose that we have a fleet consisting only of Pure-Stochasts. Despite the bad years, they seem to do very nicely, earning the payoff given in Table 4.1. Now suppose that a mutant Cartesian arises in the fleet. Being the only Cartesian, must he compete aggressively with the Pure-Stochast? No. Instead he could follow the Pure-Stochasts, letting them show him where it is best to fish. Then, by fishing at the edges of sites discovered by Pure-Stochasts, he does not irritate them too much. Remaining mostly unnoticed, he can *free-ride*, enjoying above-average returns for very little search effort.

As long as there are sites offering intermediate returns that are left untouched by Stochasts, there are niche opportunities for Cartesians to fill. Based on figures in Allen and McGlade (1986), 10-year payoffs to a mixed fleet of Stochasts and Cartesians under various levels of information exchange are given in Table 4.2.

Table 4.2. Fleet strategies and payoffs when information is shared

Fleet strategy 10-year payoff	
If catch information is shared equally:	
Stochasts	104 points
Cartesians	83 points
If Cartesians fail to inform Stochasts:	
Stochasts	95 points
Cartesians	81 points
If Stochasts fail to inform Cartesians:	
Stochasts	109 points
Cartesians	30 points

When information is shared equally between both groups, the payoff to each is higher than to a Pure-Stochast or Pure-Cartesian fleet (see Table 4.1). Thus a Pure-Stochast strategy is not evolutionarily stable, since it can be bettered by a strategy in which Stochasts and Cartesians communicate and cooperate with each other.

What happens if either group cheats by not transmitting reliable information about their catches? As Allen and McGlade noted, there is an interesting asymmetry. If Cartesians fail to inform Stochasts, there is very little effect (see Table 4.2). But if the information possessed by Stochasts is not transmitted to Cartesians, the latter perish (or are compelled to become Stochasts). Thus Cartesians will try to obtain catch information and Stochasts will try to withhold it. Such effects have evolved in real fishing fleets. For example, Vignaux (1996) found that trawlers fishing for hoki in waters off New Zealand's coastline do not share catch information, instead basing their own decisions partly on watching where other vessels fish. This amounts to spying. Other tactics include *listening in* on radios, lying about catches and spreading other misleading information.

Using agent-based simulation, a richer suite of possibilities can be explored. Recent Agent-Based Modelling (ABM) work involving Bayesian belief networks has shown that various kinds of information flow among fishing vessels has important effects on the dynamics and resource exploitation of a simulated fishery (Little et al. 2004). As stated earlier, information flow tends to benefit the Cartesians at the expense of the Stochasts. This asymmetry confirms the importance of realistically representing the rich variety of possible fisher behaviour in any modelling framework that aims to assist with integrated management of fishery resources.

A modified minority game?

The superior payoff achieved when catch information is shared reliably is in line with observed properties of the minority game. If all agents in a fleet have access to public information about catches and zones of highest return over a period of time, then the agents interact only through this public information and the system has a *mean-field* character (in the sense that no short-range interactions exist). Self-organisation in such a system is achieved by allowing each agent to have several strategies from which he selects the one that seems best (to him).

In the literature on the minority game, there is a second order phase transition between a symmetric phase (in which no information is available to agents) and an asymmetric phase (in which information is available). It is easy to see why this phase change might occur in our fisheries context. First, if reliable catch information is passed between Stochasts and Cartesians, this may ensure that a boom-and-bust series of good and bad years is avoided. Stochasts can direct some Cartesians to high-return zones and others to medium-return zones, in such a way that no high-return zone will be in great danger of over-fishing. Second, the two strategies are complementary. Cartesians make good use of information, while Stochasts generate it. Together they can exploit the resource more efficiently because of their complementarity.

Ideally, each fleet needs some Stochasts (researchers) and Cartesians (producers) who cooperate within a fleet but not with competing fleets. To survive, each needs the other. What, then, is the ideal ratio of Stochasts to Cartesians? This is a difficult question to answer without considering the dynamics of the fish population and other environmental factors. The higher payoff that a Stochast may enjoy is tempered by the higher risks involved, whereas the risks taken by a Cartesian are minimal.

In a highly dynamic world like fisheries, one may expect the ratio of Stochasts to Cartesians in a fleet to vary in response to the search success of the Stochasts and their willingness to inform their Cartesian partners. Like music lovers at the El Farol bar, the ratio of Stochasts to Cartesians can oscillate forever. But, in this case, there may be no emergent regularity as the congestion level is a dynamic variable. The situation is a self-defeating one, since the *best* thing for each vessel to do depends on what everyone else is (thinking of) doing.

Discussion

Previous simulation work referred to in this paper has shown the importance of information sharing and communication strategies (Allen and McGlade 1986; Maury and Gascuel 2001; Little et al. 2004). Irrespective of whether vessel owners within fishing fleets share reliable catch information, agent-based models can help to clarify the potential value of information-sharing. Earlier work has shown that lower returns will be experienced by Cartesians in the absence of information flow from Stochasts. In the long run, however, Stochasts suffer boom-and-bust years by fishing alone because they fish out the high-return zones in a patchwork manner. Thus the evolutionarily stable strategy for a fishing fleet is a mixed one in which Stochasts and Cartesians communicate and cooperate with each other.

When the economic viability of a fishery is under serious threat, it is worthwhile adopting a cooperative approach—not only to encourage information sharing, but also to overcome such common pool dilemmas. If conditions are suitably conducive, a sustainable solution can even be found by the participants themselves. One successful example is the inshore fishery at Alanya in Turkey (Ostrom 1990). Members of the local cooperative found an ingenious rotation system allocating fishing sites to local fishers that builds upon reliable and mutually beneficial information exchange.

As mentioned near the outset, the purpose of this paper was to look at a class of self-referential problems called *self-defeating* systems. A typical example is the bar problem, in which the simulated population of attendees oscillates in a seemingly random manner around a critical congestion level. Another self-defeating problem is urban traffic congestion. Without a degree of cooperation and information sharing between the sheep and explorers (Cartesians and Stochasts) on our roads, the collective behaviour of these complex adaptive systems will continue to defy prediction or control on a daily basis.

The fishery situation is of a similar ilk. As well as a desire to avoid congestion, fishing involves communication, adaptive learning and the emergence of role-playing and non-uniform tactics among the agents involved. Human behaviour in national parks and coral reefs raises similar collective problems. In each of these ecosystems, it is impossible to achieve an eco-efficient outcome unless a threshold level of cooperation and information sharing can be achieved. Stochasts and Cartesians need to work together in order to achieve sustainability and thus avoid over-exploiting the resources therein.

Within such socio-ecological systems, the use of agent-based simulation allows key questions about our relationship with nature and the more general problem of the management of a complex adaptive system to be explored qualitatively. Rule-change experiments—such as those that led to the successful fishing regime at Alanya—fall into the scientific domain of participatory, agent-based modelling. The great thing about this kind of simulation is that we can assess the likely outcomes of such rule changes ahead of their implementation.

As individual humans—in our roles within families, communities, firms, institutions and regions—we must decide how to divide our time and effort between doing what we know (with known values and payoffs) and searching for new opportunities and roles that may have superior payoffs in the future. We can choose to explore new pathways and connections or try to minimise such deviations from those pathways that we are accustomed to in our lives to date. Thus Stochasts and Cartesians are two important extremes in human society. The first group take more risks, venturing into the unknown, be it by hunches, tactics or entrepreneurial activities. Their discoveries nourish a society in the long run, expanding it into the adjacent possible and assuring its long-term survival in some form. The second group prefers to devote themselves to tasks already assigned to them, seeking to undertake them as quietly and efficiently as possible. They constitute the *backbone* of a society and come much closer to our definition of *normality*. The survival and sustainable prosperity of any individual, community, firm or nation, or even a human ecosystem, requires a mixture of both types of behaviour.

Acknowledgments

Helpful comments from Paul Atkins, Marco Janssen, participants at the joint CSIRO Agent-Based Modelling/Human Ecosystems Modelling with Agents (CABM/HEMA) Workshop, Canberra (17–18 May 2004) and at the IEMSS Congress, Osnabrück (14–17 June 2004) are gratefully acknowledged. The usual caveat regarding author's responsibility applies.

References

Allen, P.M. and J.M. McGlade (1986) Dynamics of discovery and exploitation: the case of Scotian Shelf fisheries. *Canadian Journal of Fisheries and Aquatic Sciences* 43: 1187–200.

Arthur, W.B. (1994) Inductive behaviour and bounded rationality. *The American Economic Review* 84: 406–11.

Batten, D.F. (2000) *Discovering Artificial Economics: How Agents Learn and Economies Evolve*. New York: Westview Press.

Beddington, J.R. and R.M. May (1977) Harvesting natural populations in a randomly fluctuating environment. *Science* 197: 463–5.

Casti, J. (1999) Would-be worlds: the science and surprise of artificial worlds. *Computers, Environment and Urban Systems* 23: 193–203.

Challet, D. and Y-C. Zhang (1998) On the minority game: analytical and numerical studies. *Physica A* 256: 514–518.

Conquest, L., J. Spyridakis, M. Haselkorn and W. Barfield (1993) The effect of motorist information on commuter behaviour: classification of drivers into commuter groups. *Transportation Research C* 1: 183-201.

Dawkins, R. (1976) *The Selfish Gene*. Oxford: Oxford University Press.

Downs, A. (1962) The law of peak-hour expressway congestion. *Traffic Quarterly* 16: 393–409.

Edmonds, B. (1999) Gossip, sexual recombination and the El Farol bar: modelling the emergence of heterogeneity. *Journal of Artificial Societies and Social Simulation* 2(3): 1–21.

Hardin, G. (1968) The tragedy of the commons. *Science* 162: 1243–8.

Hare, M. and P. Deadman (2004) Further towards a taxonomy of agent-based simulation models in environmental management. *Mathematics and Computers in Simulation* 64: 25–40.

Little, L.R., S. Kuikka, A.E. Punt, F. Pantus, C.R. Davies and B.D. Mapstone (2004) Information flow among fishing vessels modelled using a Bayesian network. *Environmental Modelling and Software* 19: 27–34.

Maury, O. and D. Gascuel (2001) Local over-fishing and fishing tactics: theoretical considerations and consequences in stock assessment studied with a numerical simulator of fisheries. *Aquatic Living Resources* 14: 203–10.

Maynard-Smith, J. (1982) *Evolution and the Theory of Games*. Cambridge: Cambridge University Press.

Ostrom, E. (1990) *Governing the Commons: the Evolution of Institutions for Collective Action*. Cambridge: Cambridge University Press.

Vignaux, M. (1996) Analysis of vessel movements and strategies using commercial catch and effort data from the New Zealand hoki fishery. *Canadian Journal of Fisheries and Aquatic Sciences* 53: 2126–36.

5. The Structure of Social Networks

David Newth

Abstract

Inspired by empirical studies of networked systems such as the Internet, social networks and biological networks, researchers have in recent years developed a variety of techniques and models to help us understand or predict the behaviour of these complex systems. In this chapter we will introduce some of the key concepts of complex networks and review some of the major findings from the field. Leading on from this background material we will show that analysing the patterns of interconnections between agents can be used to detect dominance relationships. Another important aspect of social order is the enforcement of social norms. By coupling network theory, game theory and evolutionary algorithms, we will examine the role that social networks (the structure of the interactions between agents) play in the emergence of social norms. Empirical and theoretical studies of networks are important stepping stones in gaining a deeper understanding of the dynamics and organisation of the complex systems that surround us.

Introduction

Many of the systems that surround us, such as road traffic flow, communication networks, densely populated communities, ecosystems with competing species, or the human brain (with 10^{10} neurons), are large, complex, dynamic, and highly nonlinear in their global behaviour. However, over the past 10 years it has been shown that many complex systems exhibit similar topological features in the way their underlying elements are arranged (Albert and Barabási 2002).

Underlying much of the current research is the notion that, in some way, topology affects dynamics that take place on the network, and vice versa. In this chapter I will explore some of the properties of complex social networks.

One of the most well known properties of social networks is the *small-world phenomenon*. This is something that most of us have experienced. Often we meet people with whom we have little in common, and unexpectedly find that we share a mutual acquaintance. The idea of '6 degrees of separation' is now firmly embedded in folklore, embracing everyone from Kevin Bacon to Monica Lewinsky (Watts 2003). Analysis of social networks has shown that the patterns of interactions that surround each of us, often determines our opportunities, level of influence, social circle, wealth, and even our mental well being.

In this chapter I explore some of the properties of complex social networks. The following section provides a number of examples of complex networks from a range of different contexts. Then, I provide an information theoretic-based approach to detect strong groups in social networks. It is followed by a detailed description of a game theoretic model, in which each of the agents is a decision making unit. This example shows how social network can influence the enforcement of certain behaviours within the system.

Networks

In mathematical terms, a network is a graph in which the nodes and edges have values associated with them. A graph G is defined as a pair of sets $G = \{V, E\}$, where V is a set of N nodes (vertices or points within the graph) labelled $v_0, v_1, \ldots, v_n$ and $E \subseteq V \times V$ is a set of edges (links (v_i, v_j) that connect pairs of elements v_i, v_j within V). A set of vertices joined by edges is the simplest type of network. Networks can be more complex than this. For instance, there may be more than one type of vertex in a network, or more than one type of edge. Also, vertices may have certain properties. Likewise, edges may be directed. Such edges are known as arcs. Arcs and edges may also have weights. Figure 5.1, depicts networks with various types of properties. Taking the example of a social network of people, the vertices may represent men or women, people of different nationalities, locations, ages, incomes or any other attributes. Edges may represent relationships such as friendship, colleagues, sexual contact, geographical proximity or some other relationship. The edge may be directed such as supervisor and subordinate. Edges may represent the flow of information from one individual to another. Likewise the edges may carry a weight. This weight may represent a physical distance between 2 geographical proximity, frequency of interaction, degree to which a given person likes another person.

Figure 5.1. Random graphs

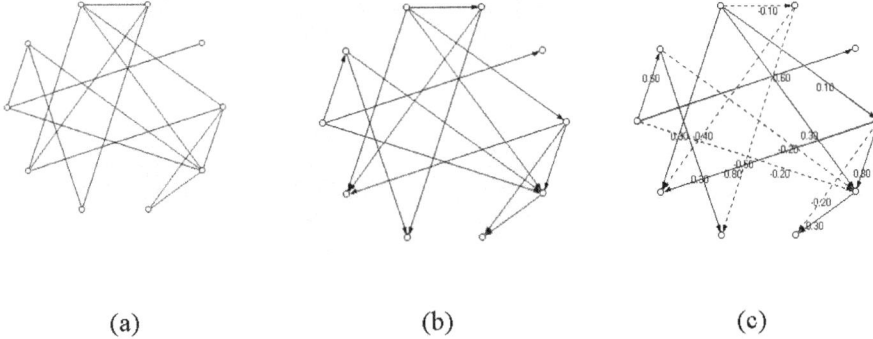

(a)   (b)   (c)

(a) Random graph. (b) Directed random graph. (c) Directed random graph with weights (random network)

Network properties

Connectivity

The degree to which the nodes of a network are directly connected is called connectivity. A network with high connectivity has a high ratio of edges to the number of nodes. To calculate a networks connectivity C, where k is the number of edges and N is the number of nodes in the network, the following equation is used;

$$C = \frac{k}{N(N-1)}$$

Degree distribution

The degree of a node in a network is the number of edges or connections to that node (Newman 2003). The distribution function $P(k)$ gives the probability that a node selected at random has exactly k edges (Albert and Barabási 2002). Plotting $P(k)$ for a network forms a histogram of the degrees of the nodes, this represents the degree distribution (Newman 2003) or the number of nodes that has that number of edges for the network.

Shortest average path length

The average path length, (l) of a network is the average number of edges, or connections between nodes, that must be crossed in the shortest path between any 2 nodes (Watts 2003). It is calculated as:

$$l = \frac{2}{N(N-1)} \sum_{i=1}^{N} \sum_{j=1}^{N} l_{min}(i,j)$$
,

where $l_{min}(i,j)$ is the minimum distance between nodes i and j.

Diameter

The diameter of a network is the longest shortest path within a network. The diameter is defined as:

$$D = \max_{i,j} l_{min}(i,j)$$.

Clustering

A common property of many social networks is cliques. Cliques are groups of friends, where every member of the group knows every other member. The inherent tendency to cluster is quantified by the clustering coefficient (Watts and Strogatz 1998). For a given node i within a network, with k_i neighbours, the degree of clustering around node i is defined to be the fraction of links the exist between the k_i neighbours and the $k_i (k_i-1)/2$ potential links. Let E_i , be the number of links that actually exist between the k_i neighbours. The clustering coefficient is then:

$$CC = \frac{1}{N} \sum_{i=1}^{N} \frac{E_i}{k_i(k_i - 1)}$$.

Subgraphs

One of the first properties of random graphs that Erdös and Rényi studied was the appearance of subgraphs. A graph G_1 consisting of the set of nodes P_1 and the set E_1 of edges is said to be a subgraph of $G = \{P,E\}$ if $P_1 \subset P$ and $E_1 \subset E$. The simplest examples of subgraphs are *cycles, trees,* and *complete subgraphs*. One of the most fundamental concepts in graph theory is a walk. Given a graph G , a walk is a sequence $(_0, _1), (_1, _2), \cdots (_{k-1}, _k)$ of vertices and edges from vertex v_0 to vertex v_k. The number of edges in the walk is its length. A walk in which all vertices are only visited once is known as a *path*. A walk in which all of the edges are distinct is a *trail*. A *cycle* is a closed loop of k edges, such that every 2 consecutive edges only have common nodes. A *tree* is of order k if it has k nodes and $k-1$ edges, and none of its sub-graphs is a cycle. The average degree of a tree of order k is $\langle k \rangle = 2 - \frac{2}{k}$, which approaches the limit 2 for large trees. A *complete subgraph* of order k contains k nodes and all the possible $k\frac{k-1}{2}$ edges, in other words, all the nodes in the subgraph are connected to all other nodes (Wilson 1994).

Criticality

Probably the most important finding from random graph theory was the discovery of a critical threshold at which giant clusters form (in this context, cluster refers to a group of connected nodes) (Erdös and Rényi 1961). Below this threshold, the network exists as a series of disconnected subgraphs. Above this threshold, the graph is one all-encompassing cluster. Figure 5.2 shows this critical phase change using a simple graph model. Initially, all nodes within the graph are disconnected. At each time step, new edges are introduced. Figure 5.2 (a) shows that as edges are added, the nodes very quickly move from being disconnected to being fully connected. While Figure 5.2 (b) shows the standard deviation of the size of the largest connected subregion, the greatest deviation occurs at the critical threshold: at either side of this point the network exists as a series of small-disconnected subregions or as one giant cluster. Finally, Figure 5.2 (c) shows the time required to traverse the graph from one node to another. Again, the maximum time required to traverse the graph exists just below this critical level. This phase change is particularly important in percolation and epidemic processes.

Figure 5.2. Example of criticality phenomena in the evolution of graphs

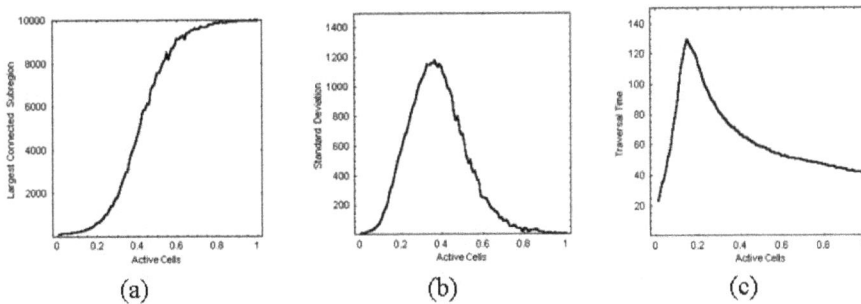

(a)   (b)   (c)

Critical phase changes in connectivity of simple random lattice, as the proportion of active cells increases (x-axes). (a) Average size of the largest connected subregion (LCS). (b) Standard deviation in the size of LCS. (c) Traversal time for the LCS. Each point is the result of 1000 iterations of a simulation. Note that the location of the phase change (here ≈ 0.2) varies according to the way we defined the connectivity within the model.

Real world complex networks

In this section we look at what is known about the structure of networks of different types. Recent work on the mathematics of networks has been driven largely by observation of the properties of actual networks, and attempts to model the processes that generate their topology. An excellent example of the dual theoretical and observational approach is presented in the groundbreaking work by Watts and Strogatz (1998). The remainder of this section examines the statistical properties of a number of complex networks. Figure 5.3 illustrates a number of different complex networks.

Ecological populations

Food webs are used regularly by ecologists to quantify the interactions between various species. In such systems, nodes represent species and the edges define predator–prey and other relationships that have positive and negative effects on the interacting species. Weights associated with the edges determines the magnitude of the relationship (May 2001). Solé and Montoya (2001) studied the topological properties of the Ythan Estuary (containing some 134 species with an average of 8.7 interactions per species), the Silwood Park ecosystem (Memmott et al. 2000) (with some 154 species with each species on average having 4.75 interactions) and the Little Rock Lake ecosystem (having a total of 182 species each of which interacts with an average of 26.05 other species). These studies found that these ecosystems were highly clustered and quite resilient against attack.

Social systems

Studies of human and other animal groups have determined that the structure of many communities conforms to the small worlds model. For example, Liljeros et al. (2001) studied the sexual relationships of 2,810 individuals in Sweden during 1996. The result was a network in which the degrees of the vertices conformed to power–law distribution. Another commonly studied class of systems consists of association networks. Given a centroid, the task is to calculate the distance (number of steps removed) one individual is from another. Perhaps the best known example is the network of collaboration between movie actors (Newman 2000), which takes the popular form of Bacon numbers (Tjaden and Wasson 1997). Other popular examples include Lewinsky numbers and scientific collaboration networks (Newman 2003). The most notable of the scientific collaboration networks are the Erdös numbers (Hoffman 1998) that use the famous mathematician Paul Erdös as the centroid.

World Wide Web and communications networks.

Computer networks and other communication networks have in recent times attracted a lot of attention. The World Wide Web (WWW), the largest information network, contained close to 1 billion nodes (pages) by the end of 1999 (Lawrence and Giles 1998). Albert et al. (2000) studied subsets of the WWW and found that they conformed to a scale free network. Other networks, such as the Internet and mobile phone networks, were also found to conform to the scale free network model (Faloutsos et al. 1999).

Neural networks

Studies of the simple neural structure of earthworms (Watts and Strogatz 1998) were able to study the topological properties of underlying networks. In Watts and Strogatz's model of these brains, nodes represented neurons, and edges were used to depict the synaptic connections. The neural structure of the earthworm was found to conform to the small world network model, in which groups of neurons were highly clustered with random connections to other clusters.

Figure 5.3. Examples of complex networks

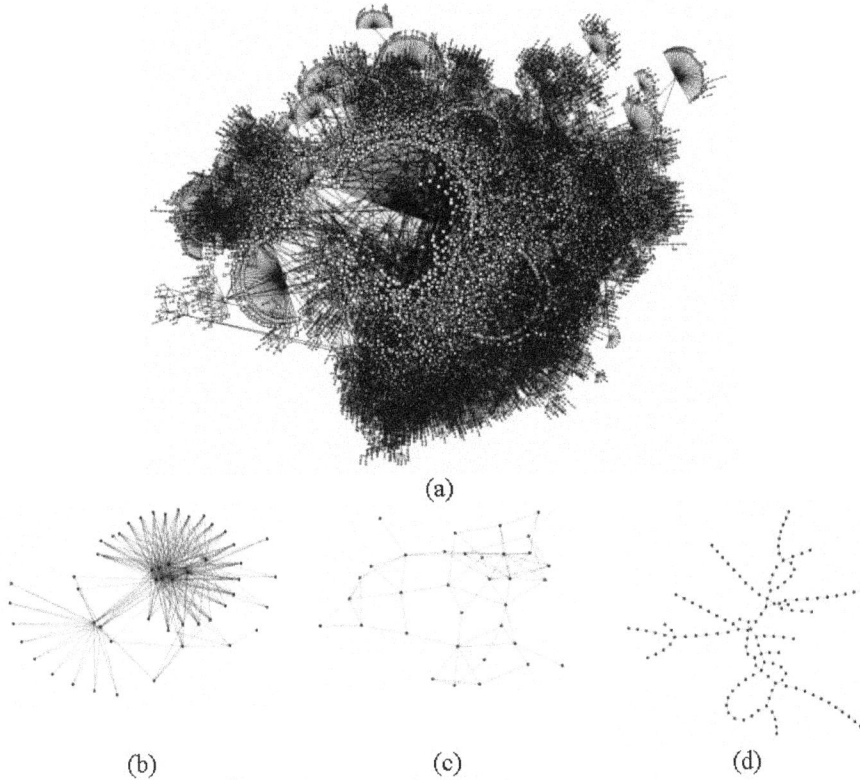

(a) the Internet, where nodes are routers and edges show physical network connections. (b) an ecosystem (c) professional collaboration networks between doctors; and (d) rail network of Barcelona, where nodes are subway stations and edges represent rail connections.

Models of complex networks

Complex networks can be generally divided into two major classes based on their degree distribution P_k, i.e. the probability that a node in the network is connected to k other nodes. This first class of graphs is referred to as exponential graphs, and the distribution of edges (i.e. the numbers per node) conforms to a Poisson distribution. This class of graphs includes the Erdös-Rényi type random graphs, and *small-world* networks.

Random graphs

Paul Erdös and Alfred Rényi (1959, 1960, 1961) were the first to introduce the concept of random graphs in 1959. The simple model of a network involves taking some number of vertices, N and connecting nodes by selecting edges from the N(N-1)/2 possible edges at random (Albert and Barabási 2002; Newman 2003). Figure 5.4 shows three random graphs where the probability of an edge being selected is *p=0, p=0.1* and *p=0.2.*

Figure 5.4. Erdös-Rényi model of random graph evolution

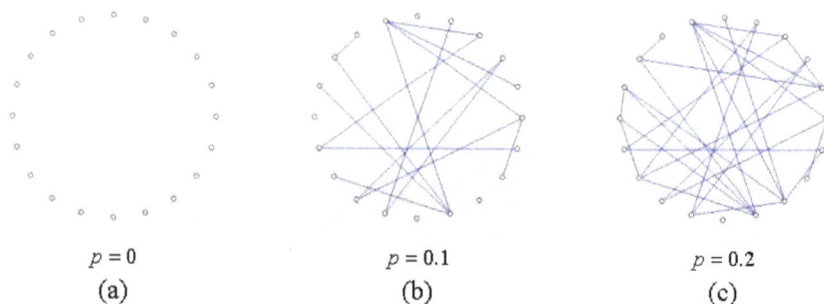

$p = 0$ $p = 0.1$ $p = 0.2$
(a)   (b)   (c)

(a) Initially 20 nodes are isolated. (b) Pairs of nodes are connected with a probability of p of selecting an edge. In this case (b) *p = 0.1* , (c) *p = 0.2,* notice how the nodes become quickly connected

The Erdös and Rényi random graph studies explore how the expected topology of the random graph changes as a function of the number of links (Strogatz 2001). It has been shown that when the number of links is below $1/N$, the graph is fragmented into small isolated clusters. Above this threshold the network becomes connected as one single cluster or giant component (Figures 5.2 and 5.4). At the threshold the behaviour is indeterminate (Strogatz 2001). Random graphs also show the emergence of subgraphs. Erdös and Rényi (1959, 1960, 1961) explored the emergence of these structures, which form patterns such as trees, cycles and loops. Like the giant component, these subgraphs have distinct thresholds where they form (See Figure 5.5).

Figure 5.5. Different subgraphs appear at varying threshold probabilities in a random graph (After Albert and Barabási 2002)

$$p \sim N^z$$

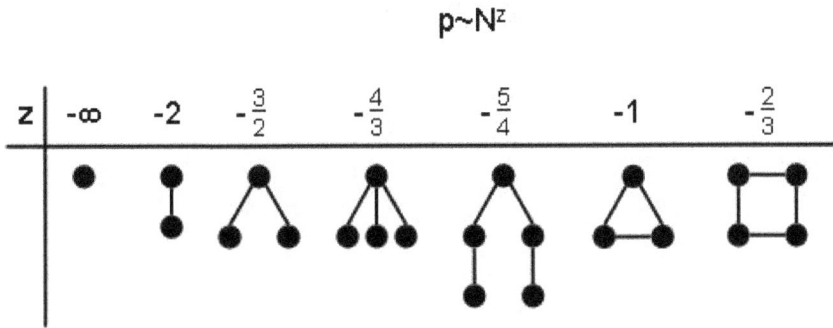

Small-world networks

In the late 1960s, Stanley Milgram (1967) performed the famous *small-worlds* experiment. While no physical networks were constructed during this experiment, the results do provide valuable insights into the structure of social networks. Essentially, the experiment examined the distribution of paths lengths in an acquaintance network by asking participants to pass a letter to one of their first-name acquaintances in an attempt to get the letter to the designated target. While most of the letters were lost, about one quarter reached the target person. One average the letter passed through the hands of between 5 and 6 people. This experiment was the source of the popular concept of *6 degrees of separation*.

The ground breaking work of Watts and Strogatz (1998) showed that many complex networks display two key features: they possessed the 6 degrees of separation phenomenon, that Milgram discovered; but locally they had many properties similar to that of a regular lattice. In an attempt to model these systems Watts and Strogatz (1998) proposed a one-parameter model, which interpolates between an ordered finite dimensional lattice and a random graph. The algorithm behind the model is as follows: start with a regular ring lattice with N nodes in which every node is connected to its first K neighbours ($K/2$ neighbours on either side); and then randomly rewire each edge of the lattice with a probability p such that self-connections and duplicate edges are excluded. This process introduces long-range edges which connect nodes that otherwise would be part of different neighbourhoods. Varying the value of p moves the system from being fully ordered ($p = 0$) to random ($p = 1$). Figure 5.6 shows steps in this transition. Small world networks have been used to describe a wide variety of real world networks and processes; Newman (2000) and Albert and Barabási (2002) provide excellent review of the work done in this field.

Figure 5.6. Progressive transition between regular and random graphs

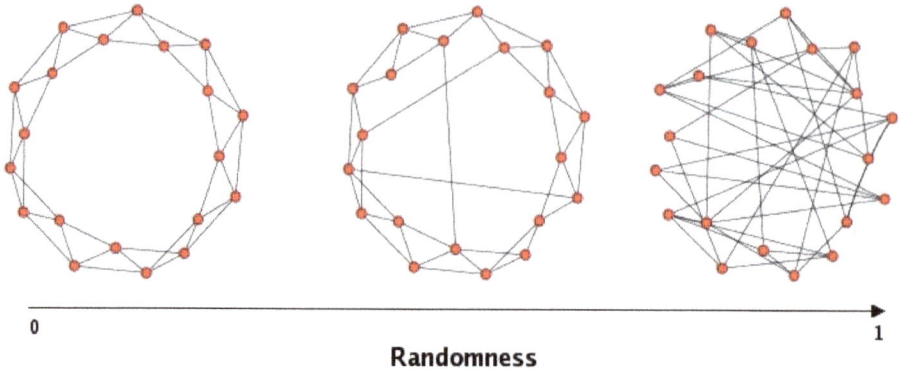

0   1

Randomness

Source: Watts and Strogatz 1998

Scale-free networks

The second class of graphs is referred to as *scale free networks*. Specifically, the frequency of nodes P_k with k connections follows a power-law distribution $P_k \approx k^{-\gamma}$, in which most nodes are connected with small proportion of other nodes, and a small proportion of nodes are highly connected (Albert and Barabási 2002). In exponential networks the probability that a node has a high number of connections is very low. In scale free networks, however, highly connected nodes (i.e. $k \gg \langle k \rangle$) are statistically significant (Albert and Barabási 2002).

Hierarchies and dominance

In competitive interactions between two individuals there is always a winner and a loser. If an individual A consistently defeats player B, then it is said that A dominates B. Such a relationship can be captured in a network in which the nodes represent players and arcs show which player is dominated. Animal behaviourists have frequently employed linear hierarchical ranking techniques to determine the dominant individuals within a community.

Dominance and linear hierarchical ranking

There are many procedures, of varying complexity, for ranking the members of a social group into a dominance hierarchy (see de Vries 1998 for a review of these techniques). In general dominance hierarchy, techniques can be divided into 2 categories. The first class of methods attempts to determine the dominance ranking by maximising or minimising some numerical criteria. The second class aims to provide a measure of overall individual success from which the rank can be directly derived. One relatively simple ranking method belonging to the second class is David's score (David 1987, 1988).

Individual ranks calculated with David's Score are not disproportionately weighted by minor deviations from the main dominance direction within dyads, because win/loss asymmetries are taken into account by the use of dyadic dominance proportions in the calculations. The proportion of wins/losses by individual i in his interactions with another individual j (P_{ij}) is the number of times that i defeats j (α_{ij}) divided by the total number of interactions between i and j (n_{ij}), i.e. $P_{ij} = \alpha_{ij}/n_{ij}$. The proportion of losses by i in interactions with j, is $P_{ji} = 1 - P_{ij}$. If $n_{ij} = 0$ then $P_{ij} = 0$ and $P_{ji} = 0$ (David 1988; de Vries 1998). The David's Score for each member, i of a group is calculated with the formula:

$$DS_i = w_i^1 + w_i^2 - l_i^1 - l_i^2$$

where w_i^1 represents the sum of i's P_{ij} values, w_i^2 represents the summed w_i^1 values weighted by the appropriate P_{ij} values, (see the worked example in the following section) of those individuals with which i interacted, l_i^1 represents the sum of i's P_{ji} values and l_i^2 represents the summed l_i^1 values (weighted by the appropriate P_{ji} values) of those individuals with which i interacted (David 1988: p. 108; de Vries 1998).

A worked example

Table 5.1 shows a worked example with calculated w, w2, l, and l2 values. Specifically for individual A, w_A^1 represents the sum of A's P_{Aj} values (i.e. $w_A^1 = 0.75 + 0.8 + 1 + 0.5 = 3.05$), and w_i^2 represents the summed w_i^1 values (weighted by the appropriate P_{ij} values) of those individuals with which A interacted (i.e. $w_i^2 = \left[(0.75 \times 2.05) + (0.8 \times 1.4) + (1 \times 0) + (0.5 \times 0.5) \right] = 2.90$. A's l^1 and l^2 values are calculated in a similar manner (David 1988; de Vries 1998).

Table 5.1. Example of ranking between players according to David's score

		A	B	C	D	E	w_i^1	w_i^2	DS
				Player Losses					
Player Victor-ies	A	—	4.5(0.75)	4(0.8)	7(1)	2(0.5)	3.05	2.90	3.20
	B	1.5(0.25)	—	4(0.8)	0(0)	5(1)	2.05	2.38	2.45
	C	1(0.2)	1(0.2)	—	0(0)	4(1)	1.40	1.52	-0.20
	D	0(0)	0(0.0)	0(0)	—	0(0)	0.00	0.00	-1.95
	E	2(0.5)	0(0.0)	0(0)	0(0)	—	0.50	1.52	-3.5
	l_i^1	0.95	0.95	1.60	1.00	2.50			
	l_i^2	1.80	1.03	1.52	0.95	3.02			

Bobby Fischer and the Ruy Lopez opening line

Bobby Fischer is probably the most famous chess player of all time and, in many peoples' view, the strongest. In the 1970s, Fischer achieved remarkable wins against top ranked grandmasters. His celebrated 1972 World Championship Match with Boris Spassky in Reykjavik made headline news all around the world. One of Fischer's favoured opening lines is the Ruy Lopez. The Ruy Lopez has been a potent weapon for Bobby throughout his career. Strategic play across the board suited Fischers talents. In his prime (and later in his career) Fischer was so proficient (dominant) in the main lines of the Ruy Lopez that many of his opponents chose irregular setups when attempting to defend their positions.

The Ruy Lopez was named after the Spanish clergyman, Ruy Lopez, of Safra, Estramadura. In the mid-sixteenth century, he published the first systematic analysis of the opening. The speed of development, flexibility and attacking nature, has seen the Ruy Lopez remain popular since its conception to the modern chess era. Figure 5.7 shows the board setup of the Ruy Lopez.

Figure 5.7. The Ruy Lopez opening line

All the games played by Fischer were converted into a network. Each player was represented as node or vertex within the system, with each game between 2 players shown as an arc. The arcs are drawn from loser to the winner, and given a weight of 1 for each victory of a given player. In the event of a draw/tie/stalemate, an arc was drawn in both directions and given a weight of $\frac{1}{2}$. In many ways, the data presented here is limited, as the dataset does not contain all top level games played within the Ruy Lopez opening system (some notable players are missing from the database), nor does it capture all the Ruy Lopez games played between players. Also, some of the games are incomplete or the result of the game is unknown. While these constraints will limit the accuracy of the result, the dataset is complete enough to detect trends and regularities, and will not greatly influence the general findings of this chapter.

Fischer's dominance

In this experiment, I examine the relationships between Fischer and his nearest neighbours. The network contains a total of 65 players. Figure 5.8 shows the network relationships between Fischer and his immediate opponents, the size of the arc represents the level of dominance of that player. Application of David's score to this network reveals that Bobby Fischer is the most successful player within his local neighbourhood. Table 5.2 lists the top 10 players in the neighbourhood.

Figure 5.8. Bobby Fischer's network of immediate opponents

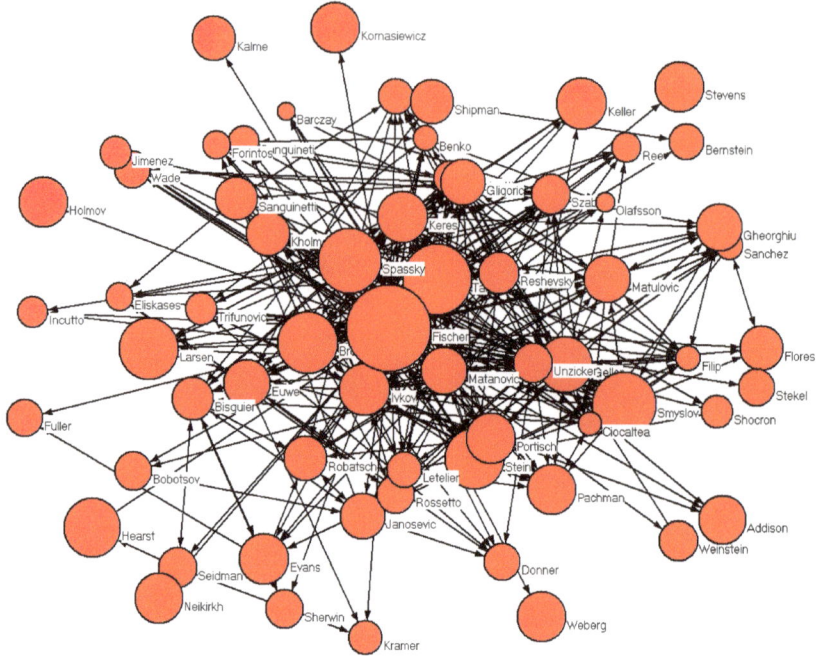

Table 5.2. Top 10 players in Fischer's gaming network

Rank	Player	Rank	Player
1.	Bobby Fischer	6.	David Bronstein
2.	Mikhail Tal	7.	Leonid Stein
3.	Vasily Smyslov	8.	Eliot Hearst
4.	Boris Spassky	9.	Yefim Geller
5.	Bent Larsen	10.	Borislav Ivkov

Some closing comments on dominance hierarchies

In this section, I have been able to show that, from the network formed by the competitive interactions of individuals, key or dominant individuals can be identified. The proposed approach can also be used to determine a 'pecking order' between individuals within a group, in which individuals are ranked from most to least dominant. While the context of illustration is the world of chess, the framework can be applied to any network in which dominance between elements can be established. One should note, however, that hidden patterns, such as incorrect weightings and missing links or nodes, can distort the final result.

Enforcement of social norms

As individuals, we are each better off when we make use of a common resource without making a contribution to the maintenance of that resource. However, if every individual acted in this manner, the common resource would be depleted and all individuals would be worse off. Social groups often display a high degree of coordinated behaviour that serves to regulate such conflicts of interest. When this behaviour emerges without the intervention of a central authority, we tend to attribute this behaviour to the existence of social norms (Axelrod 1986). A social norm is said to exist within a given social setting when individuals act in a certain way and are punished when seen not to be acting in accordance with the norm. Dunbar (1996, 2003) suggests that social structure and group size play important roles in the emergence of social norms and cooperative group behaviour.

Models of social dilemmas

All social dilemmas are marked by at least one deficient equilibrium (Luce and Raiffa 1957). It is deficient in that there is at least one other outcome in which everyone is better off. It is equilibrium in that no one has an incentive to change their behaviour. The Prisoners' Dilemma is the canonical example of such a social dilemma. The Prisoners' Dilemma is a 2×2 non-zero sum, non-cooperative game, where *non-zero sum* indicates that the benefits obtained by a player are not necessarily the same as the penalties received by another player, and *non-cooperative* indicates that no per-play communication is permitted between players. In its most basic form, each player has 2 choices: cooperate or defect. Based on the adopted strategies, each player receives a payoff.

Figure 5.9 shows some typical values used to explore the behaviour of the Prisoners' Dilemma. The payoff matrix must satisfy the following conditions (Rapoport 1966): defection always pays more, mutual cooperation beats mutual defection, and alternating between strategies doesn't pay. Figure 5.9 also shows the dynamics of this game, the vertical arrows signify the row player's preferences and horizontal arrows the column player's preferences. As can be seen from this figure, the arrows converge on the mutual defection state, which defines a stable equilibrium.

Figure 5.9. Prisoners' Dilemma

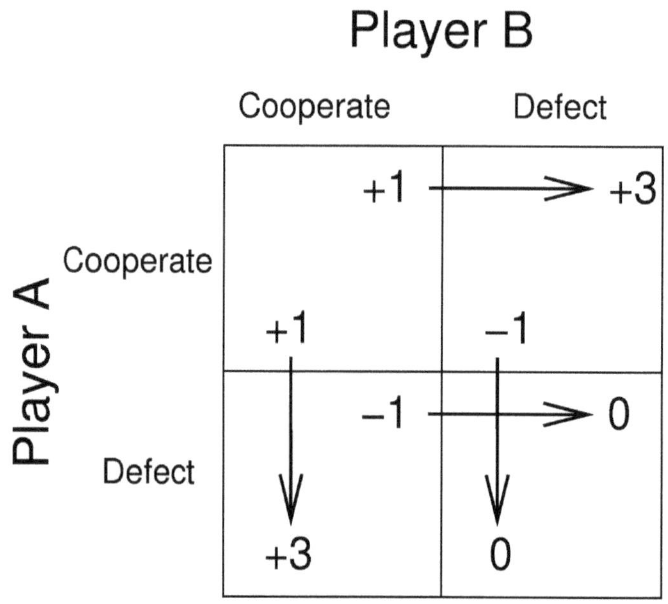

Player B

	Cooperate	Defect

The payoff structure of Prisoners' Dilemma: the game has an unstable equilibrium of mutual cooperation, and a stable equilibrium of mutual defection, this is shown by the arrows, moving away from mutual cooperation to mutual defection

While the 2-person Prisoner's Dilemma has been applied to many real-world situations, there are a number of problems that cannot be modelled. The Tragedy of the Commons is the best known example of such a dilemma (Hardin 1968). While n-person games are commonly used to study such scenarios, they generally ignore social structure, as players are assumed to be in a well-mixed environment (Rapoport 1970). In real social systems, however, people interact with small tight cliques, with loose, long-distance connections to other groups. Also, traditional n-person games don't allow players to punish individuals that do not conform to acceptable group behaviour, which is another common feature of many social systems. To overcome these limitations, I will introduce the Norms and Meta-Norms games, which are variations on the n-person Prisoners' Dilemma. They can easily be played out on a network and allow players to punish other players for not cooperating. Figure 5.10 illustrates the structure of the n-person Prisoners' Dilemma, Norms and Meta-Norms games.

Figure 5.10. The architecture of the Norms and Meta-norms games (After Axelrod 1986)

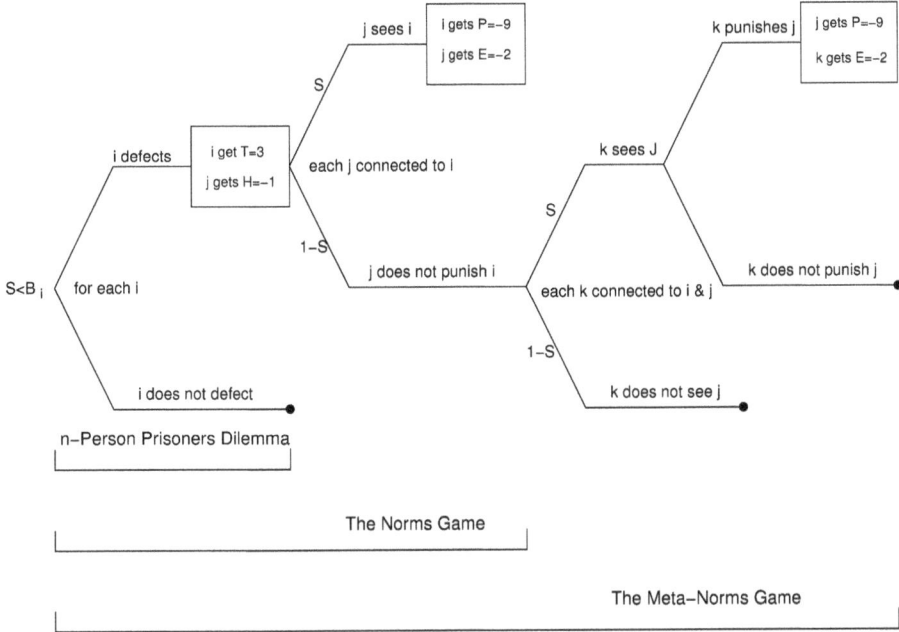

Both games start with a variation on the n-person Prisoner's Dilemma. The Norms Game allows players to punish those players caught defecting. The Meta-Norms Game allows players to punish those players who do not punish defectors.

The Norms game

The Norms game begins when an individual (i) has the opportunity to defect. This opportunity is accompanied by a known chance of being observed defecting (S) by one of i's nearest neighbours. If i defects, he/she gets a payoff T (temptation to defect) of 3, and each other player that is connected to i, receives a payoff H (hurt by the defection) of -1. If the player does not defect then each player receives a payoff of zero. To this point the game is equivalent to an n-person Prisoners' Dilemma played on a network (Rapoport 1970). However should i choose to defect, then one of his n neighbours may see the act (with probability 1-S) and may choose to punish i. If i is punished he receives a payoff of P=-9, however the individual who elects to punish i also incurs an expense associated with dealing out the punishment of E=-2. Therefore the enforcement of a social norm *to cooperate* requires an altruistic sacrifice.

From the above description it can be seen that each player's strategy has 2 dimensions. The first dimension of player i's strategy is *boldness* (B_i), which determines when the player will defect. Defection occurs when $S < B_i$. The second dimension of i's strategy is *vengefulness* (V), which is the probability that a player will punish another player if caught defecting. The greater the vengefulness the more likely they are to punish another player.

The Meta-Norms game

The Meta-Norms game is an extension of the Norms game. If player i chooses to defect, and player j elects not to punish i, and i and j have a common neighbour k, and k observes j not punishing i, then k has can punish j. Again, j receives the penalty $P=-9$, and, like the norm game, k receives an expenses $E=-2$.

Like the Prisoners' Dilemma, the Norms game and Meta-Norms game have unstable mutual cooperation equilibrium and a stable mutual defection equilibrium. The altruistic punishment is also an unstable strategy, as punishing an individual also requires a self-sacrifice. The stable strategy is mutual defection with no punishment for defectors. However, the global adoption of this strategy means that the population as a whole is worse off than if the unstable equilibrium strategy of mutual cooperation with punishment for defectors is adopted.

Model of social structure

Let us imagine 2 variables (B and V) that make up a strategy are each allowed to take on a value between [0,1]. The variables represent the probability of defecting and punishing respectively. The variables are each encoded as a 16 bit binary number (as per Axelrod 1986). The evolution of players' strategies proceeds in the following fashion: (1) A small world network of 100 players with a degree of randomness p is created; (2) Each player is seeded with a random strategy; (3) The *score* or *fitness* of each play is determined from a given player's strategy and the strategies of the players in their immediate neighbourhood; (4) When the scores of all the players are determined, a weighted roulette wheel selection scheme is used to select the strategies of the players in the next generation; (5) A mutation operator is then applied. Each bit has a 1 per cent chance of being flipped; (6) Steps 3–5 are repeated 500 times, and the final results are recorded; (7) Steps 2–6 are repeated 10,000 times. (8) Steps 1–7 are repeated for p values between 0 and 1 in increments of 0.01. The above experimental configuration was repeated for both the Norms game and the Meta-Norms game. Figure 5.11 shows the results of these simulations.

Figure 5.11. Simulation results

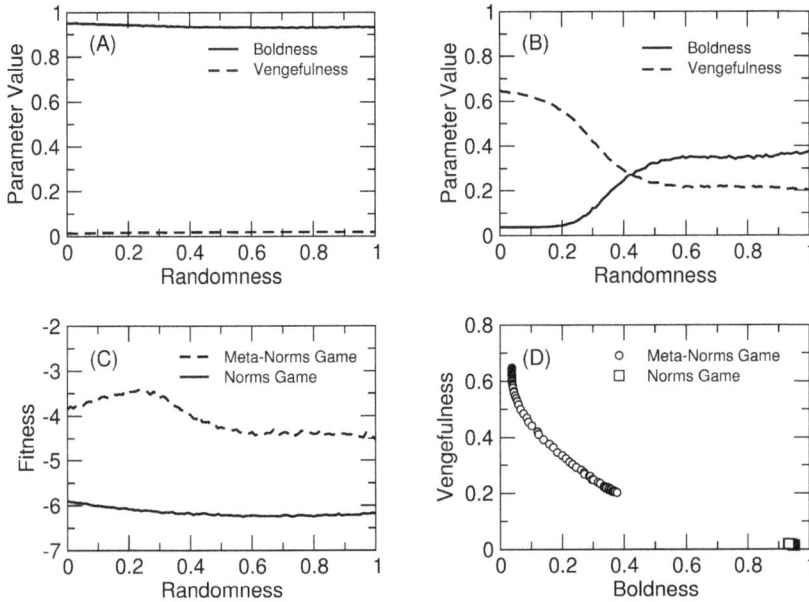

(A) Average values for boldness and vengefulness over social networks with varying degrees of randomness for the Norms game. (B) Average values for boldness and vengefulness over social networks with varying degrees of randomness for the Meta-Norms game. (C) Comparison of the fitness values for the Norms game and Meta-Norms game. (D) Trade-off between Boldness and Vengefulness

From the simulation results we can see that, regardless of the social structure, the first order altruistic punishment isn't enough to enforce the social norm of mutual cooperation. Figure 5.11(A) shows that, regardless of the social structure, the vengefulness decreases to zero, and boldness increases toward one. Essentially all players are attempting to exploit the shared resource, with no fear of being punished. However, for the Meta-Norms game, with second-order punishment, there is a distinct set of circumstances when the population as a whole will not exploit the common resource. Figure 5.11(B) shows that, when the social structure is regular and highly clustered, players boldness decreases, but as the social structure becomes more random (and clustering breaks down), the boldness of a given player increases, and each individual attempts to exploit the common resource. However, the level of exploitation is lower than that observed in the Norms game. These differences in system behaviour are also seen in the average payoff received by a player (Figure 5.11(C)). The average payoff per player in the Meta-Norms game is always higher that that received in the Norms game. The average payoff for the Meta-Norms game maximises just before the transition to a state of global exploitation. Statistical analysis of the network structure reveals that this maximum payoff point coincides with the breakdown of clustering within the network. Finally, Figure 5.11(D) depicts the trade-off between vengefulness and boldness. The Norms game (squares) converges to a strategy of low vengefulness and high boldness. While the Meta-Norms game produces a range of behaviours (circles), from the plot it can be seen that there is a trade-off between boldness and vengefulness. The Meta-Norms game produces a wide variety of strategies. These strategies are governed by the topology of the underlying social network. The trade-off surface can be thought of as the set of viable strategies, as nonviable strategies (such as high boldness and vengefulness) are selected against.

Discussion and implications

The results from the previous section provide a number of interesting insights into the emergence of social norms and group behaviour. Social structure and second order interactions seem to play an important role in the evolution of group behaviour. In the wider literature, there are many recorded instances where these 2 factors have been observed to influence group behaviour. Here I will explore 3 examples.

Animal innovation

Japanese macaque were among the first primates observed by humans to display innovation and diffusion of new novel behaviours to other group members (Reader and Laland 2004). While many individual animals invent new behaviour patterns, most new behaviours (even if they are beneficial) are unlikely to become fixed within the community. Reader and Laland (2002) have shown that there is a link between the social structure of primates and the frequency with which new technologies are uptaken. Populations that tend to be more cliquish are more likely to adopt a new behaviour as member of the clique help to reinforce the novel behaviour.

Social cohesion

Dunbar (1996) has shown that there is a correlation between neocortex size and the natural group size of primates. Also correlated with neocortex size is the cliquishness of the social structure. Dunbar (2003) conjectures that the increase in neocortex size may mean that individuals can manage and maintain more group relationships. The ability to maintain more complex relationships may allow individuals to locally enforce social behaviour. It has also been observed that, when primate groups grow too large, social order breaks down and the troop split into 2 or more smaller troops in which social order is re-established (Dunbar 2003).

Control of social behaviour

The notion of Meta-Norms is widely used in denunciation in communist societies. When authorities accuse someone of doing something wrong, others are called upon to denounce the accused. Not participating in this form of punishment is itself taken as a defection against the group and offenders are punished.

Some comments on the enforcement of social norms

In this section, I explored the emergence and enforcement of social norms through the use of 2 variations on the n-persons Prisoners' Dilemma. The simulation results suggest that a combination of second order interactions, altruistic punishment and social structure can produce coherent social behaviour. Such features have been observed to enforce norms in a number of social systems. The results from this study open a number of interesting future directions:

*   As conjectured by Dunbar (2003), social order in primate troops breaks down when the troop becomes too large. This raises the question: What is the relationship between link density, number of nodes and other network statistics, and how do these statistics influence the behaviour of evolutionary games such as those described in this chapter?
*   Coalitions and factions form and dissolve through time. How do the general results change if the underlying network is allowed to evolve?
*   Several studies (Luce and Raiffa 1957) have shown that concepts such as the Nash equilibrium don't hold when rational players are substituted for human players. Do the patterns and tradeoffs described previously hold when rational computer players are replaced by human decision makers?

All these questions require further experimentation but can be explored in the context of the framework proposed here.

Closing comments

From social networks to large scale critical infrastructure, the systems that surround us are large and complex. Despite their obvious differences these systems share a number of common regularities—such as the small world properties. In this chapter we have reviewed some recent work on the structure and function of networked systems. Work in this area has been motivated to a high degree by empirical studies of real-world networks such as the Internet, the World Wide Web, social networks, collaboration networks, citation networks and a variety of biological networks. We have reviewed these empirical studies, focusing on a number of statistical properties of networks that have received particular attention, including path lengths, degree distributions, and clustering. Moving beyond these statistical regularities, the structure and nature of the interactions between the elements within a system can provide insights into the dynamics taking place in the network and as well as the way the network is being shaped by the dynamics.

In this chapter I have explored several notions. First, by understanding the nature of the interaction between players of chess we can extract dominance hierarchies. These hierarchies can be examined through time to gain an understanding as to how the system evolves, in this case, how the dynamics affect the topology. I also explored how the structure of a network affects the dynamics taking place upon it and showed that the regularities we see in social systems may be a consequence of the dynamics taking place. These regularities are also echoed across other types of social systems, suggesting universal laws of organisation.

In looking forward to future developments in this area, it is clear that there is much to be done. The study of complex networks is still in its infancy. Several general areas stand out as promising for future research. First, while we are beginning to understand some of the patterns and statistical regularities in the structure of real world networks, our techniques for analysing networks are, at present, no more than a grab-bag of miscellaneous and largely unrelated tools (Newman 2003). We do not yet, as in some other fields, have a systematic program for characterising network structure. We need a systematic framework by which we can analyse complex networks in order to identify key dynamical and structural properties. Second, there is much to be done in developing models of networks, both to help us understand network topology and to act as a substrate for the study of processes taking place on networks (Watts and Strogatz 1998; Newman 2003). Finally, and perhaps the most important direction for future study, is the behaviour of processes taking place on networks. The work describing the interplay between social structure and game theoretic decision making is only a timid first attempt at describing such processes, and yet this, in a sense, is the ultimate goal in the field: to understand the behaviour of the network systems that surround us.

Acknowledgements

I would like to thank Pip Pattinson, Garry Robins, Steve Borgatti, Markus Brede and John Finnigan for a number of useful discussions about this chapter.

References

Albert, R., H. Jeong and A-L. Barabási (2000) Error and attack tolerance of complex networks. *Nature* 406: 378–82.

Albert, R. and A-L. Barabási (2002) Statistical mechanics of Complex Networks. *Review of Modern Physics* 74: 47–97.

Axelrod, R. (1986) An evolutionary approach to Norms. *American Political Science Review* 80(4): 1095–111.

David, H.A. (1987) Ranking from unbalanced paired comparison data. *Biometrika* 74:432–36.

David, H.A. (1988) *The Method of Paired Comparisons*. New York: Hafner.

De Vries, H. (1998) Finding a dominance order most consistent with a linear hierarchy: a new procedure and review. *Animal Behavior* 55: 827–43.

Dunbar, R.I.M. (1996) *Grooming Gossip, and the Evolution of Language*. Harvard University Press.

Dunbar, R.I.M. (2003) The social brain: mind, language, and society in evolutionary perspective. *Annual Review of Anthropology* 32: 163–81.

Erdös, P. and A. Rényi (1959). On random graphs, I. *Publicationes Mathematicae Debrecen* 6: 290–97.

Erdös, P. and A. Rényi (1960) On the evolution of random graphs. *Public Mathematical Institute of the Hungarian Academy of Sciences* 5: 17–61.

Erdös, P. and A. Rényi (1961) On the evolution of random graphs. *Bulletin of the International Statistics Institute* 38: 343–47.

Faloutsos, M., P. Faloutsos and C. Faloutsos (1999) On power-law relationships of the internet topology. *SIGCOMM* 99: 251–62.

Hardin, G. (1968) The tragedy of the commons. *Science* 162: 1243–48.

Hoffman, P. (1998) *The Man Who Loved Only Numbers: The Story of Paul Erdos and the Search for Mathematical Truth*. New York: Hyperion.

Lawrence, S. and C.L. Giles (1998) Searching the World Wide Web. *Science* 280: 98–100.

Liljeros, F ., C.R. Edling, L.A. Nunes A, H.E. Stanley and Y. Aberg (2001) The Web of Human Sexual Contacts. *Nature* 411: 907–8.

Luce, R.D. and H. Raiffa (1957) *Games and Decisions: Introduction and Critical Survey*. New York: Dover.

May, R.M. (2001) *Stability and Complexity in Model Ecosystems*. Princeton Landmarks in Biology. Princeton: Princeton University Press.

Memmott, J., N.D. Martinez and J.E. Cohen (2000) Predators, parasitoids and pathogens: species richness, trophic generality and body sizes in a natural food web. *Journal of Animal Ecology* 69: 1–15.

Milgram, S. (1967) The small world problem. *Psychology Today* 2: 60–7.

Newman, M.E.J. (2000) The structure of scientific collaboration networks. *Proceedings of the National Academy of Sciences* 98(2): 404–9.

Newman, M.E.J. (2003) The structure and function of complex networks. Physics Preprint: arXiv:cond-mat/0303516v1.

Rapoport, A. (1966) *Two Person Game Theory*. New York: Dover.

Rapoport, A. (1970) *N-Person Game Theory: Concepts and Applications*. New York: Dover.

Reader, S.M. and K.N. Laland (2002) Social intelligence, innovation, and enhanced brain size in primates. *Proceedings of the National Academy of Sciences* 99(7): 4432–41.

Reader, S.M. and K.N. Laland (2004) *Animal Innovation*. Oxford: Oxford University Press.

Solé, R.V. and J.M. Montoya (2001) Complexity and fragility in ecological networks. *Proceedings of the Royal Society* Series B 268: 2039–45.

Strogatz, S.H. (2001). Exploring complex networks. *Nature* 410: 268–76.

Tjaden, B.C. and G. Wasson (1997) The Oracle of Bacon. Available at Virginia. http://www.cs. virginia.edu/oracle

Watts, D.J. (2003) *Small Worlds: The Dynamics of Networks between Order and Randomness*. Princeton: Princeton University Press.

Watts, D. J. and S. Strogatz (1998) Collective dynamics of 'small-world' networks. *Nature* 393: 440–2.

Wilson, R.J. (1994) *Introduction to Graph Theory*. Boston: Addison Wesley.

6. Integration and Implementation Sciences: Building a New Specialisation

Gabriele Bammer

Abstract

Developing a new specialisation—Integration and Implementation Sciences—may be an effective way to draw together and significantly strengthen the theory and methods necessary to tackle complex societal issues and problems. It would place complexity science in broader context and link it to a range of complementary concepts and skills. This chapter presents an argument for such a specialisation. It outlines three sets of characteristics which will delineate Integration and Implementation Sciences. First, the specialisation will aim to find better ways to deal with the defining elements of many current societal issues and problems, namely complexity, uncertainty, change and imperfection. Second, there will be three theoretical and methodological pillars for doing this: systems thinking and complexity science; participatory methods; and knowledge management, exchange and implementation. Third, operationally, Integration and Implementation Sciences will be grounded in practical application and generally involve large-scale collaboration. The chapter concludes by examining where Integration and Implementation Sciences would sit in universities and outlining a program for the further development of the field.

Introduction

Researchers, funders and research end-users are increasingly appreciating that new research skills must be developed if human societies are to be more effective in tackling the complex problems that confront us and in sustaining the sort of world we wish to live in. Researchers must collaborate and integrate across traditional boundaries. They must bring together academic disciplines, as well as becoming more involved in the implementation of research in policy, product and action.

There is now a critical mass of researchers who have turned their efforts to meeting these challenges. By working on real-world problems they have made theoretical and methodological advances to help deal with complexity, uncertainty, change and imperfection—the primary characteristics of the vital issues modern societies face. Developments have occurred in research on the environment, public health, business and management, national security, and other applied topics, but have typically been isolated, with little interaction and communication across these areas. There have been only low levels of intellectual cross-fertilisation and learning, and limited exploitation of the significant synergies between approaches.

Mainstream research is progressively starting to embrace these new investigative imperatives, but is even more poorly connected to existing knowledge. This has led to considerable duplication of effort, reinventing, usually at less sophisticated levels, methods and frameworks that already exist.

The time is ripe for coalescence and co-ordination. An effective and efficient mechanism is to develop a new specialisation—Integration and Implementation Sciences. This involves bringing together and providing a clear identity and accepted place for a large college of peers, who can be both supportive and critical.

The vision for Integration and Implementation Sciences is to provide solid theoretical and methodological foundations to allow complex societal issues to be systematically addressed using evidence-based approaches. The three pillars are:

- systems thinking and complexity science, which orient us to looking at the whole and its relationship to the parts of an issue;
- participatory methods, which recognise that all the stakeholders have a contribution to make in understanding and, often, decision making about an issue; and
- knowledge management, exchange and implementation, which involves appreciating that there are many forms of knowledge and ways of knowing (diverse epistemologies), provides enhanced methods for accessing knowledge realising that both volume and diversity are current barriers, and involves developing better understanding of how action occurs, in other words, how policy is made, how business operates, how activism succeeds, and how action is and can be influenced by evidence.

Like statistics and epidemiology, the specialisation will advance through application to a diverse range of problems. Similarly, Integration and Implementation scientists will not necessarily have content expertise. Their work will complement, rather than replace, traditional disciplinary and specialist perspectives. Collaboration is therefore central to how Integration and Implementation Sciences operates. What Integration and Implementation scientists can contribute to these partnerships includes:

- enhanced skills in scoping problems and issues, ensuring multi-disciplinary and multi-sector involvement, and making clear where the boundaries around the problem have been set and the implications of those decisions for inclusion, exclusion and marginalisation of stakeholder groups;

- enhanced ways of thinking about integration and a range of integrative tools, including specific skills in systems-based modelling and participatory approaches;
- alternative conceptualisations of the research process, which may lead to different and innovative research approaches and the development of hybrid epistemologies;
- re-aggregation of knowledge and understanding that has been developed in separate disciplines and practice arenas;
- enhanced ability to identify and understand emergent properties, i.e., properties that disappear when a system is studied in disaggregated segments;
- enhanced understanding of policy, product development and action, and how these can be influenced by research;
- bridging between research and practice by helping develop new roles, such as boundary spanners and knowledge brokers;
- enhanced knowledge management and knowledge implementation tools;
- expanded ways of taking uncertainty into account and of managing less than perfect outcomes;
- expanded ways of encompassing change in both research and practice; and
- enhanced appreciation of how to improve collaborative processes in research, including ensuring that appropriate researchers and sectoral representatives are included, that their world-views are made explicit, that their interests are accommodated, that different strengths are harnessed, that communication mechanisms are strong, and that conflicts are appropriately mediated.

No Integration and Implementation scientist will be expert in all of these skills. However, they will have a broad framework of knowledge encompassing all these aspects and deep knowledge of some of them. They will be able to bring in colleagues to fill skill gaps. And they will be able to recognise when leading edge theory and methods are being used, when breakthroughs in thinking have been made, and when wheels are being reinvented. This is identical to how other specialisations and disciplines operate.

This chapter aims to present a broad sweep of ideas about a new specialisation. I focus on the practicalities of what the new specialisation would involve and how it would fit structurally, rather than building the case for its need, or linking the arguments to the extensive discourse on the philosophy of science, the long-standing debate about the role of scholarship and universities in society, or discussions about the future of science. Readers interested in these more philosophical issues can refer to works on critical realism (for example, Mingers 2000), post-normal science (for example, Funtowicz and Ravetz 1993), and consilience (Costanza 2003; Wilson 1998), as well as seminal works by authors such as Kuhn (1970), Ravetz (1996) and Gibbons, Nowotny and colleagues (Gibbons et al. 1994; Nowotny et al. 2001). Further, I do not cite the extensive literature underpinning each of the areas covered here. My aim is not to write a definitive treatise, but to spark discussion and stimulate action to build stronger links between the core methodologies and to embed them more firmly in academic structures.

Defining key elements of the social issues

I propose four key elements for the sorts of issues Integration and Implementation Sciences are designed to tackle.

Complexity has many dimensions, including an extensive array of factors, with both linear and nonlinear connections and interdependencies, and a range of relevant political, cultural, disciplinary and sectoral perspectives. Geographical and temporal scales can be huge. An important dimension of complexity is identifying and understanding emergence.

A necessary adjunct to complexity is *uncertainty*. In dealing with any complex issue or problem, there will always be many unknowns, including facts, causal and associative relationships, and effective interventions. Some unknowns result from resource limitations on research and some result from methodological limitations, while some things are simply unknowable. There are epistemological, ethical, organisational and functional aspects to dealing with uncertainty, ignorance and risk.

The unknowns are compounded by constant *change*, occurring on many fronts including biological evolution (for example, the development of new communicable diseases); scientific, technological and economic developments; changes in international relations; and manifold intended and unintended consequences of local, national and international policy and programs.

Perfect knowledge and solutions are impossible. *Imperfection* also has many dimensions. Dealing with complexity involves setting boundaries to the approach we take, and where we set boundaries is crucial in determining what is included, excluded and marginalised. Uncertainty and change also necessarily lead to imperfection. Further, social issues are deeply contextualised: an excellent solution in one person's eyes is anathema to another.

Theoretical and methodological pillars

The key theoretical and methodological foundations to Integration and Implementation Sciences are:

- systems thinking and complexity science;
- participatory methods; and
- knowledge management, exchange and implementation.

These provide a range of conceptual and methodological tools for dealing with complexity, uncertainty, change and imperfection, including modelling, decision and risk analyses, deliberative democracy processes, and principled negotiation processes. All are areas where considerable research has already been undertaken, but where the current situation is characterised by fragmentation and marginalisation.

Systems thinking and complexity science

While both systems thinking and complexity science concern themselves with looking at wholes, they encompass several schools of thought, which are noted for their indifference—at best—and—at worst—animosity to each other. There is not only a gulf between systems thinking and complexity science, but also within different branches of systems thinking. A major challenge would still seem to be to develop key overarching theoretical concepts that throw the commonalities and differences into sharper relief.

Participatory methods

The importance of participatory methods is based on recognition that the various stakeholders think differently about the same issue, and that exploring, sharing and synthesizing these different understandings enriches our knowledge about an issue. It can often trigger a new way to look at and contend with an issue. In addition, for some issues, an appropriate way of dealing with uncertainty and imperfection is to give the stakeholders a more direct role in making decisions.

Knowledge management, exchange and implementation

Knowledge management, exchange and implementation is a way of characterising a number of interrelated issues:

- It is a way of appreciating that there are many forms of knowledge and ways of knowing. These diverse epistemologies are important in three key ways: assisting researchers and practitioners to understand each other; appreciating that there are a number of ways in which research can be undertaken; and facilitating research that crosses disciplines.
- It provides enhanced methods for accessing knowledge. Researchers and practitioners are both confronted by and contribute to an information glut; the sheer volume of information makes it difficult to navigate. Information science is tackling this issue, for example, by improving cataloguing and search methods. These difficulties are compounded by the diversity in forms of knowledge.
- It is a way of developing better understanding of how action occurs, in other words, how policy is made, how business operates, and how activism

succeeds. This is particularly relevant to Integration and Implementation Sciences in terms of how action is and can be influenced by evidence.

Grounding in practical collaboration

The third, operational aspect of Integration and Implementation Sciences is that it has a firm footing in practical application and generally involves large-scale collaboration. As outlined above, this makes Integration and Implementation Sciences similar to disciplines such as statistics and epidemiology. The analogy with statistics, in particular, is drawn out further in examination of where this new specialisation fits in universities.

Where would Integration and Implementation Sciences sit in universities?

Although the theory and methods of Integration and Implementation Sciences are developed through engagement with practical problems, there is no home base to which breakthroughs can be reported and where they can be critically assessed. This is an important difference from disciplines such as statistics, in which such home departments play a critical role. The development of the specialisation of Integration and Implementation Sciences is a way of establishing such a home base.

The lack of a specialist or disciplinary core also means that those engaged in Integration and Implementation Sciences lack a unifying identity. As a consequence, researchers mainly characterise themselves either through their area of application (for example, as human ecologists, environmental scientists or management specialists); or through a key approach or method (such as action researcher or system dynamics specialist).

Identity as a specialist in Integration and Implementation Sciences complements, rather than replaces, these existing identities. The difference that a specialisation will make is that specialists in Integration and Implementation Sciences will be able to identify with a broader cadre of researchers and develop more rounded skill sets. For example, while there is considerable overlap in the modes of operation of researchers using soft systems methods and action researchers, there is little crossover between these groups in terms of university coursework, professional associations or even research collaboration. Soft systems researchers often have very polished systems methods, but under-developed participatory skills, with the opposite holding for action researchers. Bringing these two groups together under a unifying umbrella would increase the chances that both would bring a more highly developed set of theory and methods to bear on the problems they deal with.

Figure 6.1 illustrates the relationship between the home base (the central circle) and the key sectors in which Integration and Implementation Sciences are applied and developed. Some researchers will work predominantly in the home base, focusing on the development of theory and methods in Integration and Implementation Sciences and applying them to a broad range of problems. Some researchers (second circle) will build detailed knowledge of a single sector, such as environment or international development and use this as the basis for the development of Integration and Implementation Sciences theory and methods. A third group of researchers will be less interested in the development of theory and methods, but will focus much more on their application (outside circle).

Figure 6.1. The relationship between the home base and the key sectors for Integration and Implementation Sciences

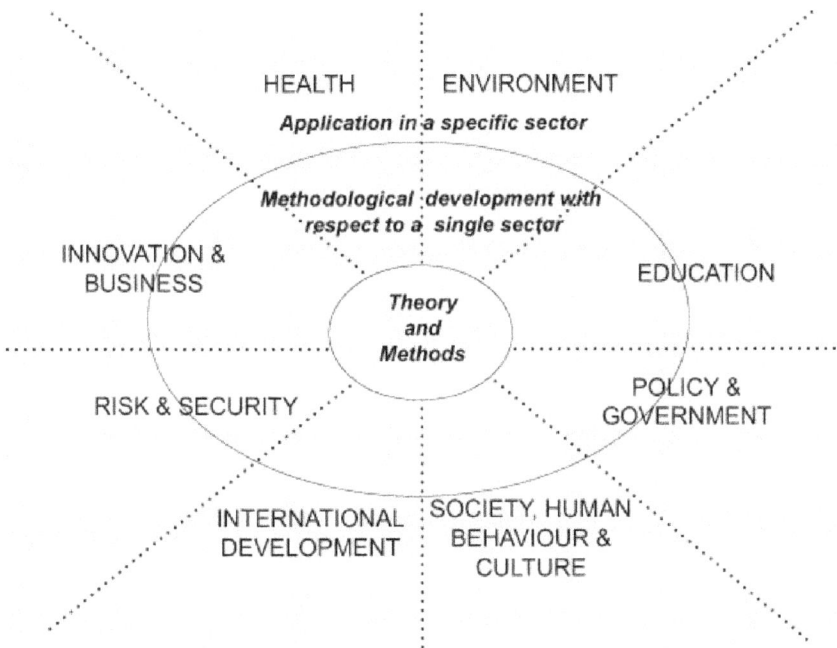

A specialisation will also provide a one-stop shop for researchers seeking access to integration and implementation skills. This will meet a growing demand, as appreciation of the need for these skills increases. Where new researchers gain a foothold currently tends to be arbitrary, as it is extremely difficult to obtain a comprehensive overview of the Integration and Implementation Sciences field, existing knowledge, and key players. Thus, researchers new to the area often spend considerable time searching for resources and key contacts and their early work often involves significant duplication of existing knowledge.

The same holds for policy makers and other practitioners seeking to link with researchers with Integration and Implementation Sciences skills. There is nowhere for them to go to receive an overview of what Integration and Implementation Sciences can offer and to match needs with available approaches. If practitioners contact universities or other public good research organisations, the aspect of Integration and Implementation Sciences they link with, and whether they indeed manage to link with any form of Integration and Implementation Sciences, is largely a matter of chance. Outside universities, there are now a large number of commercial, consultant-based packages available, but most are limited in the approaches they offer and there are no mechanisms for quality control.

This last point is not intended as a criticism of consultants practising approaches that are part of Integration and Implementation Sciences. Indeed, they have largely been responsible for the development of this field. Many have left universities to set up their own businesses because this has given them more freedom to undertake the practice-based research they care about. Further, researchers who survive in universities and other research organisations are often required to be wholly or partially self-funded, often through consultancy work. Commercially-based researchers are not in a position to develop colleges of critical peers, overarching associations, robust and comprehensive theoretical and methodological bases, or curricula for undergraduate and postgraduate education—in other words, they are not in a position to develop a specialisation. That is the role of universities. Thus, the development of a specialisation will also provide a solid underpinning for commercial consultancy practice, a place where consultants can learn new skills, or update existing ones, and where they can feed back lessons from their practice-based experience to invigorate and progress the development of theory and methods. Given that consultants rely on the methods and other intellectual property they develop to make their living, incorporating these into the academy will also be a challenge.

Statistics as a useful analogy

So far, I have dealt with the importance of a home base for Integration and Implementation Sciences. Here I will expand on this idea, using analogies between statistics and Integration and Implementation Sciences.

Statistics is embedded in the academy at three levels. First, there are home-base departments where theory and methods of statistics are developed and advanced. Second, other significant academic departments incorporate statistical training into their core curriculum and have at least some staff with a strong statistical bent. For example, disciplines such as biology, psychology, sociology and geography provide core training in statistics, particularly as it is relevant to the discipline involved, and have staff and research programs with a strong quantitative orientation. In addition, multidisciplinary departments such as public health often employ statisticians who are willing to work on public health problems. Third, there is an expectation that a large proportion of staff and students throughout the academy will have a basic level of statistical competence.

Like statistics, some elements of Integration and Implementation Sciences are already embedded in other significant academic areas. For example, many departments and centres dealing with environmental issues incorporate integrated assessment, other systems approaches and participatory approaches in their teaching and research. Public health departments often have a strong orientation to participation and implementation. However, the incorporation of Integration and Implementation Sciences is largely idiosyncratic, and there is generally little interaction between departments with different content area expertise about core or best methods. Some approaches that are key elements of Integration and Implementation Sciences have become standard in certain established academic areas. For example, most law schools now include principled negotiation (alternative dispute resolution) in their teaching, if not research.

As I have already pointed out, unlike statistics, Integration and Implementation Sciences has no home base or shared understanding of what this area encompasses. Nor is there the same level of individual competence among researchers in Integration and Implementation Sciences as there is in statistics. While many staff and students throughout the academy have basic competencies, such as building trust, thinking laterally, and seeing interconnections, (and some have very advanced abilities), these tend to be seen as personal attributes rather than academic skills. Furthermore, staff and students tend to be left to their own devices in the development of such expertise. Certainly, the building blocks for a solid home base for Integration and Implementation Sciences exist, and establishing home base departments would have positive spin-offs for established disciplines and specialisations, as well as for individual staff and students.

Statistics provides another useful analogy, namely the comfortable co-existence of diversity. Some statisticians are trained predominantly in statistics and work on a variety of problems, while others have training in statistics and another discipline and work largely in a particular area—for example, health. It is easily conceivable that some of those trained in Integration and Implementation Sciences would work on a wide range of problems, while others would work in more depth in areas such as environmental sciences or security.

The relationship between Integration and Implementation Sciences and traditional disciplines might be somewhat different from the relationship of statistics and other traditional disciplines. Those trained in Integration and Implementation Sciences and a traditional discipline might be expected to focus particularly on bringing that disciplinary perspective to the understanding of a complex problem, rather than (or in addition to) advancing the discipline. Certainly, a key task of Integration and Implementation Sciences is to harness and build on disciplinary strengths. The disciplines have developed and continue to elaborate a wealth of theoretical, methodological and content knowledge. Further, the disciplines themselves recognise the importance of building effective ways to draw together their individual strengths. There may be a case here for reinvigorating multi-disciplinary as well as inter- and trans-disciplinary approaches, with a particular focus on different methods for integrating diverse discipline-based knowledge and methods.

Statistics does not, however, provide a complete analogy. Statistics is obviously a well-developed and defined academic area, with a solid mathematical core. There are a range of widely adopted standard techniques and an array of known challenges which stimulate ongoing research. Integration and Implementation Sciences is poorly defined, with no widespread agreement about what the field does or does not encompass. As outlined above, some methods, such as principled negotiation, are relatively well defined and accepted, while others are idiosyncratically developed and applied. Even without a clear framework, the scope of Integration and Implementation Sciences is likely to be considerably broader than that of statistics. It seems unlikely that one core concept will lie at the heart of Integration and Implementation Sciences, in the same way that probability forms the nucleus for statistics. This is where the real developmental challenges for Integration and Implementation Sciences lie.

Challenges in developing a new specialisation

There are a number of key challenges in developing a specialisation of Integration and Implementation Sciences, including:

- achieving agreement on whether a specialisation is appropriate, likely to achieve the desired outcomes, and worth the downsides;
- constructing a coherent specialisation from disparate *bits*, many of which now have their own traditions. Some *bits*, such as participatory methods, principled negotiation techniques and information science, can potentially be fully encompassed within the new specialisation. Others, such as the mathematical development of complexity science, for example, may fit more comfortably within an existing discipline and might not sit well in the new specialisation. Redrawing boundaries and possibly reallocating resources are important components of this challenge;
- getting this specialisation accepted and implemented by those inside and outside the specialisation. Within the specialisation, challenges include the fact that some may not want to refocus their identity and allegiances. Others may have identified a niche in which they are doing well and may not see the need for, or be too overcommitted to contribute to a larger enterprise. Those outside the specialisation may oppose it because they fear losing resources or because they see Integration and Implementation Sciences as being about personal skills rather than academic theory, method and application;

- developing appropriate intellectual interfaces with traditional disciplines and newer multidisciplinary specialisations (such as environment studies or peace studies);
- overcoming unevenness in the development and application of approaches. For example, many of the components of Integration and Implementation Sciences are most developed in the environmental area, so that consideration needs to be given not only to further enhancing the skills that have been developed in the environmental area, but also to diffusing them into other areas (Clark and Dickson 2003);
- uniting the diverse core areas of Integration and Implementation Sciences may be extremely difficult as they have different status, require different skills and often attract different personalities. The challenge of merging model building and facilitation methods is an example; and
- finding suitable locations within universities for Integration and Implementation Sciences—locations where there is a sense of fit and where the specialisation will prosper. In order to continue to attract good people this needs to be an exciting and rewarding area for research and teaching.

Next Steps

For the specialisation of Integration and Implementation Sciences to reach its potential, considerable developmental work is required. Many of the outstanding challenges have been presented earlier. The challenges are both intellectual and practical and essentially fall into three areas:

- strengthening the intellectual base of Integration and Implementation Sciences;
- promoting networking and collaboration between researchers and practitioners interested in Integration and Implementation Sciences; and
- embedding Integration and Implementation Sciences in universities and funding programs.

An established academic specialisation can offer:

- a more clearly defined scope for Integration and Implementation Sciences and complementarities with existing disciplines and specialisations;
- a more robust theoretical base, which will be a well-spring of innovation; and
- a large and critical college of peers to evaluate current and future research and practise.

These allow for both the cross-fertilisation of ideas and advancement of knowledge, as well as opportunities for quality control. Care must be taken to ensure that the specialisation does not become too narrowly defined and lose its richness, and that it does not develop in a lopsided way, for example, that mathematical modelling takes precedence over participatory techniques.

Developing the specialisation includes:

- finding a location in the academy conducive to growth and the development of the ideas underpinning Integration and Implementation Sciences;
- developing both undergraduate and graduate curricula;
- producing textbooks and systematic reflections on case studies;
- building an overarching professional association and encouraging inter-linkage between smaller existing professional associations; and
- building up top-ranking peer-review journals.

Integration and Implementation Sciences are critical if we are serious about *integration*, *policy relevance*, *evidence-based practice*, and *innovation*, which are key concepts now driving research. The challenges are substantial, but the critical mass of researchers and approaches means that rapid development is possible. This promises intellectual excitement and fulfillment, as well as effective practical outcomes in tackling the complex social, environmental and technological issues we confront.

Acknowledgments

Lorrae van Kerkhoff made a number of astute observations that have shaped the text in various places and designed Figure 6.1. Caryn Anderson alerted me to the importance of the fields of knowledge management and information sciences. Useful comments were also received from Yoland Wadsworth, Susan Goff, Lesley Treleaven and Steve Dovers.

As an inaugural Fulbright New Century Scholar, I had the opportunity to reflect on and develop Integration and Implementation Sciences in relation to improving global health. Colleagues at NCEPH, the Hauser Center and Harvard's Center for International Development have provided stimulating discussions and feedback. The development of these ideas has also been facilitated by the following: a National Health and Medical Research Council Population Health Research Capacity Building Grant 'Environment and population health: research development from local to global', a Colonial Foundation Trust grant for the 'Drug Policy Modelling Project' and the US National Cancer Institute funded project 'Initiative on the Study and Implementation of Systems'.

References

Clark, W.C. and N.M. Dickson. (2003) Sustainability science: the emerging research program. *Proceedings of the National Academy of Sciences* 100: 8059–61.

Costanza, R. (2003) A vision of the future of science: reintegrating the study of humans and the rest of nature. *Futures* 35:651–71.

Funtowicz, S. O. and J. R. Ravetz. (1993) Science for the post-normal age. *Futures* 25: 739–55.

Gibbons, M., C. Limoges, H. Nowotny, S. Schwartzman, P. Scott and M. Trow. (1994) *The New Production of Knowledge. The Dynamics of Science and Research in Contemporary Societies.* Thousand Oaks, California: Sage.

Kuhn, T.S. (1970) *The Structure of Scientific Revolutions.* 2nd ed. Chicago: University of Chicago Press.

Mingers, J. (2000) The contribution of critical realism as an underpinning philosophy for OR/MS and systems. *Journal of the Operational Research Society* 51:1256–70.

Nowotny, H., P. Scott and M.Gibbons (2001) *Rethinking Science. Knowledge and the Public in an Age of Uncertainty.* Cambridge: Polity Press in association with Blackwell.

Ravetz, J.R. (1996) *Scientific Knowledge and its Social Problems.* New Brunswick N.J.: Transaction Publishers.

Wilson, E.O. (1998) *Consilience. The Unity of Knowledge.* London: Abacus (Little, Brown and Co.).

Part II. Exploring National Research Priorities with Agents

Viewpoint from a Practitioner

The idea that the physical environment has an impact on our behaviour is very old—there are 5,000 year old Vedic texts that talk about ways to make convivial space by adopting appropriate proportions and plan layouts. In more recent times, Winston Churchill is one of many credited with having said something to the effect that 'We shape our buildings, and then our buildings shape us'. Recent concepts of 'biophilia' and current theories of ecological design all emphasise the value of healthy, attractive environments for improving comfort and productivity. One might argue about the extent to which this architectural determinism holds true but it is hard to escape the conclusion that dysfunctional environments have a concomitant impact on human lives. However, for architects and designers there is a paucity of means to test or prove any of these assertions.

Daniell et al. note that 'current sustainability assessment tools do not adequately represent the temporal, spatial and behavioural aspects of sustainability'. Indeed, most such tools substantially neglect these issues (partly, perhaps, because time and behaviour are so hard to photograph). My architectural practice was named Ecopolis in 1990 for a theory of ecological design later pursued in my doctorate, but my own researches failed to identify an extant and effective methodology for assessing the overall sustainability of housing developments. These authors have changed that.

The authors demonstrate that the encouragement of 'community' contributes to the measurable outcomes of Christie Walk (for which I was the architect and urban designer). This is a particularly gratifying result because it is so hard to justify the Christie Walk design approach in conventional terms. There are still no comparable projects in Adelaide and developers and governments generally have great difficulty understanding that the design of the shared community realm of Christie Walk is an integral and essential part of its design for sustainability.

The authors also note that 'The results of the assessment showed that the development compared favourably to the rest of the Adelaide metropolitan area. The case study also highlighted, through behavioural scenario analyses, the importance of good infrastructure and design in reducing the impacts of human behaviour on housing development sustainability.' This is wonderful stuff! Now architects and urban designers like myself can point to a methodology that has the potential to demonstrate the value of ecological urban design strategies to the decision-makers and purse-string holders who are so critical to the realisation of any built environment.

A capacity to unequivocally demonstrate that sustainable buildings and urban environments have measurable social and material benefits would be invaluable for convincing governments, planning authorities and developers of the ecological approach to design. The complexity of human-environment systems and interactions is such that it is hard for decision-makers to grasp the issues and this research promises to fill the desperate need for quantification of the sustainability of urban developments that considers all the key dimensions: ecological, human, infrastructural and financial, in a way that can be used by policy makers as a decision support system.

Paul F Downton
BSc BArch PhD ARAIA
Principal Architect and Urban Ecologist

Ecopolis Architects Pty Ltd

7. Sustainability Assessment of Housing Developments: A New Methodology

Katherine A. Daniell, Ashley B. Kingsborough, David J. Malovka, Heath C. Sommerville, Bernadette A. Foley and Holger R. Maier

Abstract

In order to combat the rapid degradation of the world's ecosystems and depletion of natural resources, governments and planning authorities are searching for more sustainable forms of development. The need to assess the sustainability of development proposals is of great importance to policy and decision makers. However, effective methods of assessing the overall sustainability of housing developments (proposed or existing) have yet to be established. This research aims to address this problem by presenting a new methodology to assess the sustainability of housing development systems. The methodology uses a *Sustainability Scale* for indicators that are derived from percentiles of a population with resource use above a predetermined sustainable level. It has been coupled with a technique for modelling complex housing development systems using multi-agent based simulation. The methodology was shown to be operational in the case study application of the Christie Walk housing development in inner-city Adelaide, Australia. The results of the assessment showed that the development compared favourably to the rest of the Adelaide metropolitan area. The case study also highlighted, through behavioural scenario analyses, the importance of good infrastructure and design in reducing the impacts of human behaviour on housing development sustainability. It is envisaged that this new methodology of combining sustainability assessment with an integrated modelling technique will provide the basis for a solution to many of the challenges currently facing sustainability researchers, policy makers and planning authorities of urban environments both in Australia and worldwide.

Introduction

The complexity of nature-society systems such as those of urban housing developments makes the understanding and consequent sustainability assessment of these systems difficult. A large proportion of research into sustainable development over the past 15 years has attempted to assess various components of system sustainability without due respect for the complex interrelations between the components, which can have a significant effect on overall system behaviour (Clark and Dickson 2003). This has led to an incomplete understanding at government and policy making levels of what is required to achieve sustainable development for all communities. A consistent framework for sustainability assessment is therefore required for decision-making purposes (Nishijima et al. 2004).

A review of current literature into the assessment of the sustainability of housing developments (Daniell et al. 2004a), found that:

- governments and planning authorities world wide require more holistic methods for sustainability assessment in order to develop future planning strategies (Tweed and Jones 2000);
- due to the narrow focus of current assessment tools, decision makers find it difficult to make judgments which are consistent with sustainability goals for development (Macoun et al. 2001);
- current sustainability assessment tools do not adequately represent the temporal, spatial and behavioural aspects of sustainability; and
- there is no common methodology which relates measures of resource use and other variables (referred to as indicators) to a measure of sustainability; and there is a specific need for a methodology that can be used to assess the sustainability of complex housing development systems (Deakin et al. 2002).

In order to address the shortcomings outlined above, a new methodology for the assessment of the sustainability of complex housing development systems is developed in this research using multi-agent simulation. The methodology couples complex systems modelling and sustainability assessment, and provides a decision-making tool that can be used by policy makers, governments and planning authorities. The application of the methodology to a case study example, Christie Walk, an Australian eco-development, is also presented, with a special focus on determining the impacts of human behaviour on the housing development's sustainability.

Methodology

The proposed methodological framework for the sustainability assessment of housing developments is presented in Figure 7.1 and further explained throughout this chapter.

Figure 7.1. Housing development assessment methodology

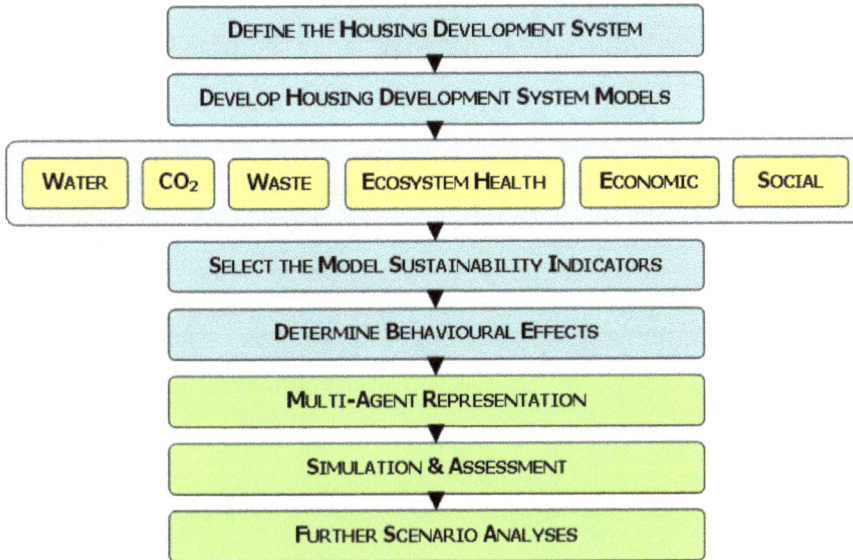

Definition of housing development systems

A housing development is a system that can be defined, and its sustainability assessed, if the definition of sustainability presented by Gilman (1992) is adopted. Gilman stated that sustainability is:

> the ability of a society, ecosystem, or any such ongoing system to continue functioning into the indefinite future without being forced into decline through exhaustion or overloading of key resources on which the system depends.

Using this Gilman definition, Foley et al. (2003) outline that for a system to be sustainable, all of the resources upon which the system relies must be managed appropriately, including natural, financial, social, and man-made (infrastructure) resources. Appropriate management requires knowledge relating to the system boundary, system resources, interactions between adjacent systems and allowable limits, or thresholds, for each resource. Each of these elements will be unique to the particular system under consideration, and each system must be assessed on its own merits. However, the process of assessment should be consistent for every system. This general systems approach to sustainability can be applied more specifically to an urban development by viewing each urban housing development as a unique system. An example of such a system with its resources and interactions is shown in Figure 7.2.

Figure 7.2. The complex housing development system

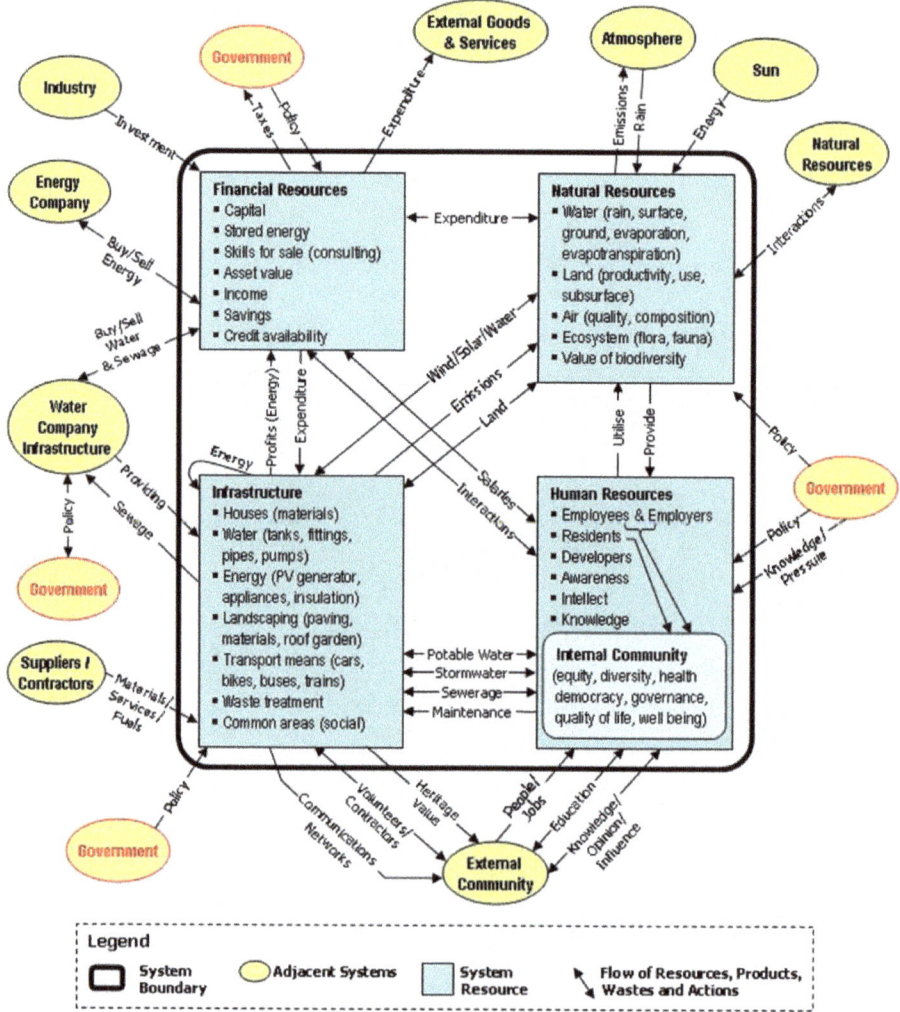

To assess the sustainability of a housing development, all of the resources and their interactions represented in Figure 7.2 (both within and external to the system) need to be determined as specified in the methodology presented in Figure 7.1. As outlined by Foley et al. (2003), if each resource in the housing development is considered as a state variable x_i, at any time t_j, the state of the system can be expressed for n state variables in vector form as:

$$\mathbf{x(t_j)} = \{ x_1(t_j), x_2(t_j), \ldots, x_{i-1}(t_j), x_i(t_j), \ldots, x_n(t_j) \} \tag{1}$$

The changes to each state variable or resource can then be modelled over each specified time interval where $t_{j+1} = t_j + t_j$.

Housing development system models

Considering a complex urban housing development system as outlined in Figure 7.2, the key resources, processes and interrelations of a housing development can be defined in terms of six interrelated models, namely water, carbon dioxide (CO_2), waste, ecosystem health, economic, and social. All of these models are affected by human behaviour and are represented in Figure 7.3. The role of human behaviour is discussed further in next section.

Figure 7.3. Framework of interrelated models for housing development sustainability assessment

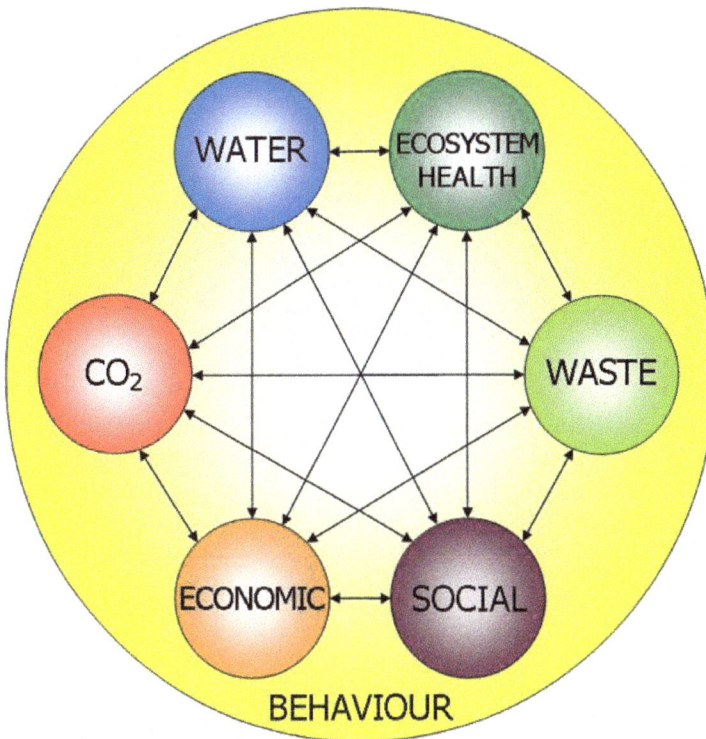

In Figure 7.3:

- the water model incorporates all the water related processes of the development, including rainfall runoff, infiltration, and potable and non-potable water use;
- the CO_2 model accounts for both embodied and operational energy use, calculated as an equivalent mass of CO_2, which incorporates the effects of building materials, infrastructure, electricity and gas use, and occupant transport use;

- the waste model accounts for all solid and liquid waste, both produced on site, and leaving the site, including sewage, compost, waste to be recycled, and waste sent to landfill;
- the ecosystem health model encompasses environmental aspects of the development such as biodiversity and land use changes, as well as air and water quality;
- the economic model accounts for both the microeconomic processes of each household based on income, expenditure and corresponding levels of debt, as well as the macroeconomic processes which affect the housing development, such as inflation and interest rates; and
- the social model incorporates levels of occupant satisfaction relating to comfort, living conditions, access to services (transport, health, education, shopping), employment, as well as equity amongst occupants.

Following the methodology presented in Figure 7.1, for each of these models an indicator representative of the model processes and consequent sustainability must be chosen for assessment purposes. [1]

Sustainability indicators and assessment

Once indicators are selected, it is important to determine the conditions under which an indicator is to be considered sustainable. Available assessment techniques for housing developments reviewed by Daniell et al. (2004a) use indicators that predominately present and collate resource use or resource quality data. There is little attempt to assess the adequacy of the data with respect to the level or condition of the resources available to the system under consideration. Foley and Daniell (2002) recognised that the use of a sustainability satisfaction scale for indicators could allow the comparison of indicators not only against each other but also against sustainability criteria. This approach, together with the System Sustainability Conditions outlined in Foley et al. (2003), was further developed by Daniell et al. (2004b) to create the Sustainability Scale for indicators, which is presented in this section.

The Sustainability Scale is based on a probability of exceeding the ultimate sustainability threshold level, *threshold* (x_{ij}), for each resource, $x_i(t_j)$, as shown in Figure 7.4.

---

[1] More than one indicator could be chosen for each model if desired.

Figure 7.4. The Sustainability Scale

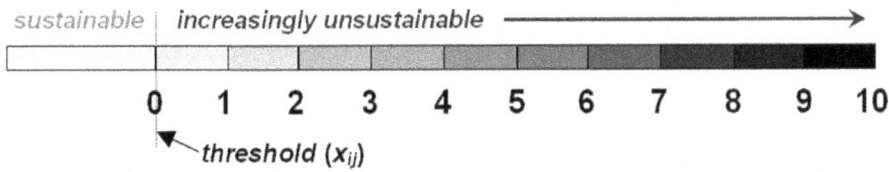

The sustainability threshold level is the resource level at which the system is deemed to be able to satisfy the following System Sustainability Conditions on an ongoing basis:

- resource levels available to the system are sufficient to meet the requirements of the system;
- resource levels within the system are maintained at levels that do not exhaust or overload the resources; and
- resources that are imported to, or exported from, the system do not compromise the ability of adjacent systems to be sustainable.

The Sustainability Scale ranges from 0 to 10, where 0 is considered as sustainable resource use, and the values between 0 and 10 represent increasingly unsustainable resource use. In other words, for a housing development system's resource use to be considered sustainable, equation 2 must be satisfied: [2]

$$x_i(t_j) \leq \text{threshold}(x_{ij}) \tag{2}$$

Individual Sustainability Scale Ratings (SSRs) for indicators are based on the cumulative probability distribution of current resource usage at a larger system scale exceeding the sustainable threshold level (i.e., a probability of threshold exceedance between 0 and 1).

The larger system chosen will depend on the purpose of the sustainability assessment. For example, a housing development might need to be compared with other developments within a local council area or to other housing developments in a larger metropolitan area. Once this larger system scale has been chosen, a distribution of the resource use of the indicator to be assessed must be developed. An example of a cumulative distribution function (in this case where the indicator is mains water use in the metropolitan Adelaide area), from which the Sustainability Scale can be derived, is represented in Figure 7.5.

---

[2] The reverse may be true when a system's resource level needs to be maximised (i.e., river flows for environmental requirements).

Figure 7.5. Cumulative distribution of mains water use exceeding the sustainability threshold level

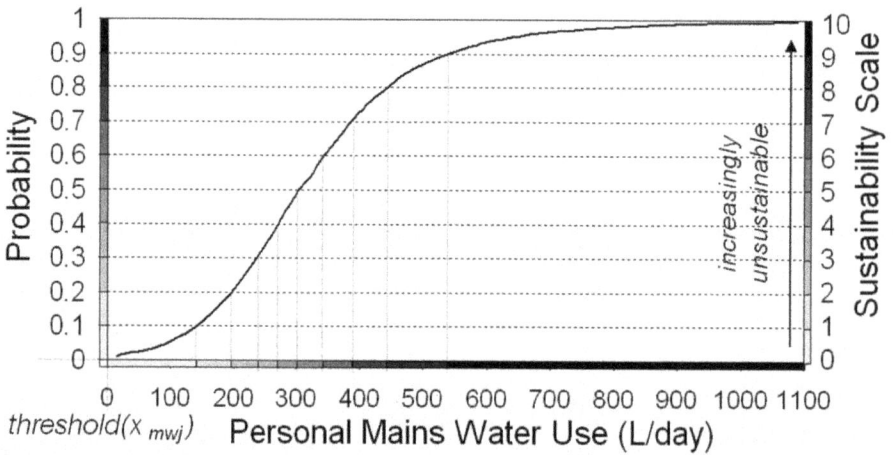

Cumulative Probability of Estimated Mains Water Use Per Person in L/day for Metropolitan Adelaide (2002-2003)

In order to create the cumulative distribution used to derive the Sustainability Scale, a number of steps need to be undertaken, dependent on the form of data available. The example of mains water use in the Adelaide metropolitan area shown in Figure 7.5 will be expanded upon here to demonstrate the process.

Step 1: The frequency of people corresponding to each level of estimated mains water usage in the Adelaide metropolitan area needs to be plotted, as shown in Figure 7.6.

Figure 7.6. Mains water use frequency distribution and the sustainability threshold level

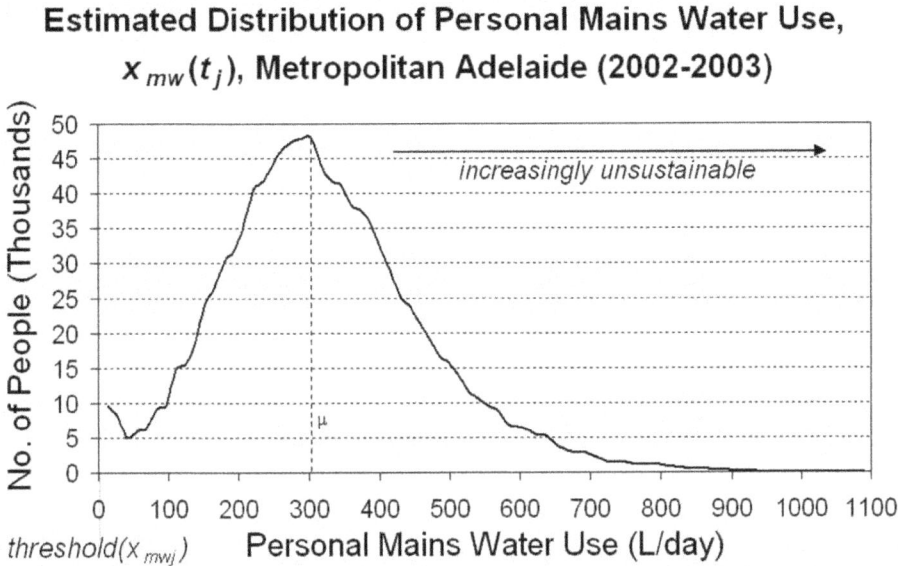

Step 2: The sustainability threshold level for the resource, in this case mains water use, needs to be defined. For a housing development scale of sustainability assessment, the threshold for mains water use is defined as 0 L/person/day, [3] i.e., $threshold\,(x_{mwj}) = 0$ as denoted in Figure 7.6.

Step 3: Now consider all people with resource usage below the threshold to be sustainable and eliminate them from the distribution, in this case all people who do not use the mains water supply. This is equivalent to removing sustainability as an outcome from the original sample space.

Step 4: The probability density function of the new sample space can be calculated (corresponding to the probability density function of mains water use of only people exceeding the sustainability threshold in the example). Since being in the sustainable state has been removed as an outcome and the distribution has then been normalised on the remaining outcomes (i.e., being in an unsustainable state), a conditional probability density function has been calculated. If this probability density function is defined as $f(x_{ij} \mid x_{ij} > threshold\ (x_{ij}))$ (for any resource), then:

---

[3] This considers that, for a housing development to be sustainable, all water used must be collected on site if the site does not contribute to the mains water supply. Considering a different example, an urban development on an isolated island with no mains water, the threshold for water use could be set at a level that would not deplete the freshwater supplies or groundwater levels.

$$P (a < X \leq b \mid X > \text{threshold}(x_{ij})) = \int_{b}^{a} f(x_{ij} \mid x_{ij} > \text{threshold}(x_{ij}))dx_{ij} \tag{3}$$

where: X is a random variable within the sample space [$threshold(x_{ij})$, ∞); and x_{ij} is the estimated value of resource usage of an unsustainable individual. [4]

Note: The order of steps 3 and 4 is very important in order to maintain the first

Axiom of Probability i.e. $\int_{-\infty}^{\infty} \text{pdf} = 1$. The relative frequencies (no. of people / total population) must be calculated using only the portion of the population exceeding the threshold required for sustainability.

Step 5: Calculate the cumulative distribution function, $F(x_{ij} \mid x_{ij} > threshold\ (x_{ij}$)), corresponding to the pdf calculated in Step 4, namely the probability density given that resource usage is in an unsustainable state:

$$F(x_{ij}) = P (X \leq x_{ij} \mid X > \text{threshold}(x_{ij})) = \int_{-\infty}^{x_{ij}} f(x_{ij} \mid x_{ij} > \text{threshold}(x_{ij}))dx_{ij} \tag{4}$$

where $x_{ij} \in [\ threshold(x_{ij})\ ,\ \infty\ )$.

For the mains water example, the resulting cumulative distribution function is shown in Figure 7.5.

Step 6: The corresponding Sustainability Scale Ratings, SSRs, for a particular resource usage can now be directly calculated from the cumulative distribution function:

$$SSR = 10 \times F(x_{ij}) = 10 \times P (X \leq x_{ij} \mid X > \text{threshold}(x_{ij}))$$

$$SSR = 10 \times \int_{-\infty}^{x_{ij}} f(x_{ij} \mid x_{ij} > \text{threshold}(x_{ij}))dx_{ij} \tag{5}$$

This can be performed without any loss of generality since a sustainable development has SSR = 0 (as in a sustainable state $x_{ij} \leq threshold(x_{ij})$) which leads to $P (X \leq x_{ij} \mid X > threshold(x_{ij})) = 0$.

This Sustainability Scale can be used to measure any indicator and thus produces a uniform method of sustainability indicator assessment. For example, waste production can be measured against water use for equivalent levels of sustainability (or, more correctly, unsustainability) or the same indicators can be compared between developments in the same larger system.

---

[4] Here an *individual* is an abstraction that can be interpreted as one individual person within a housing development or an *equivalent average individual* of a development that can be compared with other developments within the larger community or system scale.

The indicator assessment concept can also be viewed in terms of system vulnerability. If the resource or indicator level fails to be sustainable for the mains water example, if x_{mw} (t_j) > $threshold(x_{mwj})$, then the magnitude of failure or *vulnerability* of the system is quantified on the Sustainability Scale.

The proposed methodology also allows for the continual assessment of system sustainability through time. For example, if a sustainability assessment is to be carried out on a housing development every year for a number of years, and data are available to create the resource distributions required for the larger system used for comparative assessment for each of these years, then the Sustainability Scale of resource usage will change through time. The corresponding yearly resource use for the housing development being assessed can then be compared, based on the equivalent yearly Sustainability Scale. The threshold level of sustainability for a resource, $threshold(x_{ij})$, may also vary over time, depending on new scientific research or technological advances.

At the initialisation point, a sustainability assessment of the collection of 6 indicators for these models may be performed to assess the current sustainability of the housing development system. Further modelling may then be performed to assess the potential sustainability of the housing development through time.

The development of the *Sustainability Scale* for indicators is an advancement on existing assessment techniques, as the indicators provide a measure of the proximity of the indicator to the sustainable or threshold level. The methodology to this point allows the assessment of sustainability at a given point in time, but does not address other deficiencies in existing assessment techniques, such as the effect human behaviour or changes in the system over time. The last 3 steps of the methodology in Figure 7.1 address these aspects.

Effect of human behaviour

One of the criticisms of current urban simulation techniques is the lack of sufficient behavioural theory used in the modelling processes (Waddell and Ulfarsson 2004). This view is confirmed by the findings from the review of current sustainability assessment tools presented by Daniell et al. (2004a), which showed that the effect of human behaviour in relation to resource use and sustainability was not adequately included. It has commonly been reported that human behaviour has a significant impact on resource use (Georg 1999), although there has been very little research to date to quantify these effects (Jalas et al. 2001). To overcome the behavioural deficiencies of previous sustainability assessment methods, human behaviour, particularly related to resource use, has been included in the sustainability assessment framework developed as part of this research.

Behaviour relating to resource use can be analysed in many ways by studying both the causes and effects of human practices and actions. Study of behavioural patterns can be related to sociological theory, where the normative (values and preferences), cognitive (representations and beliefs), operational (practices and actions) and relational (social interaction and relationship) aspects of individuals should be analysed.

Obtaining behavioural information

Depending on the housing development to be assessed, several methods of defining behaviour relating to resource use may be applicable. If the housing development and occupants already exist, analysis of individual occupants can be performed. In such cases, questionnaires (or other forms of information gathering such as interviews) can be used to determine occupants' preferences and practices relating to their resource usage, their beliefs and goals, as well as their social practices and networks. This information can then form the basis for a behavioural typology, and social structure of the housing development's occupants can be quantified as a series of rules for modelling based on sociological or psychological decision theory (Amblard et al. 2001).

Another option of quantifying behavioural effects on resource usage is to use currently available sociological and resource use data from census collector districts or other area specific surveys (Melhuish et al. 2002). These data sets can be used to create synthetic distributions of resource use typical of the area's population. Further analysing these general resource use distributions with respect to socio-economic data can provide a good sample of what resource use relating to behavioural differences can be expected in the housing development.

The information obtained from these behavioural analyses can be integrated into the six interrelated models (water, CO_2, waste, ecosystem health, economic and social) as an initialisation point and driving mechanism to induce resource use changes for each model.

Multi-agent system modelling

In order to represent the six interrelated models in Figure 7.2, a suitable modelling platform is required. Multi-agent systems (MAS), an object-oriented programming method that was traditionally used for artificial intelligence applications, can be used to combine the water, CO_2, waste, ecosystem health, economic and social models, and their relationships to human behaviour, for temporal sustainability assessment.

It has been recognised by many authors that multi-agent based simulation has many advantages over techniques currently used to model complex systems (Huigen 2003). Multi-agent systems have the capability of explicitly incorporating human behavioural, spatial and temporal aspects into a more holistic model (Waddell and Ulfarsson 2004). They can also incorporate both qualitative and quantitative data in the same model (Taylor 2003), unlike many other modelling tools. This is of particular use in the field of sustainability assessment. The sustainability goals, or long-term objectives for resource use in housing developments, can be included in the framework of the multi-agent system as *goals* or *beliefs* of the agents (Krywkow et al. 2002). The representation of human interaction processes, such as decision-making and learning based on changes witnessed in the surrounding environment, can be programmed into the individual agents in order to allow policy makers to examine the total resulting system behaviour (Moss et al. 2000).

Housing development multi-agent representation

In the complex housing development systems to be assessed, each occupant or household can be modelled as an *agent* that uses resources in the development *environment* and can communicate with other occupant *agents*, as shown in Figure 7.7.

Figure 7.7. Multi-agent representation of a housing development

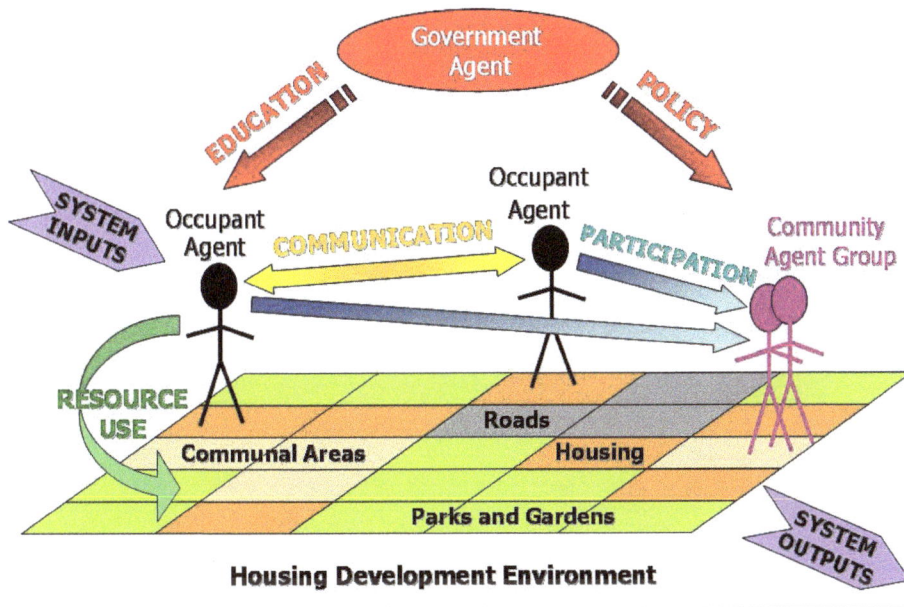

In the multi-agent representation of the housing development system shown in Figure 7.7, housing occupants can occupy the environment cells (a cell being one unit of the multi-agent model's spatial environment), interacting with the environment through resource use. The information for each house relating to its infrastructure, location and occupants can all be included in the cells' information. Using a multi-agent systems approach, occupants can communicate with each other, exchange information and learn via community participation and interaction with other occupants. The households, or occupant agents, will also be able to store specific personal information on each of the agents, such as their beliefs, needs and decision processes. The government can also be created as an agent who can exert an influence on the housing development's occupants and environment through policy change and education programs. The 5 resource use models linked to the occupant behaviours can be incorporated directly into the multi-agent model using any number of readily available multi-agent modelling platforms, including: CORMAS (Common-pool Resources and Multi-Agents Systems); REPAST (REcursive Porous Agent Simulation Toolkit); and the DIAS/FACET (The Dynamic Information Architecture System / Framework for Addressing Cooperative Extended Transactions) platforms (ECAABC 2004).

Multi-agent systems for sustainability assessment and scenario analyses

After developing a multi-agent based model of the housing development that incorporates all of the resource models and the behavioural typology and interactions of the occupant agents, the model can be run over a desired number of time-steps, potentially weekly, monthly or seasonally, following the discussion in Daniell et al. (2004b) to examine the emergent behaviour of the housing development system. Simulations of the multi-agent model can also be run to assess the impacts of various changes to the system, scenarios or policy options, and their impacts on the emergent behaviour of the housing development system and relative sustainability.

Model validation

Multi-agent models can be validated, at least in the preliminary time-steps of simulation, using a variety of methods including the use of role-playing exercises, questionnaires and forums (El-Fallah et al. 2004). Such methods can be used when the occupants of the housing development are willing to participate in the modelling process by either re-enacting the processes that take place in developments through games designed by the modeller, or by answering questions to match the processes and attributes required for the multi-agent model. Other forms of validation for multi-agent models can include more qualitative methods of assessment such as determining whether each output relates to what is seen in reality in similar housing developments, and by comparing relationships and trends obtained to trends found in literature. Considering the complex nature of the systems being examined, it is highly advantageous to validate each relationship used in the modelling process before all the relationships are combined, which has been termed 'internal verification' (Vanbergue 2003). As the use of multi-agent modelling for analysis of human ecosystems and natural resources management is still in its infancy, strict protocols for their validation are yet to be established. It is suggested that until such validation protocols are determined, modellers should use their common sense in determining the accuracy of their results, even after preliminary validation exercises have occurred (Bousquet et al. 1999).

Once validation of the multi-agent model has taken place, further simulations of the model may be performed to assess the sustainability of the housing development in its current state and in a range of other scenarios. For each simulation, the sustainability indicators chosen for each of the inner models of the multi-agent model can be obtained on a common Sustainability Scale. At this point, a total assessment of the housing development's sustainability can be determined by an analysis of the collection of indicators.

Case study application

Background

To demonstrate how the methodology proposed in this chapter can be applied in practice to assess the sustainability of housing developments, the Christie Walk housing development was used as a case study. The Christie Walk housing development, located in the central business district of Adelaide in South Australia, is a medium density urban housing development made up of 14 varied dwelling types (straw-bale cottages, aerated concrete and rammed earth construction), with other aspects of 'resource sensitive urban design' (Daniell et al. 2004c). These aspects include water sensitive urban design, passive design of buildings to maximise energy savings, an inner-city location in close proximity to services, and designated community spaces (Downton 2002).

Christie Walk is considered a leading example of sustainable development due to the innovative nature of its design. However, until now, verification of this claim has been difficult. The nature of its design, combined with the accessibility of data, made Christie Walk an appropriate case study to test the proposed methodology presented in the next section of this chapter.

Christie Walk system and multi-agent representation

The first step of the assessment requires the system and system boundaries to be carefully defined, as outlined in Figure 7.1. As with any complex system, this is not necessarily a straightforward task, since Christie Walk currently relies upon numerous systems outside its spatial boundaries. For the development to be sustainable it must be able to meet its requirements without compromising the ability of adjacent (or external) systems to be sustainable. The system boundary for the assessment of Christie Walk was considered to be its spatial boundaries (fence line). Any resource use sourced exterior to the system was considered to be unsustainable, as this allows the development to manage its own resources and minimises any impact on adjacent systems' ability to be sustainable.[5] The Adelaide metropolitan area was chosen as the larger system providing the resource use comparisons against the resource use of Christie Walk for the construction of the Sustainability Scale.

An analysis of the system, its resources and their interactions enabled detailed models representing the processes within the system to be developed as the second step of the assessment process. Models were developed for water, CO_2, waste, economic and social processes following the descriptions outlined for Figure 7.3. However, it was considered unnecessary to include an ecosystem health model. Christie Walk is located in the central business district of Adelaide and the site was previously used for commercial purposes, with little to no biodiversity on the site. Any positive impact of improving the biodiversity on the site is considered as sustainable (and thus not quantified on the Sustainability Scale). Improvements in biodiversity may be shown through increased quality of life, which is processed through the social model. Similarly, other air quality and water quality factors not included in the ecosystem health model could be partially taken into consideration in the social model. A brief outline of each of the Christie Walk models and interactions, as well as graphical interpretations, is presented in the following section.

Christie Walk conceptual models

The water model for Christie Walk was developed based on the current infrastructure, rainfall runoff qualities, occupant behaviour and garden crop production of the site. Three sources of water are available to residents: the Adelaide mains water supply, for potable uses and topping up of on-site non-potable water storage tanks; stormwater captured and stored on site for non-potable uses such as toilet flushing and garden watering; and treated wastewater, for non-potable uses (although currently not used due to the price of testing for water quality regulations compliance). The use of this water is highly dependent on occupant behaviour, which can also be potentially influenced by community interaction as represented in Figure 7.8.

[5] Although this assumption does not always hold true in reality, it is required to bound the assessment process for practicality of implementation.

Figure 7.8. Christie Walk water model conceptualisation

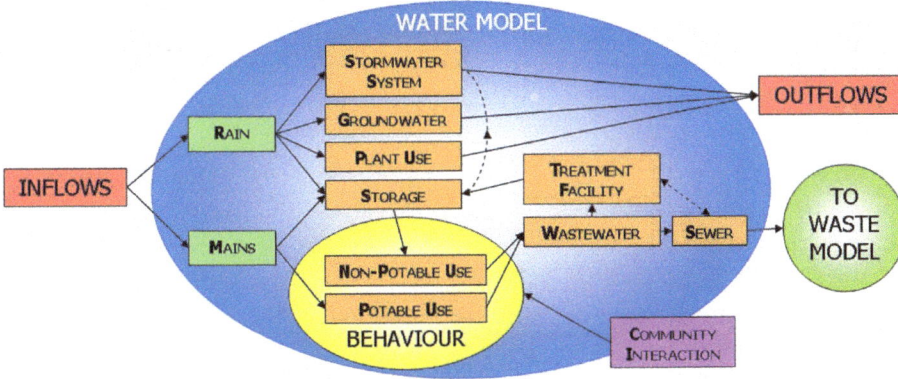

The two main inputs of CO_2 to the Christie Walk system model are embodied energy and operational energy, as shown in Figure 7.9. The only sink of CO_2 in the model is the on-site biomass. The embodied energy of the dwellings, the water, sewerage, pavement infrastructure and personal vehicles were included in the model as represented by the various categories of *infrastructure* in Figure 7.9. Operational energy requirements and equivalent mass of CO_2 for each household were calculated to include the delivered electricity and gas, fuel for transportation, an allowance for infrastructure maintenance and also conversion and transmission losses. These energy uses are affected by behaviour and subsequent community interaction as shown in Figure 7.9.

Figure 7.9. Christie Walk CO$_2$ model conceptualisation

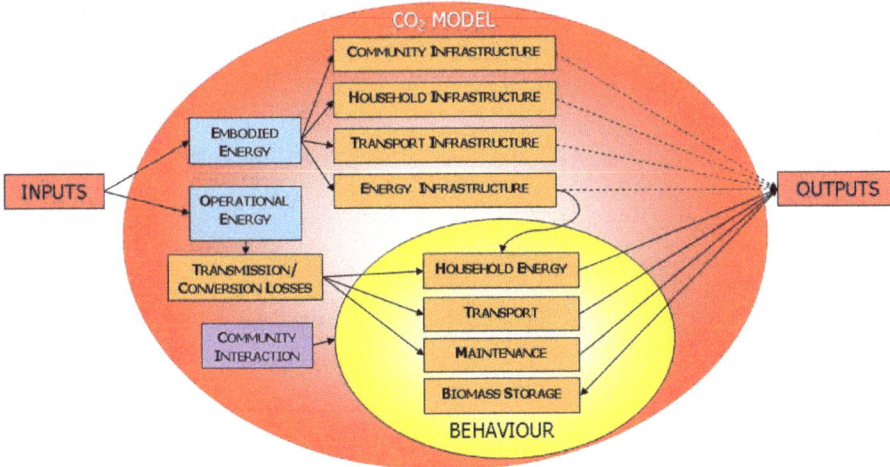

The waste model for Christie Walk was developed based on the current waste practices and facilities of the housing development. The site incorporates a community garden, which has the capability to compost all biodegradable waste, a community recycling initiative, which increases access to recycling facilities for a broad range of materials, as well as sewer mining and water recycling for liquid wastes. The waste model was divided into two main sections as shown in Figure 7.10, including solid waste, which is primarily created from general household consumption and garden production, and liquid waste, which is fed into the waste model directly from the water model. Materials use and subsequent household waste was divided into percentages of food that is composted, materials that can be recycled, and non-recyclable material—or garbage—that is sent to landfill. Each household's behavioural recycling and composting practices, as well as the effect of community interaction on these practices was determined based on survey responses of the residents relating to these issues.

Figure 7.10. Christie Walk waste model conceptualisation

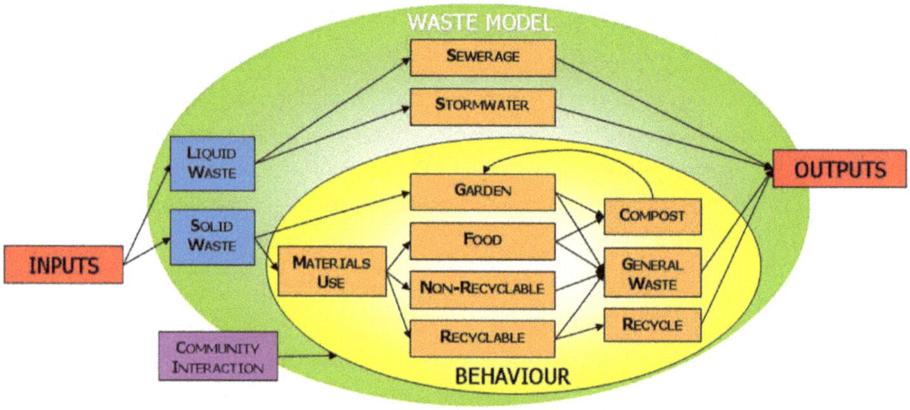

The economic model for Christie Walk has been based on cash flows for each household, all of which are influenced by the overriding macroeconomic environment as shown in Figure 7.11. The micro-economy of each household within Christie Walk is based on effective household income which occupants distribute amongst a range of mandatory expenditures such as rates (based on resource usage from the water, waste and CO_2 models and current South Australian pricing structures), rent or mortgage payments, development strata fees and essential goods and services. Remaining disposable income use is then dependent on occupant behaviour, with money being able to be borrowed or saved at interest rates affected by inflation and government policy in the macro-economy.

Figure 7.11. Christie Walk economic model conceptualisation

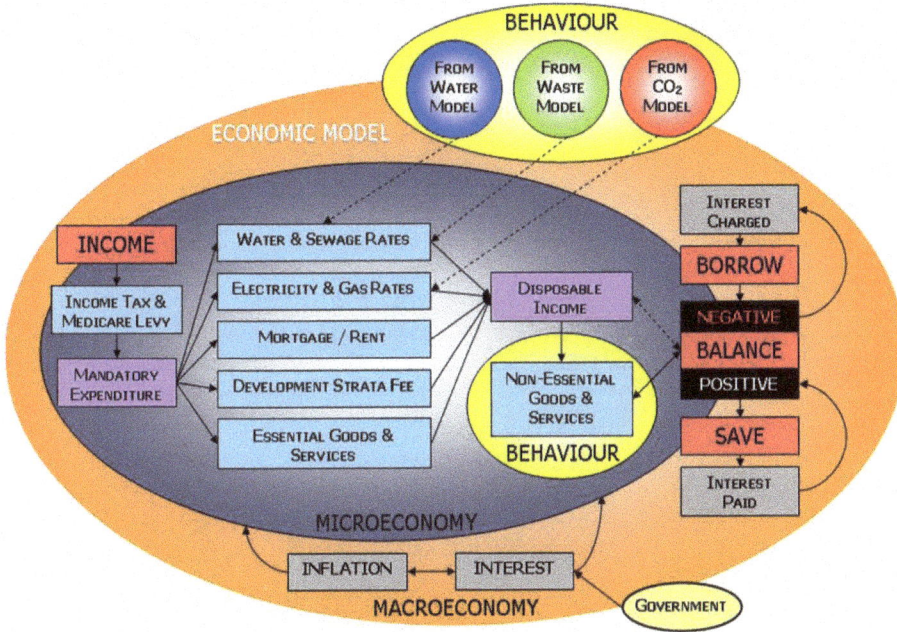

The social model for Christie Walk has been devised based around 5 components of social sustainability: equity, diversity, quality of life, interconnectedness, and democracy and governance (McKenzie 2004) . The model allows for the assessment of individuals' *social sustainability* through the 7 key areas of the European Sustainable Cities program *citizen satisfaction* indicator. These 7 areas of satisfaction were measured for each of the Christie Walk occupants as the difference between the current and desired state of each attribute as shown in Figure 7.12. These satisfaction levels, as well as their relative importance, were obtained from occupant survey results. The range of individual occupants' satisfaction levels was then used to quantify the state of equity in the development as represented in Figure 7.12.

Figure 7.12. Christie Walk social model conceptualisation

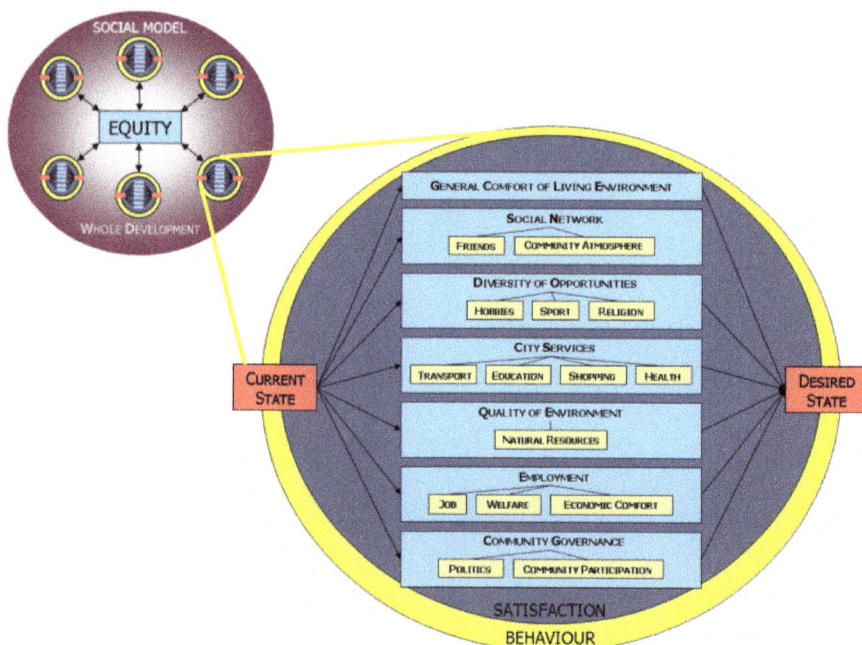

A sustainability indicator for each of the water, CO_2, waste, economic and social models created for the Christie Walk housing development was then chosen as the next step in the methodology, as shown in Figure 7.1. These indicators and the equivalent measures used in the construction of the Sustainability Scales are presented in Table 7.1.

Table 7.1. Model sustainability indicators

Model	Sustainability Indicator	Equivalent Measure
Water	mains water use, $x_w(t_j)$	litres / person / day
CO_2	mass of equivalent CO_2 , $x_c(t_j)$	tonnes / person / year
Waste	waste sent to landfill, $x_l(t_j)$	kg / person / week
Economic	average % usage of available household debt, $x_d(t_j)$	% usage of available debt
Social	equitable satisfaction level, $x_s(t_j)$	equitable satisfaction level

Defining behaviour

Behaviour for the Christie Walk occupants was defined using 2 methods based on those previously described. The first method was to examine the occupants' preferences and practices relating to resource usage, their beliefs and goals, and their social practices and networks through the use of a survey which was distributed to all Christie Walk residents. The second method was based on examining currently available sociological and resource use data in order to quantify the behavioural aspects of the Christie Walk occupants with respect to their resource use patterns, in comparison with the resource use of residents in the larger metropolitan Adelaide area.

The surveys distributed to the Christie Walk residents were used to obtain behavioural data regarding the occupants' values, preferences, practices and social interactions in a range of domains related to resource utilisation and other aspects of the 5 models described previously. For the water, CO_2 and waste models, the resource use distributions created for the metropolitan Adelaide area were used in conjunction with the survey information to assign general behavioural classifications to each household. In order to demonstrate the procedure for behavioural classification of occupants, the example of waste production will be outlined.

From the responses to the waste related questions in the survey, the behavioural profile of the occupants, relating to both quantity of waste produced and the amount of this waste they are likely to recycle, were assessed. From these profiles, each occupant was assigned a grouping from the cumulative distributions of waste production and percentage waste diversion in the metropolitan Adelaide area. To simplify the process used in this case study, each distribution was only broken down into three levels of behavioural classification of resource usage (33 percentile sections of the population). The cumulative distribution for total waste production in the Adelaide metropolitan area is shown with these waste production level groupings in Figure 7.13.

Figure 7.13. Occupant behaviour categories for total waste production

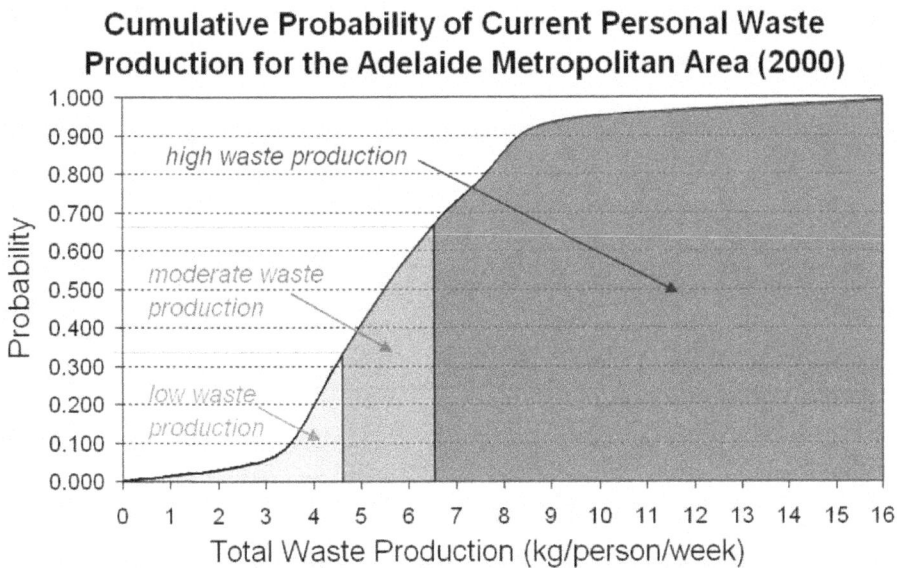

Cumulative Probability of Current Personal Waste Production for the Adelaide Metropolitan Area (2000)

This division into behavioural categories enables the occupants' waste production to be assigned randomly in the multi-agent model from the appropriate category at each iteration. This allows the waste produced to fluctuate much like weekly household waste production in reality. Behavioural classifications for the other resource practices (for recycling, energy use, transport, water use and financial use) were formulated in the same way for the Christie Walk residents compared with the larger metropolitan Adelaide area.

Multi-agent model implementation

The multi-agent modelling platform, CORMAS (COmmon-pool Resources and Multi-Agent Systems), which uses the SmallTalk programming language, was used in order to combine all the required models and behavioural information for the sustainability assessment of Christie Walk. This modelling platform was developed by the Centre de Coopération Internationale en Recherche Agronomique pour le Développement (CIRAD) in Montpellier, France, specifically for modelling the relationships between societies and their environments for natural resources management (Bousquet et al. 1998). This particular platform was chosen as it is freely available and has a strong support network to help with any problems experienced with programming or the software.

Representation of the Christie Walk system in CORMAS was performed in several stages: representing the *environment*, representing the *agents*, and representing the models and interactions. After the model was created, it was calibrated, tested and validated before operational use.

The CORMAS platform allows for a spatial representation of a system as a grid of cells or *spatial entities*. Each of these spatial entities can have attributes such as area or land use, as well as processes such as rainfall or vegetation growth. For the purposes of the model, a grid was devised to mimic the land use pattern of the Christie Walk development, including the land use types of unit (house), garden, path/vegetation, car park, waste treatment plant and bike shelter. From the architectural plan drawings of Christie Walk, the conversion to the model environment is shown in Figure 7.14.

Figure 7.14. Architectural plan of the Christie Walk development and the CORMAS model environment representation

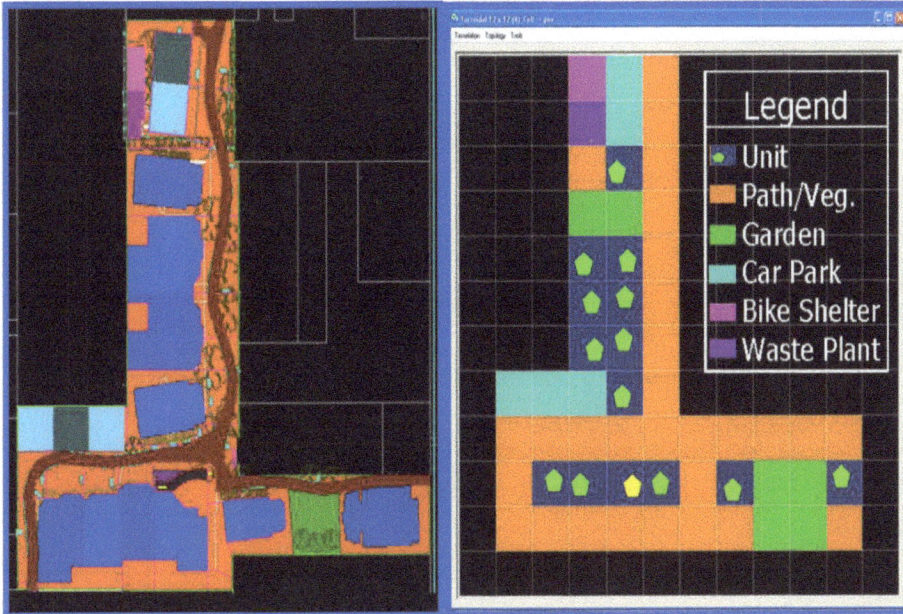

One agent was initialised per household with behavioural categories as outlined in the previous section, which could be graphically represented on the CORMAS model environment (in Figure 7.14 as pentagons in each home unit). Each agent was also pre-programmed with a specific behavioural category of community interaction which was determined from the surveys and interviews with residents.

The 5 models previously defined (water, CO_2, waste, economic and social) and other processes for the Christie Walk model were written on several levels within the CORMAS platform. Methods relating to household resource use were written at the household level, for example, in-house water use, energy use, transport use, financial use and individual social sustainability. Other methods that related to the overall housing development situation were run at the main model level, for example, the development's water use and the 5 sustainability indicators. At the government level, the methods for updating the interest rates and corresponding inflation rates, CPI, wages and tax brackets were performed.

A seasonal time-step (3 months) was chosen for the model in order to ensure reasonable computational efficiency, as well as allowing seasonal variation to be gauged.

The model was validated to the greatest possible extent using several of the methods outlined in the first part of this chapter, including internal verification (checking individual model outputs such as water and energy use against known meter readings), survey responses and general matching of the model outputs with observations of real-world housing developments.

Results of sustainability assessment simulations

Following the construction and validation of the CORMAS multi-agent Christie Walk model, simulations were run over a 30-year period with the sustainability indicators prescribed in table 1 for the water, CO_2, waste, economic and social models being rated using the Sustainability Scale framework previously outlined.

A simulation of the Christie Walk multi-agent model over a 30-year simulation period assuming relatively stable economic, political and climatic conditions is illustrated in Figure 7.15, showing the 5 model indicators against the SSRs.

Figure 7.15. Sustainability Scale Ratings for Christie Walk for the 5 indicators

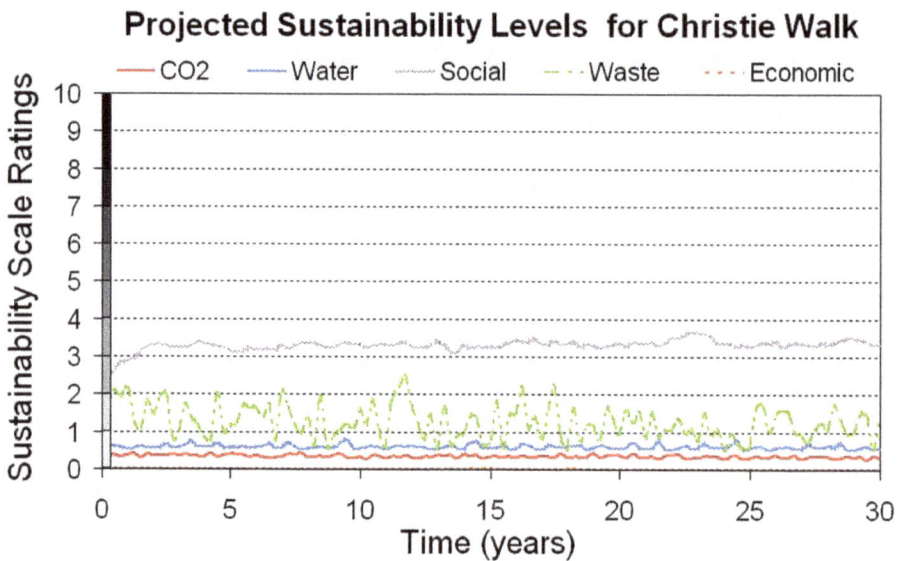

Sustainability indicators of mains water use, waste sent to landfill and CO_2 production, are all based on distributions of the indicator variables within the metropolitan Adelaide area, according to the framework presented in the first part of this chapter. It can be seen that for each of these indicators in Figure 7.15, Christie Walk is well within the lower 30th percentile of resource users in the Adelaide metropolitan area that exhibit unsustainable resource use. This demonstrates the effectiveness of some of the components integrated into the Christie Walk housing development design compared with other Adelaide residential developments.

As the modelling process adopted is stochastic, every simulation performed will vary within reason. Outputs from each of the simulations give an overview of possible trends in the variation of the sustainability indicator ratings. These indicator ratings may be used to target specific areas for improvement and further analysis by government and planning authorities.

This simulation of the Christie Walk model through time reinforces the notion that the sustainability of a housing development will vary due to many parameters, including behavioural and climatic variation. For future government planning and assessment applications of this framework, both extreme and average Sustainability Scale values should be considered when such results are to be used for decision making.

Scenario analyses

One of the greatest advantages of multi-agent modelling is the capability to perform *What if?* scenario analyses. Using the Christie Walk model, a variety of scenario analyses were undertaken during the course of this research, including assessment of the effects of droughts, changes to building materials, location, behaviour and community interaction on the sustainability of the Christie Walk development. The full analysis of these scenarios is given in Daniell et al. (2004b). For this chapter, the question of whether occupant behaviour is closely linked to the sustainability of housing developments will be examined.

Behaviour scenarios

The effect of human behaviour on natural resources utilisation is largely recognised, but rarely quantified (Curwell and Hamilton 2003). Decision makers such as governments and planning authorities have the ability to influence people's behaviour through legislation, education, increased awareness, information sharing, and price manipulation. However, the effectiveness of such campaigns has been difficult to analyse.

In order to analyse if occupant behaviour has a significant impact on the sustainability of Christie Walk, several scenarios were run, specifically focusing on waste production, recycling, water and energy use. In each case, high, moderate and low levels of each resource behaviour (see Figure 7.13) were initialised for all residents in the Christie Walk model and run for a 30-year simulation. These results were also compared with results for the current Christie Walk occupants (shown by 'CW' on the graphs). All other behavioural patterns were kept constant at the Christie Walk levels when an individual behavioural characteristic was analysed.

Figure 16 shows the effect of different behavioural levels of total waste production (total quantity of waste that needs to be reused, recycled, composted or disposed of to landfill) on the waste sustainability of a housing development (quantity of waste sent to landfill).

Figure 7.16. Effect of waste production behaviour on waste sustainability

Projected Waste Sustainability Rating Time Series
(Behavioural Effect of Waste Production)

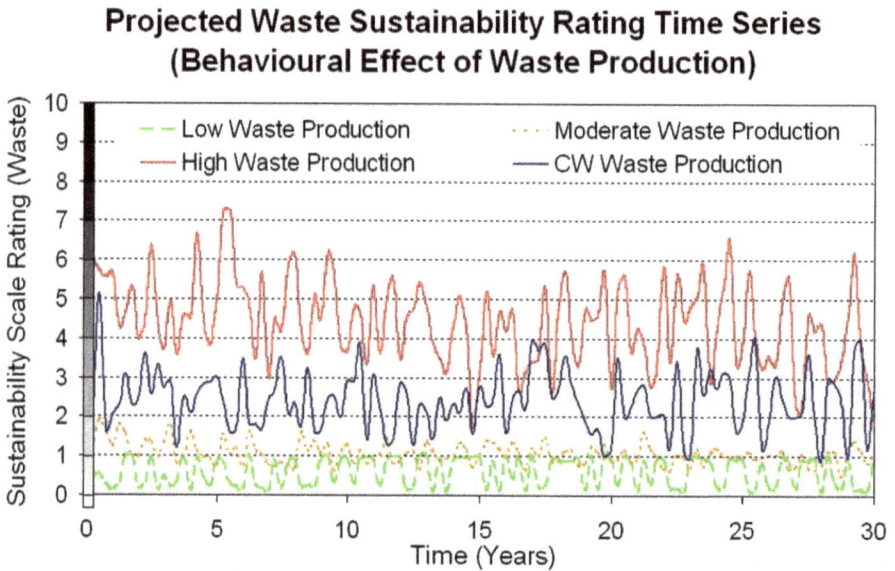

It can be seen from Figure 7.16 that the effect of total waste production behaviour on the waste sustainability of the housing development is substantial. Occupants with high waste production have Sustainability Scale Ratings as high as 7.3, while occupants with low waste production approach the threshold level for waste sustainability. A similar pattern, although with an improvement in ratings, is also observed for the behavioural effect of recycling on the amount of waste sent to landfill in Figure 7.17.

Figure 7.17. Effect of recycling behaviour on waste sustainability

Projected Waste Sustainability Rating Time Series
(Behavioural Effect of Waste Recycling)

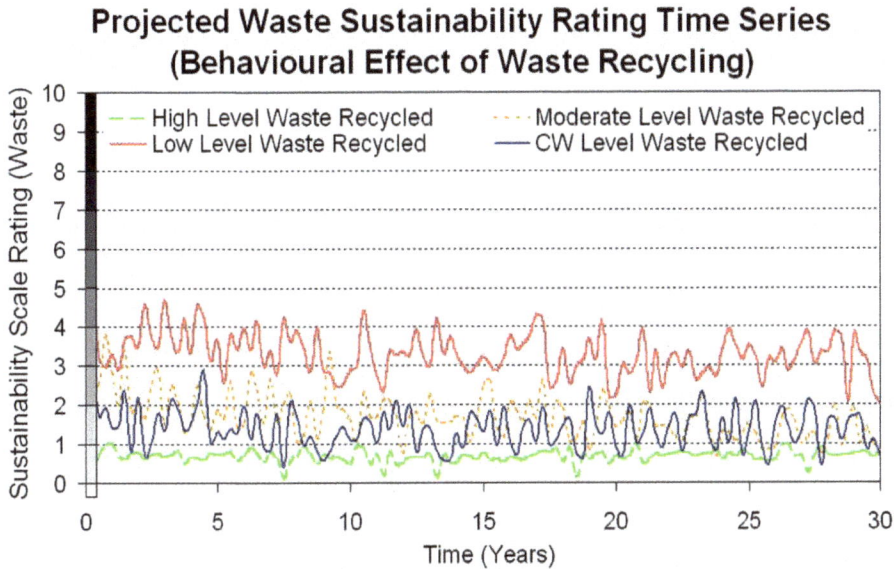

These results show that minimising waste at the source and recycling behaviour can have a significant effect on waste sustainability. It must also be noted that these results show that even with the worst recycling behaviour, the infrastructure at the Christie Walk development, including composting and recycling services, helps to reduce the effect of behaviour to below the 50th percentile of the greater Adelaide population.

A significantly different pattern is seen in Figure 7.18 for the effect of water use behaviour on water sustainability (based on mains water use) in Christie Walk. In this case, Sustainability Scale Ratings (SSRs) for water use show the effects of behaviour to be quite insignificant within the Christie Walk development. Figure 18 shows an average difference in water SSRs of less than 0.4 between low and high water use. This indicates that, in the Christie Walk development, behaviour has very little effect on water sustainability compared with the rest of metropolitan Adelaide. This is thought to be predominately due to the inclusion of water saving devices and the use of stormwater for toilet flushing and garden watering, which reduces the overall mains water use in the housing development. Changes in behaviour therefore do not have the effect they might have in developments without water saving infrastructure or small gardens.

Figure 7.18. Effect of water use behaviour on water sustainability

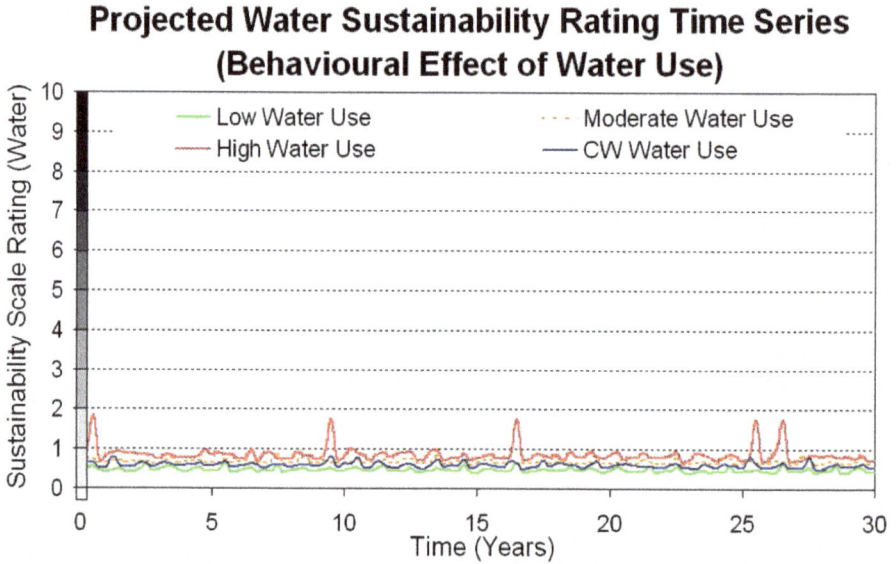

Projected Water Sustainability Rating Time Series (Behavioural Effect of Water Use)

The behavioural effects relating to in-house energy use on the CO_2 Sustainability Scale Ratings in Figure 7.19 show similar results to the water use behaviour example. Once again, the small spread of SSRs indicates that in the Christie Walk environment, behavioural influence on energy use has very little effect on sustainability. Due to the inclusion of energy efficient appliances, solar hot water heaters and building materials with high thermal efficiency, which lead to a lack of air conditioners, energy use in Christie Walk remains at very low levels for a range of behavioural types compared with metropolitan Adelaide.

Figure 7.19. Effect of energy use behaviour on CO_2 sustainability

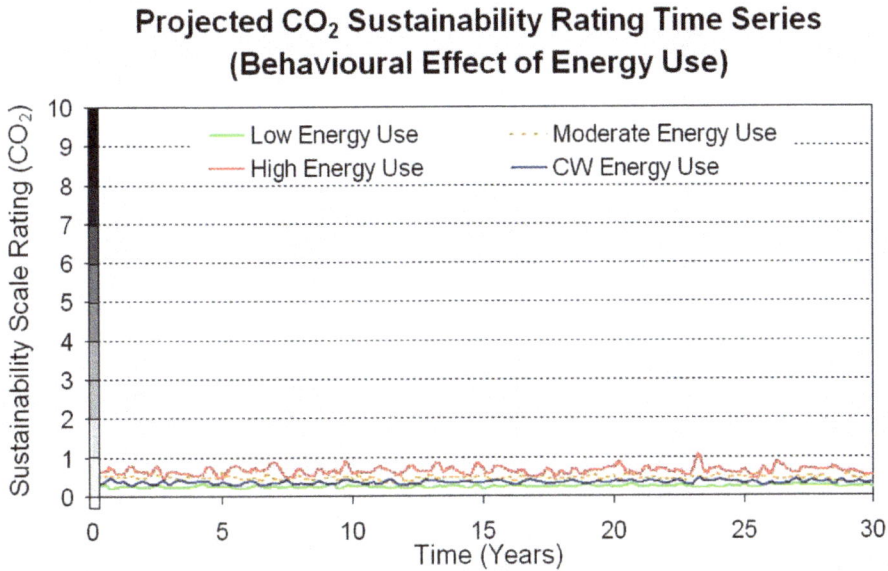

Projected CO_2 Sustainability Rating Time Series
(Behavioural Effect of Energy Use)

Discussion of behavioural results

The results of the behavioural analyses showed that, for certain household factors such as total waste production and waste recycling, changes in behaviour could significantly influence sustainability. Christie Walk includes basic infrastructure to encourage occupants to recycle and compost waste. However, because of their latent effectiveness, the infrastructure and design of other components of the development such as water infrastructure and passive building design, are seen to drastically reduce the potential impacts of residents' behaviour on Christie Walk's sustainability.

These findings highlight the importance of good design and infrastructure in achieving sustainability in built environments. Although it is possible for governments and planners to potentially have an influence on occupant behaviour, it has been shown that attempts to change behaviour can be very difficult, time consuming, and often met with extreme opposition. It could therefore be more effective for governments and planning authorities to concentrate on the improvement of infrastructure that will lead to a reduction in the effect of human behaviour.

From the information obtained from the survey and the knowledge of the practices at Christie Walk, it was found that most residents had a high waste diversion to recycling (they recycled large percentages of the waste they produced, rather than disposing of it to landfill). Many residents had a low waste production, although it was interesting to note that quite a number of residents exhibited a moderate waste production, seeming to focus more on the reduction of their waste going to landfill through recycling, as opposed to the minimisation of waste generation (buying foods with less or no packaging, etc.). It could therefore be advantageous to focus any future education campaigns on buying produce with less packaging rather than focusing on the end solution of recycling.

From the results of the survey of Christie Walk residents, it was also found that community interaction can have a significant impact on improving resource use behaviour (although not comparatively as large as the impacts of infrastructure and design), especially on recycling and energy usage, and to a lesser extent on water usage and waste minimisation. These findings highlighted the fact that if it is not possible to improve infrastructure, then improvements to resource use through certain community programs can be achieved. It has been stated by Marks et al. (2003) that the most effective way to introduce behavioural change is by including communities in the decision-making processes from the beginning of any plans for change. In this way, resistance can be reduced and the eventual uptake of practices and changes in behaviour can thus be made more quickly and smoothly.

Conclusion and future directions

The new methodology presented in this chapter allows a housing development to be examined as a complex system rather than being broken down into components. It also provides a method of assessing the sustainability of a housing development through time using multi-agent-based simulation. The multi-agent framework allows the integration of interrelated models including water, CO_2, ecosystem health, waste, economic, social, as well as their respective sustainability indicators. The methodology assists policy and decision making within governments and planning authorities by examining and comparing the quantitative sustainability of housing developments using Sustainability Scale Ratings. The assessment process allows different indicators of sustainability to be compared on a common scale (e.g., water and waste can be directly analysed for their comparative sustainability) or indicators in different housing developments can be compared. Furthermore, simulations of the multi-agent-based housing development model can be used to examine the emergent behaviour of the housing development system for various system changes and *What if?* scenarios along with the corresponding sustainability assessment of the indicators on the Sustainability Scale. This methodology would provide an ideal decision support system for stakeholders interested in urban developments, particularly in policy and planning applications.

The methodology, shown to be operational in the case study application of the Christie Walk housing development, simultaneously overcomes identified deficiencies in existing assessment tools, specifically, the inadequate inclusion of behavioural, spatial and temporal aspects within the sustainability assessment of complex housing development systems.

Results from the Christie Walk case study, with its many components of resource sensitive urban design, showed that the development compared favourably to the rest of the Adelaide metropolitan area. The importance of good infrastructure and design in reducing the impacts of human behaviour on housing development sustainability were also highlighted through the behavioural scenario analyses.

From the success of this study, it is considered that the methodology outlined in this research could be applied to assess other developments throughout the world for comparative purposes or as part of planning and policy assessment. It could equally be used as part of a participatory framework for decision-making and to stimulate stakeholder discussions relating to sustainability agendas. Future directions for this research are numerous and could including the following:

- further analysis of methods to model occupant behaviour based on more complex decision theory, game theory or other sociological and psychological theory;
- further analysis of the impacts of resource pricing on usage and behavioural changes;
- studies of behaviour relating to the uptake of sustainable technologies and practices, and how policy makers can better work with communities to ensure a successful uptake of such technologies and practices; and
- expansion of the methodology to assess the sustainability of rural systems, companies, countries, or any other complex system, potentially with the integration of Graphical Information Systems (GIS).

Acknowledgements

Thank you to Ariella Helfgott, Paul Downton, the residents of Christie Walk and everyone else who has helped and supported this work.

References

Amblard, F., N. Ferrand and D. Hill (2001) How a conceptual framework can help to design models following decreasing abstraction, Paper prepared for the 13th SCS-European Simulation Symposium, 18–20 October, pp. 843–47. Marseille: France.

Bousquet, F., I. Bakam, H. Proton and C. Le Page. (1998) Common-Pool Resources and Multi-Agent Systems (Cormas). *Lecture Notes in Artificial Intelligence* 1416: 826–37.

Bousquet, F., O. Barreteau, C. Le Page, C. Mullon and J. Weber (1999) An Environmental Modelling Approach: the Use of Multi-Agent Simulations,

Advances in Environmental and Ecological Modelling, pp. 113–22. Paris: Elsevier.

Clark, W.C. and Dickson, N.M. (2003) Sustainability Science: The Emerging Research Program. Proceedings from the National Academy of Sciences, Vol. 100, No. 14, 8 July, pp. 8059–61.

Curwell, S. and A. Hamilton (2003) *Intelcity Roadmap–Version 4*, University of Salford, United Kingdom. Available at http://www.scri.salford.ac.uk/ic-part/documents/Docs/Roadmap/Roadmap%204.doc [09/10/04].

Daniell, K.A., A.B. Kingsborough, D.J. Malovka, H.C.Sommerville, B.A. Foley and H.R. Maier (2004a) A review of sustainability assessment for housing developments, Departmental Report R175, School of Civil and Environmental Engineering, The University of Adelaide, Australia. Available at http://www.civeng.adelaide.edu.au/research/reports/R175.pdf [03/02/05].

Daniell, K.A., A.B. Kingsborough, D.J Malovka and H.C. Sommerville (2004b) Assessment of the sustainability of housing developments, Honours Research Report, School of Civil and Environmental Engineering, The University of Adelaide, Australia.

Daniell, T. M., B.A. Foley and K.A. Daniell (2004c) Sustainability in water resources management. How can it help?, *Proceedings of the 2004 International Symposium on Hydrological Environment*, Daegu, Korea, International Hydrologic Society, 8–9 October, pp. 7–26.

Deakin, M., S. Curwell and P. Lombardi (2002) Sustainable urban development: the framework and directory of assessment methods. *Journal of Environmental Assessment Policy and Management* 4(2):171–98.

Downton, P.F. (2002) Ecopolis: towards an integrated theory for the design, development and maintenance of ecological cities. Thesis (Ph.D.), Department of Geographical and Environmental Studies, University of Adelaide, Australia.

El-Fallah, A., I. Degirmenciyan and F. Marc (2004) Modelling, control and validation of multi-agent plans in dynamic context, Paper prepared for the Third International Joint Conference on Autonomous Agents and Multiagent Systems (AAMAS'04), 19–23 July, New York, USA.

European Co-ordination Action for Agent-Based Computing (2004) AgentLink – Agent Software (website). Available at http://www.agentlink.org/resources/agent-software.php [09/07/04].

Foley, B.A. and T.M. Daniell (2002) A sustainability tool for intrasectoral and intersectoral water resources decision making, Paper prepared for the

27th Hydrology and Water Resources Symposium, Melbourne, Australia, 20–23 May.

Foley, B.A., T.M. Daniell and R.F. Warner, (2003) What is sustainability and can it be measured? *Australian Journal of Multidisciplinary Engineering* 1(1):1–8.

Georg, S. (1999) The social shaping of household consumption. *Ecological Economics* 28:455–66.

Gilman, R. (1992) Sustainability, from the 1992 UIA/AIA. Call for Sustainable Community Solutions. Available at http://www.context.org/ICLIB/DEFS/AIADef.htm [04/02/03].

Huigen, M.G.A. (2003) Agent based modelling in land use and land cover change studies, Interim Report IR-03-044. Available at http://www.iiasa.ac.at/Publications/Documents/IR-03-044.pdf [12/05/04].

Jalas, M., A. Plepys and M. Elander (2001) Workshop 10 – Sustainable consumption and rebound effect, Presented at the 7th European Roundtable for Cleaner Production, 2–4 May, Lund, Sweden.

Krywkow, J., P. Valkering, J. Rotmans and A. van der Veen (2002) Agent-based and integrated assessment modelling for incorporating social dynamics in the management of the Meuse in the Dutch province of Limburg. In: *Integrated Assessment and Decision Support*, Proceedings of the First Biennial Meeting of the International Environmental Modelling and Software Society IEMSS, 24–27 June, University of Lugano, Switzerland, Vol. 2, pp. 263–68.

Macoun, T., G. Mitchell and P. Huovila (2001) Measuring urban sustainability: the challenge of integrating assessment methods and indicators, BEQUEST Lisbon Symposium, 26–27 April 2001, Lisbon, Portugal.

Marks, J., N. Cromar, H. Fallowfield and D. Oemcke (2003) Community experience and perceptions of water reuse. *Water Science and Technology: Water Supply* 3(3): 9–16.

McKenzie, S. (2004) Social sustainability: some definitions. Hawke Research Institute Working Paper Series, No. 27. Available at http://www.hawkecentre.unisa.edu.au/institute/resources/working%20paper%2027.pdf [11/10/04].

Melhuish, T., M. Blake and S. Day (2002) An evaluation of synthetic household populations for census collection districts created using spatial microsimulation techniques, Paper prepared for the 26th Australia and New Zealand Regional Science Association International (ANZRSAI) Annual

Conference, Gold Coast, Queensland, Australia, 29 September–2 October 2002.

Moss, S., T.E. Downing and J. Rouchier (2000) Demonstrating the role of stakeholder participation: an agent based social simulation model of water demand policy and response, Centre for Policy Modelling Discussion Papers, Manchester. Available at http://cfpm.org/~scott/water-demand/demand-pilot1.pdf [12/05/04].

Nishijima, K., D. Straub and M.H. Faber (2004) Sustainable decisions for life cycle costs based design and maintenance, Draft Paper. Submitted for the IFED Workshop on Consequence Modelling, Stoss, Switzerland, December.

Taylor, R. (2003) Agent-based modelling incorporating qualitative and quantitative methods: a case study investigating the impact of E-commerce upon the value chain, prepared for the 1st International Conference of the European Social Simulation Association, Gronigen, the Netherlands, September. Available at http://cfpm.org/papers/Taylor-abmiqaqm/Taylor-abmiqaqm.pdf [12/05/04].

Tweed, T. and P. Jones (2000) The role of models in arguments about urban sustainability. *Environmental Impact Assessment Review* 20: 277–87.

Vanbergue, D. (2003) Conception de simulation multi-agents: application à la simulation des migrations intra-urbaines de la ville de Bogota. Thèse de doctorat, Université Pierre et Marie Curie.

Waddell, P. and G.F. Ulfarsson (2004) Introduction to urban simulation: design and development of operation models. In D.A. Hensher, K.J. Button, K.E. Haynes, P. Stopher (eds), *Transport Geography and Spatial Systems, Handbook in Transport, Volume 5*. Amsterdam: Elsevier. Available at http://www.urbansim.org/papers/waddell-ulfarsson-ht-IntroUrbanSimul.pdf [12/05/04].

Viewpoint from a Defence Expert

The development of future war fighting concepts and capabilities involves the exploration of complex and multi-dimensional military domains. The effective use of traditional modelling and simulation approaches in this area is often constrained by the time and resources required to prepare, run and process high resolution combat models. Consequently, studies are often based on a limited set of runs that restricts examination of the wider problem space.

The use of agent-based simulation models has been adopted as a means of circumventing these shortcomings, enabling the modelling of complex military systems and thus enhanced scope for studies and exploratory experimentation. Typically, agent-based simulations are easy to use, flexible and have fast runtimes. The characteristics of agent-based simulations provide a tool that is well suited for the initial exploration of advanced war fighting concepts. This ability to explore the function landscape quickly is a major benefit of agent-based simulations and can be exploited by efficient experimental design that exhibits good space filling properties, allowing subsequent statistical investigation. Such tools can supply the analyst with a vehicle to explore the problem landscape, determine the opportunities and limitations of a new concept, gain insights into critical issues and thus identify areas requiring more detailed investigations. Importantly, military activities can be investigated at the effects level, without recourse to modelling of actual hardware and systems. Thus more abstract issues can be explored and this provides an independent force development process that could lead to disruptive rather than incremental changes in development trajectories.

Employing agent-based simulations as a exploratory tool provides a twofold benefit. They can provide the means to minimise resource costs, in terms of information, manpower and time, and increase flexibility. They also allow more creative investigation of new concepts without the constraints imposed by the properties of existing military systems. Critically, current agent based distillation studies have been found useful in that the game play and outputs are thoroughly compatible with the perspectives of military practitioners, and thus the technique is gaining ready acceptance.

John Hall
Land Operations Division—Canberra
Defence Science and Technology Organisation (DSTO), Australia

8. WISDOM-II: A Network Centric Model for Warfare

Ang Yang, Neville Curtis, Hussein A. Abbass, Ruhul Sarker
and Michael Barlow

Abstract

With recognition of warfare as a complex adaptive system, a number of Agent-Based Distillation Systems (ABDs) for warfare have been developed and adopted to study the dynamics of warfare and gain insight into military operations. These systems have facilitated the analysis and understanding of combat. However, these systems are unable to meet the new needs of defence, arising from current practice of warfare and the emergence of the theory of Network Centric Warfare (NCW). In this chapter, we propose a network centric model which provides a new approach to understand and analyse the dynamics of both platform centric and network centric warfare.

Introduction

Simulation is an important part of the Operations Researcher's tool box in assisting decision makers. Traditionally, Defence uses human-based warfare simulation to assess risks, optimise missions, and make tactical, operational and strategic decisions on change. However, simulation can be expensive and may not explore enough aspects of the problem or provide statistical validity.

Recent research (Ilachinski 1997; Lauren 2000) shows that warfare is characterised by nonlinear behaviours and that combat is a Complex Adaptive System (CAS). This has attracted attention from researchers and military analysts and a number of ABDs have been developed to understand and gain insight into military operations. They help the military analysts explore *What if?* questions that allow investigation of concepts and development of capability options. However, with recent practice of warfare and the emergence of the theory of NCW (Alberts 1999; Wilson 2004), analysts have found that it is hard to use these systems to understand and verify the new theory and concepts in warfare (Yang 2004). The military analysts often ask questions such as: how can we reach certain goals; why certain strange behaviours have emerged; and why the same setup has produced different results. Because almost all existing ABDs were built on platform centric warfare and existing agent architectures, it is very difficult for them to answer these questions. In this chapter, we propose version II of the Warfare Intelligent System for Dynamic Optimization of Missions (WISDOM)

(Yang et al. 2004, 2005b), which is built on a novel agent architecture, called the Network Centric Multi-Agent Architecture (NCMAA) (Yang et al. 2005a). With such agent architecture, WISDOM version II (WISDOM-II) allows analysts to study the dynamics of warfare easily, especially for NCW.

In this chapter we identify the key limitations of existing ABDs in the following section, then the NCMAA architecture and WISDOM-II are described. After that, scenario analysis is conducted before conclusions are finally drawn.

Limitations of existing agent-based distillation systems

Based on the understanding of warfare as a complex adaptive system (Ilachinski 1997, 2000, 2004; Lauren 2000; Scherrer 2004), a number of ABDs have been developed. These include the Irreducible Semi-Autonomous Adaptive Combat (ISAAC) (Ilachinski 1997, 2000) and the Enhanced ISAAC Neural Simulation Toolkit (EINSTein) (Ilachinski 2000, 2004) from the US Marine Corps Combat Development Command, the Map Aware Non-uniform Automata (MANA) (Lauren 2000; Lauren and Stephen 2002; Galligan et al. 2003; Galligan 2004) from New Zealand's Defence Technology Agency, BactoWars (White 2004) from the Defence Science and Technology Organisation (DSTO), Australia, the Conceptual Research Oriented Combat Agent Distillation Implemented in the Littoral Environment (CROCADILE) (Barlow and Easton 2002) and the Warfare Intelligent System for Dynamic Optimization of Missions (WISDOM) (version I) developed at the University of New South Wales at the Australian Defence Force Academy (UNSW@ADFA) (Yang et al. 2004).

ISAAC and EINSTein created a new era for warfare analysis. They are the first 2 systems that modelled warfare as a CAS. Almost all latter ABDs were inspired by them. MANA introduced the concept of way-points, internal situational awareness (SA) map and event-driven personality changes. The latest version of MANA (released at the end of 2004) concentrated on the model of communication, including the reliability, accuracy, capacity and latency of each communication channel. BactoWars focused on problem representation and tried to provide a simple framework that allows analysts to model the real world problems more adaptively and flexibly. CROCADILE used a 3D continuous environment with a higher fidelity than ISAAC, EINSTein and MANA. Version I of WISDOM improved the movement algorithm used by the above systems and adopted evolutionary computation techniques to search for optimal strategies for pre-defined scenarios (Yang et al. 2005b).

These systems have facilitated the analysis and understanding of combat, for example, using MANA to explore factors for success in conflict (Boswell et al. 2003). They offer an opportunity to analyse the behaviours that we would intuitively expect on the battlefield. Through the use of these systems, people are able to gain understanding of the overall shape of a battle and what factors are playing key roles in determining the outcome of a battle. However, in our opinion, current ABDs are facing several shortcomings:

- Hard to validate and verify. System behaviours emerge from simple low level rules in any CAS. In current ABDs, agents are programmed without an underlying theoretically sound software architecture. Therefore, it is very difficult to validate and verify them.
- No reasoning during the simulation. Due to the high degree of nonlinear interaction between agents, it is impossible to reason at the agent level, which makes it hard to understand the results of the whole simulation.
- Can be a computationally expensive exercise in some systems. This is either because of a bad design, unnecessary fidelity, or fancy tools without proper modelling.
- No connection between tactic and strategy. Existing ABDs are developed either on the reactive agent architecture, which focuses on tactics, or on the BDI (Belief-Desire-Intention) architecture, which focuses on strategies (Wooldridge and Jennings 1995; Nwana 1996; Sycara 1998; Wooldridge 1999). There is almost no interaction between tactics and strategies being modelled by existing ABDs.
- Hard to capture the underlying structural interaction between agents. Although existing ABDs embed the structural interaction between agents, there is no explicit model for such interactions. It is hard for the user to capture these interactions during the simulation, which is a crucial point of a CAS.
- Difficulty in application to complexity. Current ABDs are based on conventional military tactics and tend not to be approached from an overarching systems view. Concepts such as NCW, with its inherent complexity and interdependency, present challenges to identifying correct inputs at the entity level. Thus, techniques addressing higher level manipulations must be employed.

To address the limitations above, WISDOM-II, which is re-designed and re-implemented based on the NCMAA architecture, is proposed.

Network Centric Multi-Agent Architecture

The limitations of existing ABDs and current agent architectures motivate the appearance of the new agent architecture of NCMAA, which is purely based on network theory (Wasserman and Faust 1994; Dorogovtsev and Mendes 2002). The system is designed on the concept of networks, where each operational entity in the system is either a network or a part of a network. The engine of the simulation is also designed around the concept of networks. Each type of relationship between the agents forms a network. We distinguish between a static network, where the topology does not change during the course of the simulation, and a dynamic network, where the topology changes. An example of the former is a network of families, while an example of the latter is a communication network. It is important to emphasise that the definition of static or dynamic may vary from one application to another.

The properties of each network may change because the agents take actions that may trigger changes in agents' states, environmental states, or simply simulation clock advances. These triggers affect the dynamic relationships and the cycle continues. Figure 8.1 depicts a coarse-grained view of the system.

Figure 8.1. A coarse-grained view of NCMAA

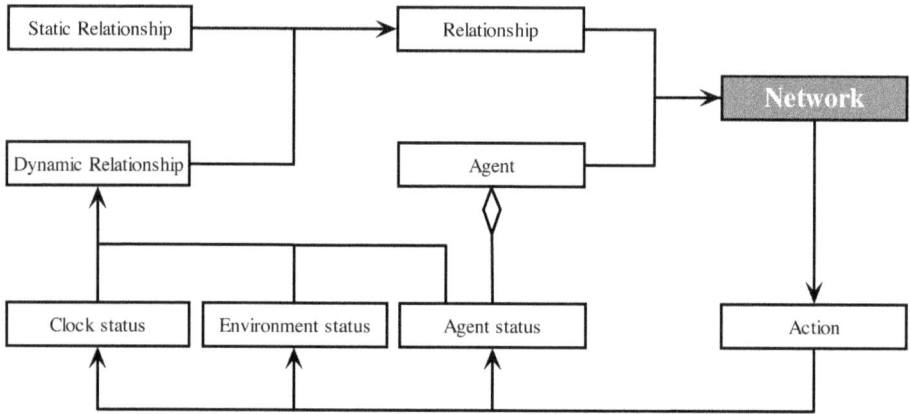

To design a system based on NCMAA, the developer first needs to define an influence diagram of concepts, and then develop a finite state machine to control the simulation engine. The influence diagram is a directed graph of concepts defining the interdependency of concepts in the concept space. It provides the basis for establishing a meta-level reasoning system. The finite state machine is a collection of states, each representing the state of a network in the system. The finite state machine represents the sequence of executing each network in the system and the control of the system clock.

NCMAA adopts a two-layer architecture (Figure 8.2). The top layer, which we call the influence network, based on the influence diagram, defines the relationship types and how one type of relationship influences other types. Each of these relationship types is reflected in the bottom layer by a set of agents who interact using that relationship. For example, agents that can see each other would be connected in the lower layer, and agents that can communicate with each other would also be connected in the lower layer. The influence of vision on communication would form a connection in the top layer.

Figure 8.2. wo-layer architecture in NCMAA

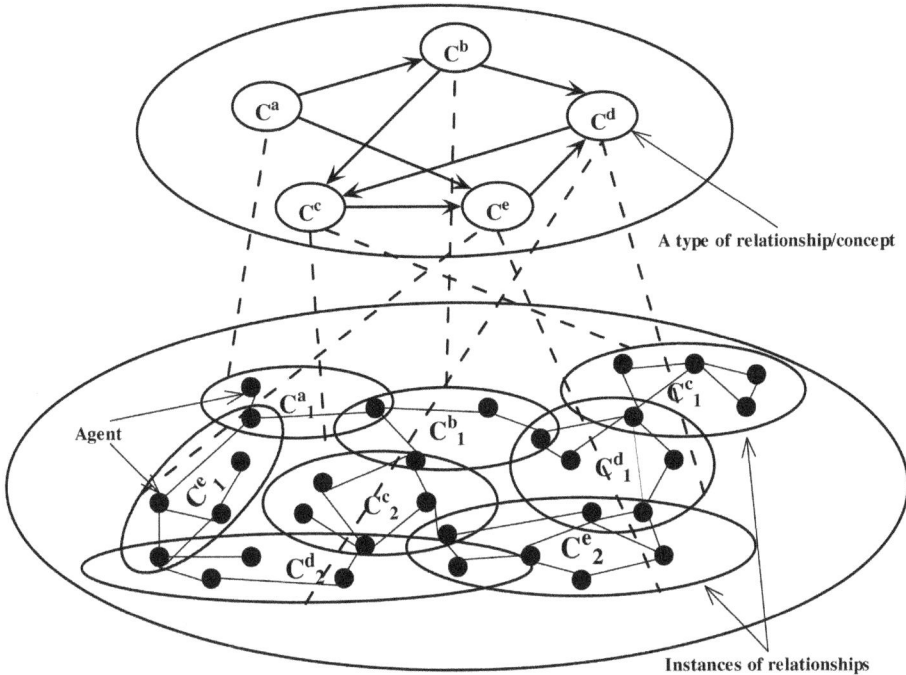

WISDOM Version II

WISDOM (Yang et al. 2004) was first proposed in 2004 and is similar to other existing ABDs. Although it improved the movement algorithm and used the relational database to store information during simulations, it faced the same problems as other ABDs. In order to overcome these problems, WISDOM-II has been developed based on NCMAA. It not only uses the spirit of CAS in explaining its dynamics, but also centres its design on fundamental concepts in CAS. In WISDOM-II, concepts have been abstracted at a level that balances efficiency and depth of understanding. It is implemented based on the concept of keeping it simple, informative, efficient but realistic. Generally speaking, there are 5 components in WISDOM-II. The first 3 components are used to model the internal behaviours of warfare, and the last 2 components are analysis tools:

- *the C3 component*—including both Command and Control (C2), and communication. C2 is a key concept in military operations. It is the process where the commander directs and coordinates the assigned forces to achieve certain missions according to their own knowledge and information collected. Allard (1996) defined C2 as:

  'The exercise of authority and direction by a properly designated commander over assigned forces in the accomplishment of the mission.

Command and control functions are performed through an arrangement of personnel, equipment, communications, facilities, and procedures which are employed by a commander in planning, directing, coordinating, and controlling forces and operations in the accomplishment of the mission.'

Dissemination of information in a command and control system requires secure and robust communication network. The C3 component in WISDOM- II has 2 computer models, one for C2 and the other for the communication system;

- *the sensor component*—retrieving information from environment;
- *the engagement component*—including firing and movement activities;
- *the visualisation component*—presenting various information with graphs; and
- *the reasoning component*—interpreting the results in natural language during the simulation process.

Architecture

There are 5 concept networks—relationships between agents (Figure 8.3)—that form the influence network in WISDOM-II and make up the top layer of the NCMAA. These 5 concept networks are:

- *The C2 network* defines the command and control hierarchy within one force. Each force has its own commander and control structure. Figure 8.4 depicts the C2 hierarchy in WISDOM-II. Each force may have several teams, each of which may include several groups. Each group may have a number of agents with different characteristics. We first introduce heterogeneous agents at the group level in WISDOM-II. Since the commands can only be sent from agents at the higher level to agents at the lower level, the C2 network is a directed graph.
- *The vision network*—if agent A can see agent B, then there is a link from Agent A to Agent B. The vision network is also a directed graph.
- *The communication network*—communication only occurs within the same force in WISDOM-II. Therefore, there are 2 instances of the communication network within the system. These communication networks could carry 2 types of information: situation information and commands. In a traditional force, the situation information typically flows from an agent to other agents in the communication channel, from the group leader to the team leader, and from the team leader to the general commander. In a networked force, the situation information flows directly from the agent to its general commander. In either a traditional force or a networked force, a Common Operating Picture (COP) is developed at the headquarter based on the collected information. Based on the COP the general commander makes decisions for each group in

the battlefield and sends commands to the team leader, and then the team leader sends commands to the group leaders. However, in a networked force, all agents in the battlefield can access the COP through the communication channel. Therefore, each agent has a global view of the battlefield, while in a traditional force each agent only has its own local view of the battlefield. Since we use network to model communication, it is easy for WISDOM-II to support various types of communications: Point to Point directly (P2Pdirect), Point to Point indirectly (P2Pindirect), and Broadcast (BC). Because the information flows from source to sink, the communication network is obviously a directed graph.

- *The information fusion network*—defines current knowledge about friends and enemies through vision and communication. We fuse the information collected by vision and communication and then develop this network. Since both vision and communication are direction dependent, the information fusion network is a directed graph too.
- *The engagement network*—defines the agents being fired at, based on the firing agent's current knowledge about its enemies and friends. This network is also a directed graph.

Figure 8.3. Influence Network for NCMAA in Warfare

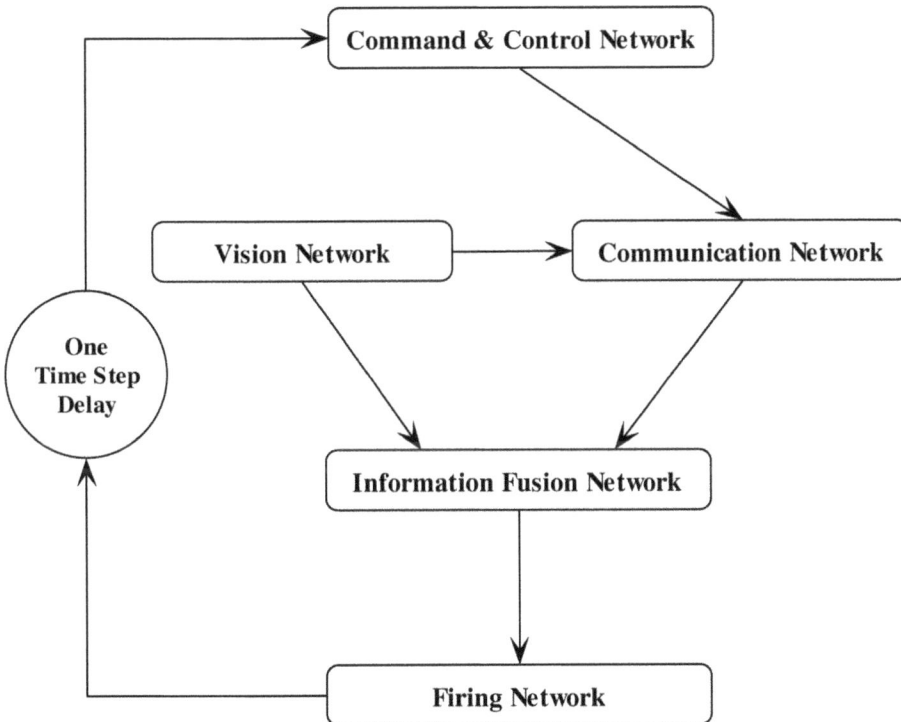

Figure 8.4. Command and Control Hierarchy in WISDOM-II

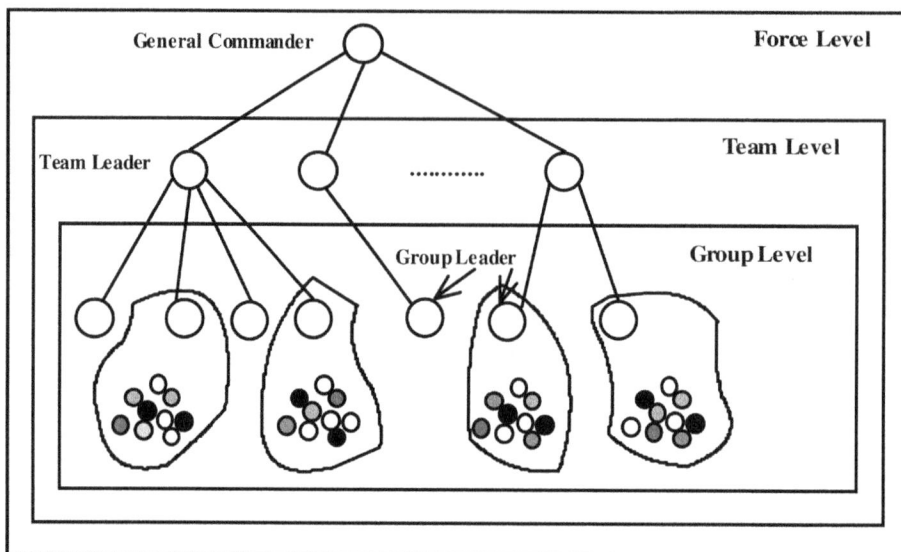

Each of these networks influences each other. The C2 network specifies how the command is flowing, therefore it influences the communication network. Both vision and communication network define the information fusion network, based on which agents decide which agent they may fire at. Agents may die through firing, so the engagement network will influence the C2 network in the next time step. Each of the above concept networks may have one or more instances defined for the blue and red agents with their interactions. These instances make up the lower layer of the NCMAA. Since different forces have different C2 structures, and communication can only occur within the same force, there are 2 instances of the C2 network and 2 instances of each communication network, while there is only one instance for each other network in WISDOM-II.

Characteristics of agents

4 types of agents are supported in WISDOM-II: combatant agent, group leader, team leader and general commander. Agents are defined by their characteristics and personalities. Both team leader and general commander are virtual agents that exist in the force headquarters. They only have one capability: communication. Each combatant agent and group leader has 8 characteristics: health, skill, probability to obey command, invisibility, vision, communication, movement and engagement. The initial levels of health, skill, visibility and vision are set by the user. They may differ based on different agent types. The health defines the level of energy for an agent. With user defined wounded threshold and immobile threshold, each agent may have 4 health statuses: healthy, wounded, immobile, and dead.

The skill level of each agent is used to set the hit probability, which is the probability of the enemy being hit. High skill level means high hit probability. The probability to obey command is the probability to follow the command sent by its leader. When an agent receives a command from its commander, it may obey the command or take actions based on its own knowledge.

Invisibility defines how hard it is to be detected by other agents if it is within their vision range. The value of invisibility should be between [0, 1], with 1 meaning invisible to all. Each combatant agent has its own sensor that is defined by sensor range and detection. Detection defines what kind of agents can be detected by using this sensor. If the detection of agent A is equal to or larger than the invisibility of agent B, and agent B is within agent A's vision range, then agent A may detect agent B. The value of detection should be between [0, 1], with 1 representing able to detect all.

Combatant agents can communicate with other agents linked directly to them through the communication network. This communication occurs through a communication channel, which is modelled by noise level, reliability, latency and communication range. The agent may only communicate with the agents within the range of that communication channel. We also adopted a probabilistic model to implement the noise level and reliability of a communication channel. Each communication channel has 2 probabilities corresponding to noise level and reliability. At each time step the message can only be transferred from one agent to another agent. The message will permanently be lost if it is older than a number of time steps predefined by the user.

How agents move is determined by their speed and personality. WISDOM-II supports 4 kinds of speeds: still, low speed, medium speed and high speed. Agents with a low speed must wait for 2 time steps until they can move, and agents with a medium speed must wait for 1 time step until they can move, while agents with a high speed can move every time step. The movement algorithm is described in the section about tactic decision making.

Engagement in WISDOM-II is determined by what kind of weapon the agent uses. The weapon is defined by the weapon power, fire range and damage radius. Based on the damage radius, 2 types of weapon are supported in WISDOM: point weapon, the damage radius of which is zero, and explosive weapon, the damage radius of which is larger than zero. WISDOM-II also supports direct and indirect fire. Indirect fire can fly over obstacles.

Personalities of agents

Existing ABDs use personality to define the tendency of an agent to move close to or far from certain types of agents, while WISDOM-II uses it to define the influence of other agents on this agent. The influence of other agents may attract the agent to move closer to them or repulse the agent far from them. The personality in WISDOM-II is a vector quantity specified by its magnitude and its direction. The magnitude is the strength of the influence which is between 0 (no influence) and 1 (strongest influence). The direction defines at which direction the influence occurs. There are eight directions, defined by a value between 0 and 1 (Figure 8.5).

Figure 8.5. Influence directions

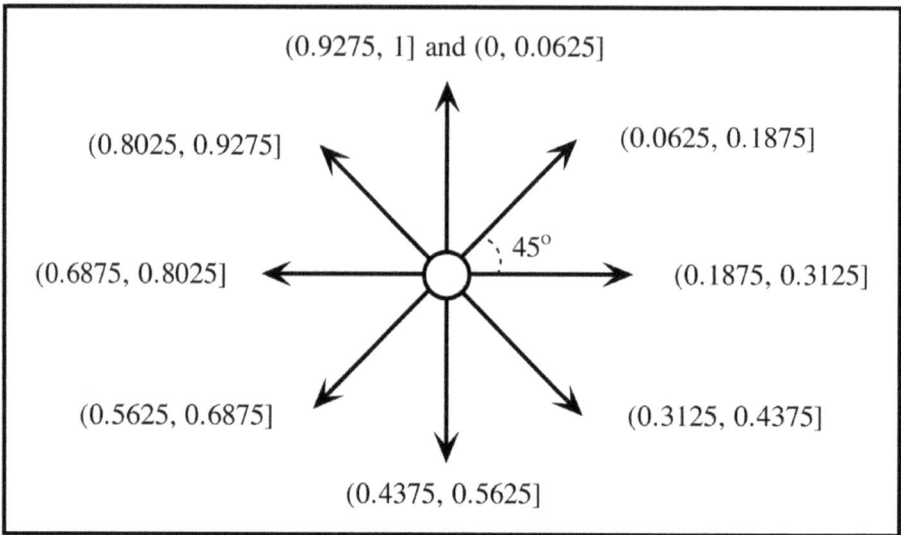

The influence of agent A on agent B may be different from that of agent B on agent A. This difference can be reflected in either the value of the influence, or the direction of the influence, or both. Similar to existing ABDs, the personality is used to determine the new location the agent should move to. For an agent, its movement is influenced by all the agents which it can detect or communicate with. The algorithm to aggregate and resolve movement is described later. The personalities for each type of movement influence are:

1. Influence vector of healthy friend agent via sensor.

2. Influence vector of wounded friend agent via sensor.

3. Influence vector of immobile friend agent via sensor.

4. Influence vector of healthy hostile agent via sensor.

5. Influence vector of wounded hostile agent via sensor.

6. Influence vector of immobile hostile agent via sensor.

7. Influence vector of healthy friend agent via communication.

8. Influence vector of wounded friend agent via communication.

9. Influence vector of immobile friend agent via communication.

10. Influence vector of healthy hostile agent via communication.

11. Influence vector of wounded hostile agent via communication.

12. Influence vector of immobile hostile agent via communication.

13. Influence vector of group leader.

The group leader has the same number of personalities as a combatant agent. However, the group leader has the influence vector of the target flag assigned by the higher level commander instead of the influence vector of the group leader. Since they are virtual agents and do not move, the team leader and general commander do not have personalities.

Decision-making mechanism

There are 2 decision making mechanisms in WISDOM-II: tactical and strategic. The former is used by each agent to determine where it should move to and is based on the agent's current knowledge and personalities, while the latter is used by each general commander to determine the way point for each group and is based on the group's mission type and the firepower of each force in the battlefield.

Tactical decision-making mechanism

The movement of each agent is determined by its knowledge and personality. Only a healthy or wounded agent in the battlefield can move to a new location. In each time step, the agent can only move to its neighbour cell at the direction of the overall influence of all influencing agents. A movement function, as in Equation 1, is constructed on the influence vectors, and an agent moves in the direction of the resultant influence vector. The sum of the influence vector in the equation is a vector summation. Calculations are done synchronously with the moves. This process is repeated for each time step in the simulation. This movement algorithm is totally different from that implemented in any other ABDs.

$$RF_k = \frac{\sum_i^n \vec{F}_i^v + \sum_j^m \vec{F}_j^c + \vec{F}^t}{D}$$

(1)

where:

RF_k denotes the resultant influence of an evaluated agent k;

n denotes the total number of agents perceived by agent k through vision in the information fusion network;

$\vec{F}_i^v$ denotes the influence vector from agent i, who is perceived by agent k through vision in the information fusion network;

m denotes the total number of agents perceived by agent k through communication in the information fusion network;

$\vec{F}_j^c$ denotes the influence vector from agent j, who is perceived by agent k through communication in the information fusion network;

$\vec{F}^t$ denotes the influence vector from agent k's target. If the evaluated agent is a combatant agent, then the target is its group leader. If the evaluated agent is the group leader, then the target is the group waypoint.

D denotes the distance between the agent and the influencing agent.

In WISDOM-II, each cell can only accommodate one live agent. If more than one agent would like to move to the same cell, a collision occurs, then the collision resolution mechanism is used to remove it. The collision resolution mechanism is defined by a set of rules:

- the agent that occupied this cell in the previous time step has the highest priority to stay in this cell;
- the group leader has the second highest priority to move to this cell;
- the wounded agent has the third highest priority to move to this cell, the wounded agent normally goes back to its hospital for recovery; and
- if multiple agents with the same priority wish to occupy the cell, one is uniformly randomly chosen to move and the others stay in their original cells.

Strategic decision-making mechanism

A strategic decision is made for each group by the general commander of each force based on the COP, which is the global view of the battlefield for that force, and the mission type. Each team is assigned one mission in the input XML configuration file by the user. Then all the groups in that team have the same mission as the team mission. Three types of missions are supported in WISDOM-II: defend, occupy and surveillance. Defend means that the group is trying to protect a certain area; occupy means that the group is trying to occupy a certain area; and surveillance means that the group is trying to collect all possible information about the battlefield.

Three types of decisions can be made for each group: advance, defend and withdraw. Advance means that the group needs to go forward; defend means that the group needs to protect the current location; and withdraw means that the group needs to escape from the current location. After making decisions, the high level commanders send the decision/command/intention to the lower level agents. To simplify the implementation of the command, the command sent to the group leader is the location of the waypoint.

The general commander abstracts the whole environment into $n \times n$ (n is predefined by the user) hyper cells and calculates the total firepower of the hostile and own force for each hyper cell. The total firepower is a function of individual agent weapon power, fire range, damage radius and health level (see Equation 2).

$$P = \sum_{i}^{n} (f_i^P \times f_i^R \times r_i \times h_i)$$

(2)

where:

P denotes the total fire power of one force in this hyper cell;

n denotes the total number of agents of one force in this hyper cell;

f_i^P denotes the weapon power of agent i in this hyper cell;

f_i^R denotes the fire range of agent i in this hyper cell;

r_i denotes the damage radius of agent i in this hyper cell;

h_i denotes the health of agent i in this hyper cell.

For each group with the mission of occupy, the general commander calculates the force power ratio defined in Equation 3 for each of the surrounding hyper cells.

$$R_p = \frac{P_o}{P_h}$$

(3)

where R_p is the fire power ratio, P_o is the total fire power of own force in that hyper cell, and P_h is the total fire power of hostile force in that hyper cell. The waypoint for each group is generated based on Equation 4.

$$WP = \begin{cases} C_i, & \text{if } \exists R_p^i \leq \theta_1, C_i \in C^b; \\ C_0 & \text{else if } \exists R_p^i > \theta_1, R_p^i \leq \theta_2, C_i \in C^b; \\ C_j & \text{else } \forall C_k \in C^s, R_p^j = \min(R_p^k) \end{cases}$$

(4)

where:

θ_1 is the advance threshold while θ_2 is the defend threshold. Both of them are defined by the user in the input configuration XML file;

C^b is the set of hyper cells which is between the group and its goal. It is defined in Table 8.1. The maximum number of the hyper cells in this set is 3;

C^s is the set of hyper cells which are surrounding the hyper cell of that group. The maximum number of the hyper cells in this set is eight;

C_0 is the hyper cell which is occupied by that group;

C_i is a hyper cell in C^b ;

C_j is a hyper cell in C^s ;

R_p^i, R_p^j, R_p^k is the force power ratio in the hyper cell i, j or k respectively.

Table 8.1. The set of hyper cells in between

Goal relative position to group position	The set of hyper cells in between (C^b)
Top-left	Left, top-left and top hyper cell
Top	Top-left, top and top-right hyper cell
Top-right	Top, top-right and right hyper cell
Right	Top-right, right and bottom-right hyper cell
Bottom-right	Right, bottom-right and bottom hyper cell
Bottom	Bottom-right, bottom and bottom-left hyper cell
Bottom-left	Bottom, bottom-left and left hyper cell
Left	Bottom-left, left and top-left hyper cell

If the firepower ratio of one hyper cell in C^b is less than or equal to the advance threshold θ_1, the centre of this hyper cell will be assigned to the group as the waypoint. If the firepower ratios of all hyper cells in C^b are between the advance threshold θ_1 and the defend threshold θ_2, the group will stay in the original hyper cell. Otherwise the commander selects the centre of the hyper cell which has the minimal firepower ratio among all surrounding hyper cells as the waypoint for the group.

If more than one hyper cell meets the condition, the general commander will randomly select one. When the command is sent to the group, the group leader may misunderstand it. We model such misunderstanding by the probability and the variance of misunderstanding. The probability of misunderstanding stands for the chance that a group leader may misunderstand the command. The variance defines the degree of misunderstanding. In other words, it defines the degree of a received waypoint deviating from its correct location. After receiving the command, the group leader may or may not follow it, based on the probability of following the command mentioned above.

For a group with the mission of surveillance, the waypoint is the surrounding hyper cell with highest hostile force power. In order to maximise the information collected, no more than one group will be assigned to the same hyper cell. Since the area the team needs to defend is predefined by the user in the input configuration XML file, no command will be sent to the group with the mission of defend. The frequency of the general commander sending the command is also defined by the user in the input configuration XML file.

Recovery

One of the most important aspects of military operations is logistics, and the medical treatment system is one of its key components. The model of the artificial hospital is first introduced in WISDOM-II. Each team may have a hospital in the team base, which is defined by the number of doctors and the recovery rate. If the team has a hospital, the wounded agent will move back to the hospital for treatment. Each doctor can treat only one wounded soldier at each time step and the health of that treated soldier will be increased by the recovery rate at each time step. If all doctors are already treating, the wounded soldier will be put in the queue to wait for treatment. When the agent is fully recovered, it will move back to the battlefield near its group leader. If its group leader is in the hospital, or if it is the group leader, it will be positioned in a cell around the team base.

The waiting queue and the number of doctors in the hospital are used to model the limited resources available and the recovery capability of a force. The capability to make a wounded soldier fully recovered is a key issue in warfare. With this kind of recovery model, analysts may search for the minimal resources needed to maintain a minimal level of recovery capability for a force.

Visualisation

One drawback of current ABDs is that they only provide limited information to analysts during simulations. WISDOM-II fills this gap and visualises many types of information. The visualised information in WISDOM-II can be in two categories: information about entities in the battlefield, and information about the interaction between entities. The entity information includes the agent status (position and health), the number of dead agents, the number of agents in hospital, the number of wounded agents, the total number of agents in each force, and the total damage for each force. With such information the user may quickly identify the damage, casualty and firepower for each force. The interaction information between agents includes: C2 structure, information flow chain, vision, knowledge, and engagement. By spatially visualising the interaction information, WISDOM-II provides a graphic view for each type of interaction between agents. It offers a window for the user to know what kind of interaction occurs between agents, and to identify the role of each interaction during the simulations. For example, visualising the C2 and communication network for each force may allow the user to see how information is transferring, how the C2 structure is evolving, and if it is efficient for a certain communication structure. Combining with visualisation of the entity information, the user may develop the relationship between the damage or casualty of a force with the structure of C2 or communication network.

Spatially visualising the information also makes it possible to validate and verify the system itself. If something happens, it will immediately be reflected on the visualisation. The developer or the user can then identify and capture it quickly.

Finally, combining with the real-time reasoning component described in the following subsection, the analysts may gain deeper understanding of what happens in each time step during the simulation and how the simulation is progressing.

Reasoning

The architecture of NCMAA makes it possible and easy to conduct real-time reasoning during the simulation. BDI agents do not scale well because all reasoning is at the individual level. The nonlinear interaction between agents makes it almost impossible to reason at the individual level. To overcome this, WISDOM-II conducts reasoning at the network (group) level. We define the reasoning component through 2 stages: establishing causal relationships, and establishing degree of influence. The first stage is taken care of through the influence diagram, which defines the causal relationship. The second stage is undertaken through statistical inference.

We define 5 interactions (concept networks) between agents in WISDOM-II and the influence network identifying the influence between them (Figure 8.3). In order to see which network (interaction) among these 5 concept networks is the key aspect leading to the damage of the enemy and to see why certain behaviours emerge during the simulation, 2 kinds of reasoning are conducted in each time step: time series analysis, and correlation analysis.

Time series analysis allows the analysts to capture the dynamics over time during the simulation. It includes:

- damage of each force over time;
- average degree of each network over time;
- average path length of each network over time;
- clustering coefficient of each network over time;
- spatial entropy at the coarse, medium and fine level over time; and
- combat entropy over time.

When any network collapses, the user may immediately capture it according to these network measures.

Correlation analysis may allow the analysts to understand which network is playing the key role in damaging the enemy. We don't randomly select two parameters and conduct the correlation analysis between them. Correlation analysis is based on the influence network (Figure 8.3). For example, in the influence network, the information fusion network directly influences the firing network, while the information fusion network is based on the communication network and the vision network. In order to see that the damage of the enemy is mainly caused by the activity in the vision network or in the communication network during a certain time frame, we may conduct correlation analysis of some network measures of the vision and communication network with the damage of the enemy.

The window and data-streaming techniques are used in the correlation analysis. All the data of the parameters, which are used to conduct the correlation analysis, are streamed and filtered. Only the valid data within the window size, which is defined by the user in the input configuration XML file, are captured for the correlation analysis. For example, if the window size is set to 5 and we would like to see if the communication plays a key role in damaging the enemy, we might calculate the correlation coefficient between the average degree of communication network and the damage of the enemy. Only the data of the average degree of communication network and the damage in the last 5 time steps are captured and used for the calculation. Following are 11 examples of correlation coefficients calculated between the *damage of the enemy* and the:

- average path length of communication network;
- average degree of communication network;
- ratio of the number of links from vision to the total number of links in information fusion network;
- ratio of the number of links from communication to the total number of links in the information fusion network;
- correctness of the knowledge about their enemy at the force level;
- correctness of the knowledge about their enemy at the agent level;
- correctness of the knowledge about their friend at the agent level;
- correctness of the knowledge about their friend at the force level;
- spatial entropy at the coarse level;
- spatial entropy at the medium level; or
- spatial entropy at the fine level.

In WISDOM-II, two kinds of knowledge correctness are measured: correctness of knowledge about the friend (C_f) and the enemy (C_e), which can be calculated as:

$$C_f = \frac{N_f^a}{N_f}$$

(5)

$$C_e = \frac{N_e^a}{N_e}$$

(6)

where N_f^a denotes the number of friends perceived by a single agent or a force;

N_f denotes the total number of friends; N_e^a denotes the number of enemies perceived by a single agent or a force; N_e denotes the total number of enemies.

Based on the real-time reasoning, WISDOM-II provides an English-like interface to interpret the dynamics during the simulation to the user. Since the command sent by the general commander for each group is very important too, we integrate the command log to this English-like interface.

Terrain feature

WISDOM-II supports impassable objects. Agents cannot see or travel through impassable blocks. With the indirect weapon, the agent can shoot its enemies through them. The impassable blocks can be defined in the input configuration XML file by specifying each block's coordinates and the corresponding length and width.

Interactive simulation

Any simulation system is designed and implemented based on the domain knowledge of its developers. The limitation of the domain knowledge of the system developers may lead to misunderstanding of the simulations. This requires the simulation system to be able to import the domain knowledge from the domain experts during the simulation. Most existing ABDs generally don't allow interaction between the system and the user during the simulation.

WISDOM-II supports interactive simulation. In any time step, the users may send a command to any group based on their knowledge of the military domain. Therefore, the simulation may automatically run based on a predefined scenario or be guided by the user.

Scenario analysis

A simple scenario has been built to exemplify the use of our model. The red force is a traditional force with a large number of soldiers and traditional weapons while the blue force is a networked force with a small number of soldiers and advanced weapons. There are two surveillance agents in the blue force, they do not have weapons but are invisible to the red force. The scenario settings for each force are shown in Table 8.2. Figure 8.6 depicts the initial position of each force. The blue force is initialised at the top left corner of the environment while the red force is initialised at the bottom right corner.

Table 8.2. Scenario settings

	Blue Force	Red Force
Number of Agents	11	50
Vision	9 short	50 medium
	2 long	
Communication	Networked	Medium range
	2 no weapon	
Weapon	8 P2P	50 P2P
	1 explosive	

Figure 8.7 presents the damage of each force over time. From this figure, it is easy to see that the engagement started around time step 15. Although the red force has over 4 times number of agents than the blue force, the losses of the red force are much higher than that of the blue force. This is consistent with the importance of communication in combat. Two blue surveillance agents collect information and send it back through communication. Based on the COP developed in the headquarter of the blue force, the blue agent with the long range, explosive weapon may then shoot enemies. However, the red agents only have local information and they do not know where their enemy is, so they always get fired upon.

Figure 8.6. Initial position

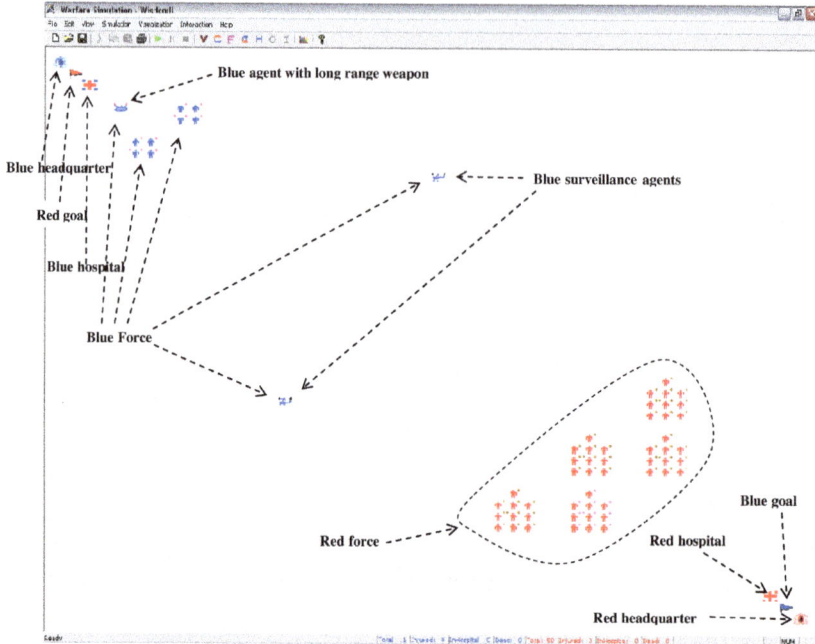

Figure 8.7. Damage of each force over time

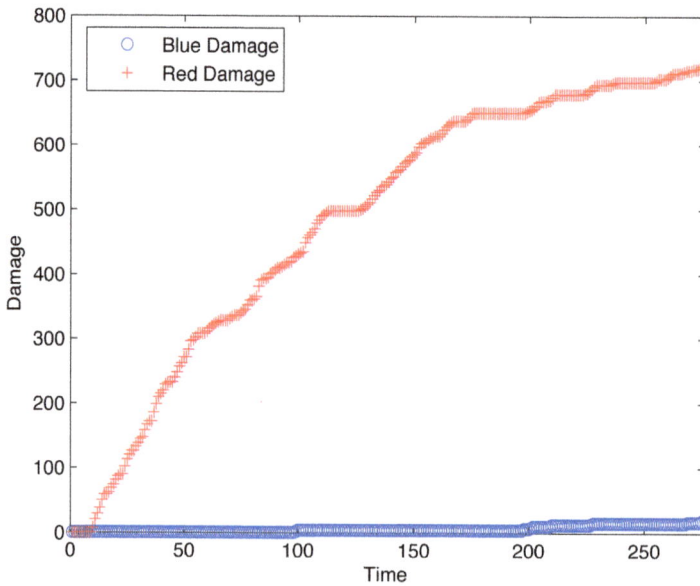

Figure 8.8 presents the average degree and average path length of the blue and red communication network over time. It supports our intuitive view above. The average path length of the blue communication network is always longer than that of the red communication network. The average degree of the blue communication network is larger than that of the red communication from about the time step of 35. This implies that information can be transmitted to most of the blue agents in the battlefield. We also note that the average degree and path length of the blue communication network are almost unchanged over time, while those of the red communication network are continuously going down. This may imply that the red communication network has broken and that they cannot exchange information properly. They know much less than the blue force and cannot win the game.

Figure 8.8. Average degree and average path length of the blue and red communication network over time

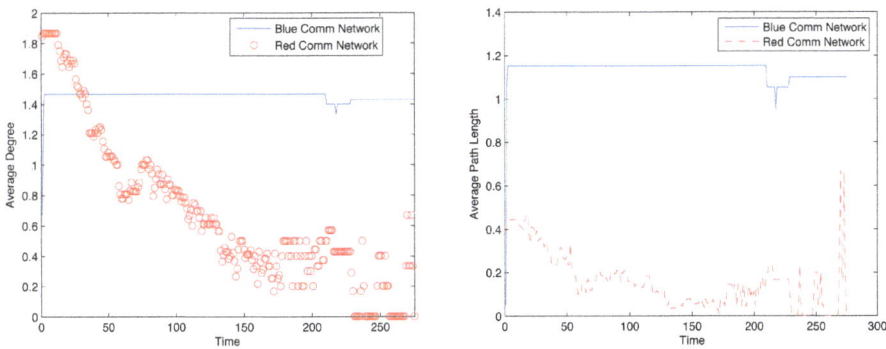

Figure 8.9 depicts the correlation coefficients between the damage of the red force and the knowledge of the blue force. It is obvious that the red damage is strongly related to the knowledge about both friend and enemy at the agent level, while at the force level it is more related to the knowledge about enemy than friend. At the agent level, agents may get more information about their enemies through their friends, so the knowledge about friend is of similar importance to that about enemy. However, at the force level, the only thing agents need to know is where their enemies are. Therefore, here the knowledge about enemy is more important than that about friend.

Our scenario demonstrates that a lesser number of networked blue agents can overtake a large number of red agents. Obviously, one cannot generalise from a single run when using a stochastic simulation. However, the objective of this chapter is to introduce a new model for warfare and it is not our objective to provide a detailed analysis of warfare simulations.

Figure 8.9. Correlation coefficient between red damage and the knowledge correctness at agent level (left) and force level (right) (window size is 5)

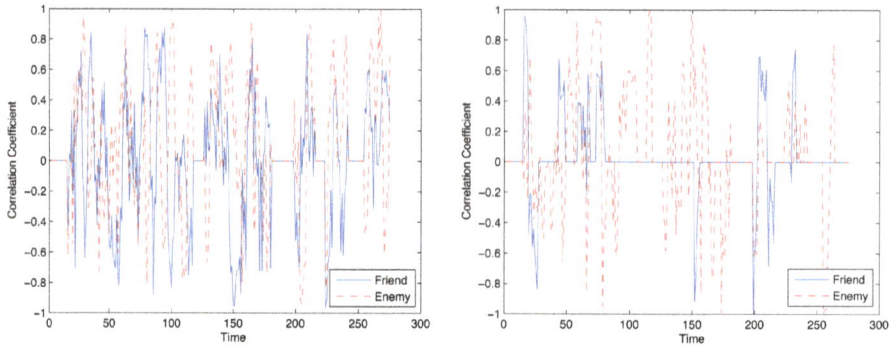

Conclusion

WISDOM-II is different from other existing ABDs in the sense of architecture, functionality and capability. Since it is developed on NCMAA architecture, it can overcome the limitations discussed above to some degree. With the help of network measures, WISDOM-II may easily validate the underlying structure. If a certain network collapses, the user can immediately detect it through network measures and visualisations. WISDOM-II conducts the real-time reasoning at the network (group) level. Such reasoning captures the domain specific interaction between networks (interactions), and a natural language interface provides a real-time interpretation of the simulation process in natural language for the analyst. This may overcome the limitations of the reasoning at the individual agent level. Since everything in WISDOM-II is a network or part of a network, the theory of NCW can be easily modelled, analysed and verified by using WISDOM-II.

The use of a network as the representation unit in WISDOM-II also facilitates efficient parallelism based on network structure and grounded modelling. In WISDOM-II, a rule based algorithm is used to make strategic decisions that guide a semi-reactive agent to make tactical decisions. Therefore, the interaction between tactics and strategies is easily captured. Concepts such as information misunderstanding, communication, and level of information fusion can now be studied in a unified framework. Overall, WISDOM-II is a promising ABD system that creates a new approach for analysts to understand the dynamics of and gain insight into warfare.

Acknowledgement

This work is supported by the University of New South Wales grant PS04411 and the Australian Research Council (ARC) Centre of Complex Systems grant number CEO0348249.

References

Albert, R. and A.L. Barabási (2002) Statistical mechanics of complex networks. *Reviews of Modern Physics* 74: 47–97.

Alberts, D.S., J.J. Garstka and F.P. Stein (1999) Stein Network centric warfare: developing and leveraging information superiority (2nd ed.). CCRP Publication Series.

Alberts, D.S. and J.J. Garstka (2001) Network centric warfare: Department of Defense report to Congress. Technical report, Department of Defense, United States of America.

Allard, K. (1996) *Command, Control, and the Common Defense*. (rev. edn.) Washington, D.C.: National Defense University, Institute for National Strategic Studies.

Barlow, M. and A. Easton (2002) Crocadile: an open, extensible agent-based distillation engine. *Information and Security* 8: 17–51.

Boswell, S., N. Curtis, P. Dortmans and N. Tri (2003) A parametric study of factors leading to success in conflict: potential warfighting concepts. In *Proceedings of the Land Warfare Conference*, pp. 369–81. Adelaide, Australia.

Dorogovtsev, S.N. and J.F.F. Mendes (2002) Evolution of networks. *Advances in Physics* 51: 1079–187.

Galligan, D.P., M.A. Anderson and M.K. Lauren (2003) *MANA: Map Aware Non-uniform Automata Version 3.0 Users Manual*. Devonport, Auckland: Defence Technology Agency. (Draft version).

Galligan, D.P. (2004) Modelling shared situational awareness using the mana model. *Journal of Battlefield Technology* 7: 35–40.

Ilachinski, A. (1997) Irreducible semi-autonomous adaptive combat (isaac): an artificial life approach to land combat. Research Memorandum CRM 97–61. Alexandria: Center for Naval Analyses.

Ilachinski, A. (2000) Irreducible semi-autonomous adaptive combat (isaac): An artificial life approach to land combat. *Military Operations Research* 5: 29–46.

Ilachinski, A. (2004) *Artificial War: Multiagent-Based Simulation of Combat*. Singapore: World Scientific Publishing Company.

Lauren, M.K. (2000) Modelling combat using fractals and the statistics of scaling systems. *Military Operations Research* 5: 47–58.

Lauren, M. and R. Stephen (2002) Mana: Map-aware non-uniform automata: A New Zealand approach to scenario modelling. *Journal of Battlefield Technology* 5: 27-31.

Newman, M.E.J. (2003) The structure and function of complex networks. *SIAM Review* 45: 167–256.

Nwana, H.S. (1996) Software agents: an overview. *Knowledge Engineering Review* 11: 205–44.

Rao, A.S. and M.P. Georgeff (1995) Bdi agents: from theory to practice. In *Proceedings of the 1st International Conference on Multi-Agent Systems*, pp. 312–319, San Francisco, USA.

Sycara, K.P. (1998) Multiagent systems. *AI Magazine* 19: 79–92.

Scherrer, J.H. (2003) *Risks and Vulnerabilities of Network-Centric Forces: Insights from the Science of Complexity*. Newport, R.I.: Naval War College.

Wasserman, S. and K. Faust (1994) *Social Network Analysis: Methods and Applications*. Cambridge, UK: Cambridge University Press.

White, G. (2004) The mathematical agent a complex adaptive system representation in BactoWars. In First workshop on complex adaptive systems for defence, Adelaide, Australia.

Wilson, C. (2004) Network centric warfare: background and oversight issues for Congress. Report for Congress, CRS.

Wooldridge, M.J. and N.R. Jennings (1995) Intelligent agents: theory and practice. *Knowledge Engineering Review* 10: 115–52.

Wooldridge, M.J. (1999) Intelligent agents. In G. Weiss (ed.), *Multiagent Systems: A Modern Approach to Distributed Artificial Intelligence*, pp. 3–44. Cambridge, MA:MIT Press.

Yang, A. (2004) Understanding network centric warfare. *ASOR Bulletin* 23: 2–6.

Yang, A., H.A. Abbass and R. Sarker (2004) Landscape dynamics in multi-agent simulation combat systems. In *Proceedings of 17th Joint Australian Conference on Artificial Intelligence*, LNCS. Cairns, Australia: Springer-Verlag.

Yang, A., H.A. Abbass and R. Sarker (2005a) Wisdom-II: A network centric model for warfare. In *Ninth International Conference on Knowledge-Based Intelligent Information and Engineering Systems,* KES, LNCS. Melbourne, Australia.

Yang, A., H.A. Abbass and R. Sarker (2005b) Characterizing warfare in red teaming. *IEEE Transactions on Systems, Man, Cybernetics, Part B*, (forthcoming).

Viewpoint from a Policy-Maker

The Australian and state/territory governments in partnership with Australia's plant industries established Plant Health Australia (PHA) as a public company in April 2000 with the challenge of taking a partnership approach to key plant health issues and enhancing Australia's ability to respond to emergency plant pests. PHA aims to improve Australia's plant health systems and processes in order that the impact of pests on Australian plant industries—which are worth approximately $21.4bn per year in terms of gross value of production, and provide $5.5bn in export income to Australia—is decreased.

One of the 4 priority areas for PHA is the development of an enhanced emergency plant pest emergency response system. This priority has been addressed through 2 large projects; the development of national industry biosecurity plans and the development of the Emergency Plant Pest Response DEED (EPPRD). The industry biosecurity plans have enabled the key threats for each industry to be identified along with the development of appropriate response strategies. The EPPRD, once enacted, will define the process by which future incursion response is undertaken and will determine the relative cost shares for both government and industry in undertaking the response strategies.

It became evident early on in the development of these projects that there were some gaps in our capacity to ensure the most effective response strategies were undertaken. While contingency (response) plans could be developed for specific pest threats, there was no capacity to evaluate and validate what was proposed. Evidence from past incursion responses had highlighted that often the strategies defined prior to the incursion were not appropriate and most decisions were made 'on the run'. In addition, no cost benefit analysis had been undertaken on any contingency plans.

The other issue that arose was the inability to determine the regional economic impact of an emergency plant pest incursion. Economists could determine the likely impact at the farm gate or at a national level but there was no capacity to measure at the regional level. Given the strong regional focus of Australia's plant industries and the large variation in the size of industries, it was considered essential that Australia's plant health system could determine the impact at a farm gate, a regional, a state and a national level. An accurate measure of this is required by the EPPRD to ensure that the appropriate cost sharing between public and private is achieved.

The bio-economic modelling approach using cellular automata has enabled both of these gaps to be addressed. PHA has invested and worked in partnership with ABARE to develop a proof of concept incursion management model. Further resourcing for the full development of the incursion management predictive simulation system (model) is presently being sought through the Cooperative Research Centre for National Plant Biosecurity. The successful undertaking of this project requires a mix of researchers with a range of expertise in scientific (biology/complex system modelling/climatology/terrestrial data), economic and public policy fields. The system will enhance the strategic assessment of emergency plant pest incursions (proactive assessment) and contribute significantly to the rapid assessment of appropriate control measures during an incursion. These assessments cannot be feasibly done by experimentation. Australia's capacity to successfully control or eradicate an EPP would be greatly enhanced if those managing an incursion response had ready access to appropriate modelling and simulation technology.

Simon McKirdy
Program Manager
Plant Health Australia

9. Managing Agricultural Pest and Disease Incursions: An Application of Agent-Based Modelling

Lisa Elliston and Stephen Beare

Abstract

An incursion management model was developed to estimate the regional economy effects of a potential exotic pest or disease incursion in the agricultural sector. By developing an agent-based spatial model that integrates the biophysical aspects of the disease incursion with the agricultural production system and the wider regional economy, the model can be used to analyse the effectiveness and economic implications of alternative management strategies for a range of different incursion scenarios. A case study application of the model investigates the impact of a potential incursion of Karnal bunt in wheat in a valuable agricultural producing region of Australia.

Introduction

Australia has a valued reputation for supplying high quality agricultural products with pest free status to export markets (a pest being any species, strain or biotype of plant, animal or pathogenic agent injurious to plants or plant products). Pest incursions pose a serious threat to both consumer confidence and Australia's agricultural production. Incursions can impose significant costs on the economy in the form of lost production, income and trade, as well as damaging unique natural environments.

A wide range of strategies can be put in place to reduce the incidence and effects of pest incursions within Australia. These include border control and surveillance measures that focus on preventing a pest from entering the country or a particular region. In the event of an incursion being identified, containment strategies—such as the establishment of quarantine boundaries—as well as measures to eradicate the pest can be employed. Often a rapid response to an exotic pest incursion is required in order to limit the extent of the incursion and to maximise the likelihood of successful eradication. While real time modelling and evaluation of alternative management strategies has, to date, been infeasible, advance assessments of potential response options have been able to increase preparedness.

With responsibility for coordinating and designing policies to improve the ability of Australian agriculture to respond to pest incursions, organisations such as Plant Health Australia and Animal Health Australia require tools to evaluate the effectiveness of potential incursion management strategies for key pests that pose the greatest threat to Australian agriculture.

This chapter documents the development of ABARE/PHA's agent-based Exotic Incursion Management (EIM) model and its application to a disease that affects wheat. The results of several simulation scenarios are presented together with a discussion on the implications for government policy and management of agricultural pest incursions in Australia.

Exotic incursion management model

The EIM model is an integrated bio-economic model that integrates the biophysical aspects of a pest incursion with the agricultural production system and the wider regional economy. While biophysical models and economic models in their own right are capable of offering insights into the impact of a pest incursion, they often miss the complex interaction and feedback effects that exist between agricultural production, the characteristics of the pest, and economic returns.

The multi-agent system approach is ideal for developing a simulation framework to model pest incursions. The agent-based nature of the model enables the simulation of numerous *What if?* scenarios, investigating the sensitivity of both the extent of an incursion and the overall effect on the economy to different assumptions. The results of such simulations then become valuable input into the design of incursion management and response plans.

The EIM model was developed using CORMAS, a spatial natural resource and agent-based simulation modelling framework running on the VisualWorks platform (CIRAD 2003). The spatial component of the modelling framework was particularly important to capture the physical process of a pest moving across the landscape. A cellular automata process is used to drive the spread of the pest across neighbouring paddocks, and a range of potential transmission pathways are modelled explicitly. At the same time, numerous agents including farmers, contractors and quarantine officers—each with their own specific patterns of behaviour and movement—are also interacting in the spatial environment.

The model consists of 4 main components that represent: the pest, the farm system, the incursion management system, and the regional economy. The pest module captures the unique characteristics of the particular pest, including transmission pathways and estimates of the rate of movement or spread. The farm system is modelled on a weekly time-stepped basis and includes production choices and estimates of financial returns. The incursion response and management module incorporates the methods by which the pest is first identified, the process by which potential incursions are investigated, and any subsequent containment and eradication measures that are put in place. A stylised representation of the regional economy is also included to enable calculation of the flow-on effects of an incursion to the wider community.

Karnal bunt case study

The EIM model was used to analyse a hypothetical incursion of the wheat disease Karnal bunt in a region of south eastern Queensland. A number of scenarios were constructed to analyse the ability of specific quarantine measures to contain and eradicate Karnal bunt from the region, and to estimate the likely impact of an incursion on the regional economy.

Karnal bunt of wheat

Karnal bunt of wheat is caused by the fungal pathogen *Tilletia indica* Mitra (Bonde et al. 1997). Karnal bunt teliospores, spread by air currents, vehicles and farm machinery, infect developing wheat heads at the time of flowering. Despite the strong fishy odour of infected grains, only a portion of an infected kernel is replaced with a mass of teliospores with only a few kernels in each head usually infected, making detection of the disease difficult.

Karnal bunt teliospores have proven resistant to a number of adverse environmental conditions, remaining viable for up to 5 years in contaminated soils (Bonde et al. 1997). Despite some recent advances in the development of fungicide treatments and resistant wheat varieties, the primary means of containing the disease is to ban the planting of wheat in affected paddocks for at least 5 years.

Case study region

The south eastern Queensland case study region was chosen because it is a significant producer of high quality export wheat in Australia. The region relies heavily on the agricultural sector, with almost 60 per cent of all businesses belonging to the agriculture, forestry and fishing industry (OESR 2003). Around 90 per cent of all land in the region consists of agricultural holdings. In 2001 more than 1.1 million hectares of crops were grown in the case study region, over half of which was sown to wheat. The wheat produced in this region is generally of a high quality and attracts a significant price premium on world markets. In the event of a Karnal bunt incursion in Australia, wheat producers, including those in south eastern Queensland, would likely lose immediate access to valuable international markets due to the quarantine restrictions imposed by wheat importing countries.

Customised model

The generic EIM agent-based modelling framework was customised to analyse the impact of Karnal bunt in the case study region (Figure 9.1). A range of transmission pathways were explicitly incorporated into the model. These included the wind, farmers and their machinery, contractors and their machinery, and farm inputs such as seed and fertiliser. Each disease pathway interacts with the spatial environment, with its own patterns of behaviour or movement. For example, contract workers and farmers are able to spread Karnal bunt teliospores across and between farms as they move throughout the region. Subsequently, the disease may spread across neighbouring paddocks by wind or the movement of machinery on-farm.

Figure 9.1. Structure of the EIM model

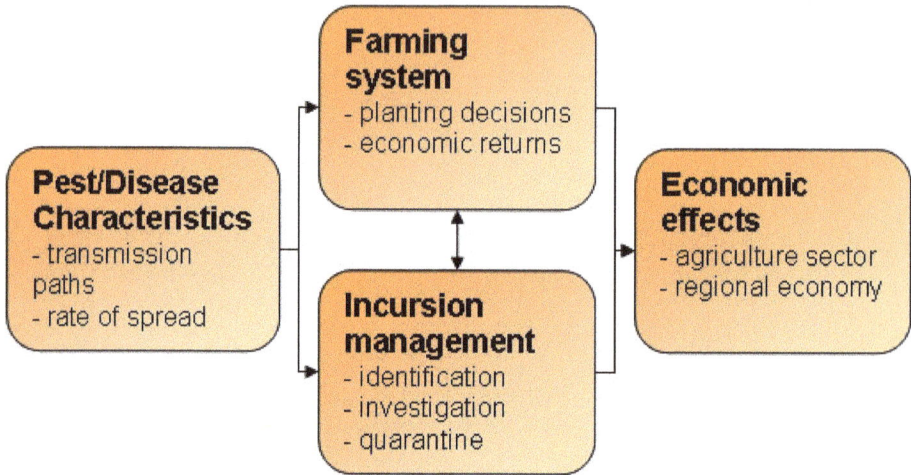

Farm Component

The farm system captures the decisions made by farmers to plant different crops. 4 major agricultural activities are included in the model: wheat for export, wheat for domestic animal feed requirements, sorghum and an aggregate representing all other agricultural activities. The timing of planting, spraying for weeds, harvesting and the use of contract labour is all determined within this module. These key events are included in the model because they capture the most likely ways the disease can spread. For example, during spraying and harvesting, Karnal bunt teliospores can be spread across and between farms in the region. At other times of the year farmers move around their farm, potentially spreading the disease if it is present on their property. Based on average yields reported in the region, the total volume of wheat, sorghum and other commodities produced on each farm is calculated. Based on average prices, gross receipts for each of the major commodities are calculated.

The model allows for the identification of a Karnal bunt incursion in one of two ways. First, farmers can identify signs of diseased crops on their property. Second, teliospore infested grain can be identified when harvested grain is sent to the silo and inspected for quality.

Quarantine Component

When diseased crops are identified, a quarantine response is triggered to invest-igate the extent of the incursion and attempt to contain it so that it cannot spread further. The farm from which the infected grain came is immediately quarantined and a tactical response officer is dispatched to the property. All neighbouring properties are placed in a buffer quarantine zone (Figure 9.2). Tactical response personnel visit each of the neighbouring properties and search for signs of Karnal bunt. If the disease is identified on any of the properties, these properties are upgraded to full quarantine status and all properties neighbouring the newly identified farm are then searched. Where signs of infestation are not found on neighbouring properties, those properties remain in the buffer quarantine region and the search of other properties is stopped.

Figure 9.2. Spatial visualisation of the environment: Quarantine response

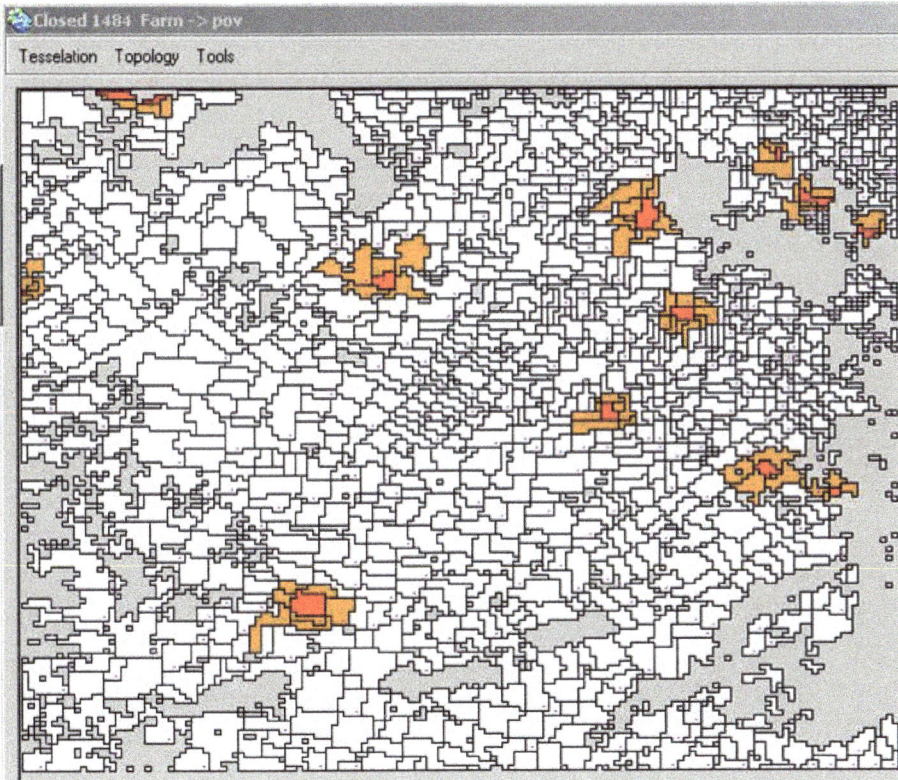

At the same time, any contractors that have visited Karnal bunt infested farms that are now fully quarantined are identified to trace back the source of the in-cursion and limit its spread. In the first instance, contractors identified in this process are asked to provide a list of farms that they have visited during the year. Tactical response personnel are then dispatched to each of these farms in order to identify the extent of the incursion. Where an infestation is identified on a property, that property is fully quarantined and the search through all neighbouring properties begins. Any contractors contacted in this trace back process who were found to be carrying teliospores on their machinery are disinfected before the next season begins.

In the event of an identified disease incursion, farms can be classified as:

- identified as having an infestation and fully quarantined;
- identified as not having the infestation but in a buffer quarantine zone because neighbouring properties have the infestation; or
- not quarantined, either clear of the infestation or not yet identified.

The planning options available to farmers are then restricted for a period of years based on their level of quarantine. Farmers identified as having the disease are quarantined and are unable to grow grain crops for 5 years. Neighbouring farmers that form the buffer zone are unable to grow grain crops in the first year after an incursion is identified. In the remaining years of the 5 year quarantine period, wheat grown on these farms can only be used to feed livestock within the region.

In all scenarios considered, when the disease is first identified the price for wheat produced in the region, even wheat that is free of the disease, receives a lower price. The price for wheat only returns to the higher export price after the disease is deemed to have been eradicated from the region on the basis that no new incursions are identified for at least one year.

Regional Component

A 12 sector input-output model represents the regional economy of south eastern Queensland. Input-output tables contain the supply and demand of goods and services in an economy over a particular period, along with the interdependencies between the industries and associated primary factors of production. Changes in the value of agricultural production as a result of a disease incursion can therefore be traced through the rest of the regional economy in terms of output, employment, income and imports. The input-output analysis provides estimates of both the direct and indirect impacts of a change in agricultural production resulting from a disease incursion. The direct—or initial—impact captures the changes in the value of production and any associated changes in employment and income in the directly affected agricultural industries, as well as any changes in imports required by these industries. Subsequent changes in all other industries and the directly affected industries form indirect or flow-on impacts.

Scenario-based results

Incursion scenarios

2 different incursion scenarios, with different levels of farmer detection and reporting were analysed to investigate the importance of early detection on the likelihood of eradicating the disease and the overall economic cost of a Karnal bunt outbreak. The first scenario involved a limited and slowly expanding incursion with Karnal bunt introduced into the case study region by contractor equipment. The incursion begins with just 2 contractors and spread across the region by the movement of farmers and contractors, as well as the wind. The second scenario represented a diffuse starting point with potentially rapid expansion, with a load of fertiliser contaminated with Karnal bunt sold throughout the region at the beginning of the simulation.

Figure 9.3. Area infested, contractor scenario

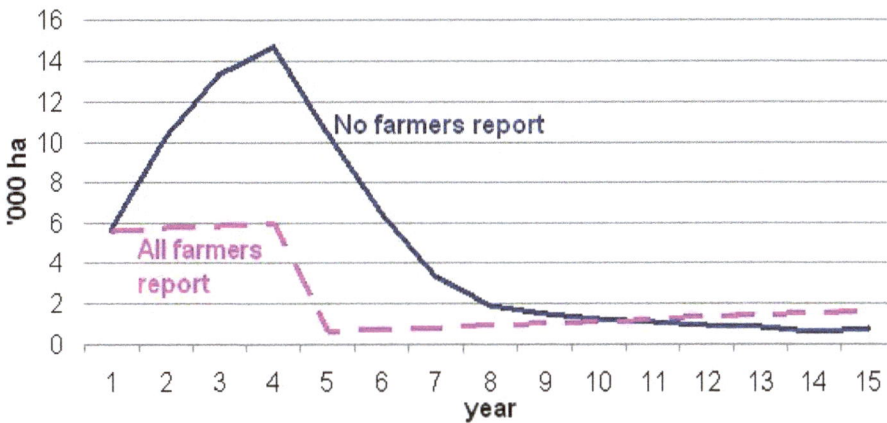

Contractor based incursion

2 contractor-based incursion scenarios were investigated. In the first scenario, the likelihood of teliospore infested grain being identified at the silo was assumed to be 50 per cent, and farmers did not report signs of the disease on their property. In the second scenario, the likelihood of detection at the silo remained at 50 per cent and all farmers reported signs of the disease on their property. When farmers do not report signs of the disease on their property the area infested increases from around 5,500 hectares in the first year to more than 14,000 hectares by the fourth year of the 15-year simulation (Figure 9.3). At this point in time, almost 80 per cent of all infested land had been identified and quarantined. As a result, the extent of the incursion is curtailed and the area infested reduces to negligible levels by the end of the planning horizon.

In contrast, when farmers report signs of the disease on their property the extent of the incursion is reduced, with a maximum of 5,900 hectares infested by the fourth year of the simulation. More than 95 per cent of all infested land is quarantined in the fourth year and the ability of the disease to spread is reduced. The area infested is reduced significantly. Despite a slightly higher level of infestation in this scenario compared with the scenario where no farmers report in the latter years of the simulation, more than 95 per cent of the infested land is quarantined and the likelihood of the disease being eradicated within another 5 years is high.

The combined effect of the low level of infestation in the region and the ability of quarantine measures to contain an outbreak caused by contractors, means that the adverse economic effects of this hypothetical Karnal bunt incursion are relatively minor. Over the 15-year planning horizon, an incursion, even when no farmers report signs of the disease on their property, results in a net loss of production valued at around $58 million in net present value terms (Table 9.1).

Table 9.1. Regional economy effect of alternative incursion scenarios

	Initial	Flow-on	Total
	Direct	Indirect	
	$m	$m	$m
Contractor incursion, no farms report	−58.0	−22.3	−80.4
Contractor incursion, all farms report	−55.6	−21.5	−77.1
Fertiliser incursion, no farms report	−430.2	−165.3	−595.5
Fertiliser incursion, all farms report	−368.9	−141.5	−510.3

Over the 15-year planning horizon, the indirect effect of the hypothetical incursion on all industries is estimated to be around $22 million (in 2003 prices). The total industry and consumption effects, reflecting the indirect effects along with the initial (direct) effects, capture the overall impact of this particular Karnal bunt incursion. It is estimated that over the 15-year planning horizon, the decline across the case study region is around $80 million (in 2003 price).

When farmers report signs of the disease on their property the incursion is contained in a shorter period of time and the overall economic effects of the outbreak are reduced. Over the 15-year planning horizon, the loss in value of production is estimated at under $56 million. When the direct and indirect effects of changes in production are aggregated across the region, the decline in economic performance is $77 million. The $3.3 million difference in the economic performance of the region under these two contractor based incursion scenarios provides an indication of the value associated with improving the likelihood of detection by farmers on their property. This in turn can provide a benchmark against which expenditure aimed at improving farmer awareness of the disease, and therefore the likelihood of detection, can be assessed.

Fertiliser based incursion

2 fertiliser based incursions, with the same likelihood of detection at the silo and on-farm, were also investigated. When farmers do no report signs of the disease on their property, the area infested increases from around 76,000 hectares in the first year to more than 300,000 hectares by the fourth year of the simulation (Figure 9.4). Unlike the contractor scenario, only around two-thirds of all infested land has been identified and quarantined at this point in the simulation. As a result, despite the reduction in infestation between the fifth and eighth years of the simulation, the disease fails to be contained and the area infested continues to increase throughout the remainder of the planning horizon.

Figure 9.4. Area infested, fertiliser scenario

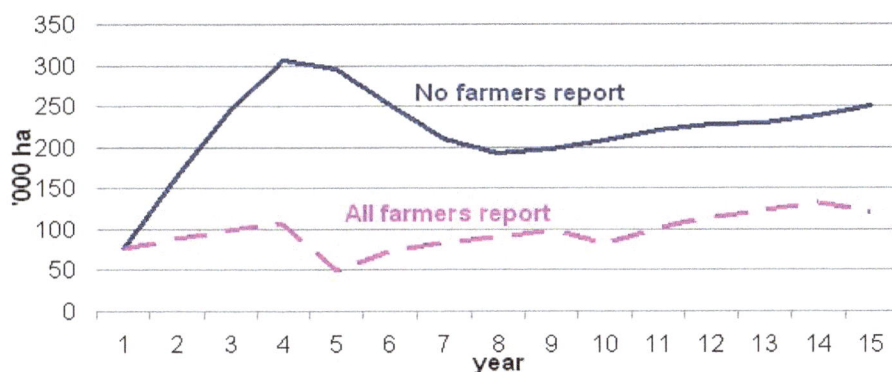

When farmers report signs of the disease on their property, the extent of the incursion is reduced significantly, but still fails to be eradicated. The area infested follows a cyclical pattern and at the end of the 15-year planning horizon almost 120,000 hectares of land is infested. The much larger incidence of infestation across the region and the failure of quarantine measures to adequately contain the disease when it is brought into the region via contaminated fertiliser results in the economic impact of this scenario being much larger than the contractor based incursion scenario.

Over the 15-year planning horizon, a fertiliser based incursion where farmers do not report signs of the disease results in a net loss of agricultural production valued at around $430 million, compared with a reference case of no disease (Table 9.1). When the indirect effects are added to this, the economic impact of the disease incursion is estimated at more than $595 million over the 15-year planning horizon. When farmers do report signs of the disease on their property the extent of the incursion is reduced and the economic impact of the incursion is correspondingly reduced. The net loss in agricultural production that results from the fertiliser based incursion falls to around $369 million when farmers report signs of the disease on their property. This converts to an overall loss in regional income of $510 million.

Once again, the difference in the economic performance of the region under the two fertiliser based incursion scenarios provides an indication of the value associated with improving the likelihood of detection by farmers on their property.

Incentives to self report

The results presented in the previous section indicate that the region as a whole is better off when farmers report signs of the disease on their property. However, an analysis of the gross receipts earned by farmers identified as having the disease suggests that at least in the initial years of the simulation, farmers are unlikely to have an incentive to report signs of the disease on their property.

Figure 9.5. Receipts per infested farm, contractor scenario

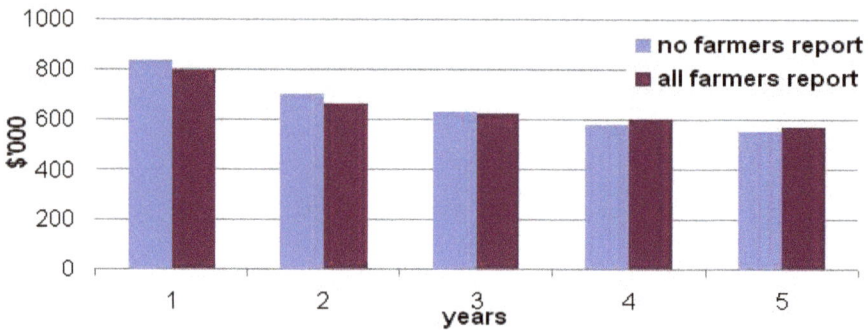

In the contractor based incursion scenario the receipts per infested farm for the scenario where no farmers report signs of the disease on their property are higher than the scenario where all farmers report for the first 2 years (Figure 9.5). In the third year the receipts per farm are on average equivalent. In subsequent years, when the quarantine measures take effect and the disease is contained, the receipts per infested farm are higher under the scenario where farmers report signs of the disease on their property.

Similar results were observed for the fertiliser based incursion scenario. Average receipts per infested and quarantined farm were higher in the first 3 years of the simulation under the scenario where no farmers report signs of the disease on their property (Figure 9.6). In the final years of the simulation this trend reverses and the average receipts per infested farm are higher when farmers report signs of the disease on their property.

Figure 9.6. Receipts per infested farm, fertiliser scenario

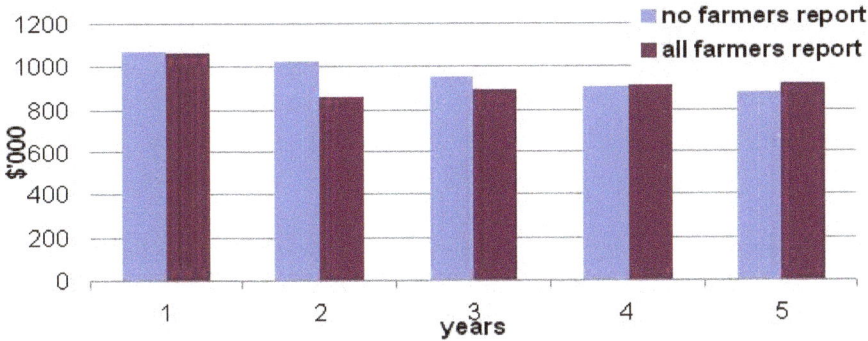

These results indicate that while the region as a whole benefits from an early reporting of the disease, individual farmers have little incentive to report signs of infestation, at least during the early years of the simulation. This is an important result, because the ability to successfully contain a disease such as Karnal bunt is likely to depend critically on it being identified within the first few years of its introduction to Australia. Further, the estimated economic costs of a Karnal bunt incursion presented in this chapter include only the costs incurred within the case study region. If Australia's major wheat export markets were to refuse all wheat produced within Australia, the additional cost of the incursion across the rest of the country is likely to be significant, particularly if the disease cannot be eradicated.

Government policy implications

The findings from these scenarios suggest that successful eradication is possible if a Karnal bunt incursion was to begin with the importing of contaminated machinery by a contract worker and be identified early. In contrast, a widespread and diffuse incursion, for example as a result of the sale of contaminated fertiliser, is likely to lead to a rapid expansion of the disease that fails to be eradicated despite a range of containment and eradication strategies. These results have important implications for the allocation of biosecurity resources to reduce the economic impact of a Karnal bunt incursion. Given the apparent low likelihood of successful eradication, consideration should be given to allocating resources to border control activities that reduce the likelihood of the disease entering Australia. Further, given the importance of early detection to successful eradication, resources allocated to regular surveillance and monitoring may also be warranted. In contrast, if the disease is well established before being identified, the economic return on investment in eradication activities is likely to be negligible.

The results from this analysis also demonstrate the ability of improved farmer detection and reporting to reduce the overall economic costs associated with a potential Karnal bunt incursion. The results highlight the need to provide appropriate incentives to farmers in order to obtain the estimated benefits to the regional economy. For example, to ensure self reporting, farmers with the disease may require financial payments to offset the disadvantages associated with reporting the disease and being placed under quarantine restrictions.

This ability to investigate the incentives to engage in strategic behaviour that farmers face is likely to greatly enhance the accuracy with which the effect of proposed containment strategies can be assessed. Further, it can assist in the development of any incentive structures put in place to encourage farmers to report signs of infestation early.

Conclusion and perspective

The initial case study application of the EIM model indicates that it can be used for assessing the likely effectiveness of alternative management strategies in the event of an exotic plant disease incursion such as Karnal bunt. The simulation environment allows decision makers to better understand the key characteristics likely to influence the economic assessment of alternative management strategies. This in turn provides an indication of where resources should be optimally allocated to reduce the impact of pest incursions, and be incorporated into biosecurity planning processes such as the development of contingency management plans.

The EIM modelling framework is generic in nature and is capable of being adapted to analyse a wide range of incursion scenarios, as well as the incursion of other plant pests. A case study of *fireblight*, a bacterial disease that affects apples and pears, is currently being undertaken to assess the likely effectiveness and impact of proposed management strategies in the event of an incursion. Additional development of the agent-based modelling framework is also being undertaken to develop a rapid incursion management simulation model to assist in the development of optimal control strategies for emergency plant pests as part of the Co-operative Research Centre for National Plant Biosecurity.

References

Bonde, M.R., G.L. Peterson, N.W. Schaad, and J.L. Smilanick (1997) Karnal bunt of wheat. *Plant Disease* 81(12):1370–7.

Centre de Coopération Internationale en Recherché Agronomique pour le Développement (2003) *Cormas: Natural Resources and Agent-Based Simulations,* Montpellier, France (cormas.cirad.fr/indexing.htm).

Office of Economic and Statistical Research (2003) *Local Government Area Profile*. Report generated by the Queensland Government Office of Economic and Statistical Research, Brisbane, September. (www.oesr.qld.gov.au).

Viewpoint from a Policy Adviser

'If I have $100m and want to spend it on reducing drug trouble, what is the best investment? Is it prevention, or treatment for drug dependent people, or interdiction and more law enforcement?'

So began a conversation, with questions that cannot be readily answered. Yet this question or its equivalent underpins daily decision making in drug policy, priority setting and programme selection. In an era calling for such decisions to be research-evidence-based, it is initially embarrassing to admit that we don't know. Upon reflection, however, it proves to be a challenge.

As a person committed to finding answers using research evidence where possible, the response, 'it's complicated', is inadequate. The researcher's more precise articulation is often characterised by qualifications: "well, on the one hand...but on the other...'; or: 'the research suggests...but more research is needed before I could confidently say...". These responses are often hard to apply or utilise in decision making. Furthermore, it's rare that choices in a social and health policy context are clearly yes or no choices, simple, or ever fully informed by data and evaluations. Usually, choice of one strategy to the exclusion of others is highly risky and politically unsustainable. In this case, it is also likely that all of the nominated responses to drugs have potential merit and the question then becomes not which to fund but what mix or proportions will produce the best outcome?

So the challenge becomes one of exploration and experiment. But this involves people and large amounts of money, and social experiments of this sort are not tenable. To abandon any effort to inform our answers because it's hard, complicated or because we don't have all the answers or data to inform us leaves policy makers victims of belief, opinion, general attitudes and simplistic guestimates, prejudice or passion. Many find the necessary mental juggling or efforts to analyse multifactored phenomenon overwhelming, and, as this question involves human behaviour, uninformed or simple mathematical formula do not suffice. Long ago, one of my early professional mentors introduced me to the notion of 'partialising' complexity when faced with problems that might seem overwhelming.

The development of SimDrug is an example of pooling such early career lessons, showing some tenacity in approaching questions posed but not yet answered, and utilising the experience of real world policy development, unpicking or partialising the pieces and putting them together again. Choices will be made—the issue becomes the extent to which we can inform them. This exciting new opportunity is not a mere game. It is an opportunity to 'conduct' or, in this case, model social experiments. It is grounded in real world experience, data, evaluation and feedback.

In my experience, a model such as this offers enormous potential to facilitate thoughtful policy debate, to improve the chance that evidence will be a fundamental ingredient in drug policy decision-making, and be a driver for improved data collection, analysis and utilisation. It is only when we try to utilise data and experience that we provide an incentive for improvement.

SimDrug will not and does not intend to make decisions for us. But it reveals and 'manages' a level and extent of interactivity that goes beyond any currently available tool—including my own brain/capacity—to juggle and hold together the dynamic interconnected threads that influence who will inject what drug, where, when and with what consequence, and what response, and more importantly, what mix of responses might be optimum—when dealing with illicit drug use. It's exciting. It's a powerful tool. It's clever. It's adaptable to a range of conditions. It's novel and revealing. It's great!

Margaret Hamilton
School of Population Health, University of Melbourne
Chair, Multiple and Complex Needs Panel

10. SimDrug: A Multi-Agent System Tackling the Complexity of Illicit Drug Markets in Australia

Pascal Perez, Anne Dray, A. Ritter, P. Dietze, T. Moore and L. Mazerolle

Abstract

Complexity of illicit drug markets mirrors the complexity of illicit drug use itself. The intricacy of multiple interactions between individuals, the various time lines linked to different aspects of harm reduction, and contrasted social rationalities observed among field practitioners (prevention, law enforcement, harm reduction) contribute to the creation of complex and unpredictable systems. In order to explore this complexity, an Agent-Based Model (ABM) called SimDrug was designed. The prototype model includes users, dealers, wholesalers, outreach workers and police forces. The model is focused upon local drug market and the hot spots in Melbourne. The time span for the model is 4 years, and the team endeavoured to replicate the period of the Australian heroin drought. One interesting question to answer was whether the drought was an emergent phenomenon or caused by externalities. Preliminary results tend to provide comprehensive answers to the question and meaningful insights for further developments of the model.

Introduction

Complexity of illicit drug use and markets

Ritter (2005) provides meaningful examples of inherent complexity of illicit drug markets in her comprehensive review. First, she mentions work from Dorn et al. (1992) who describe a qualitative research study of drug markets in the UK. They found that the drug markets are constantly fluid and changing. Some of the variables that may drive this diverse and messy phenomenon include social background, resources and cultures. The researchers describe people weaving in and out of the trade, with constant interactions with law enforcement resulting in market changes. In the USA, South (2004) describes two case studies of heavy recreational drug users. In this context of small-scale dealers, selling drugs becomes a norm with its inherent rules. The author emphasises the fact that better understanding the epistemology of these markets challenges existing notions of drug dealers. May and Hough (2004) describe trends in the American drug market over 10–15 years. They note the change in the market from an open street-based market to a closed market, and associate this with the widespread introduction of mobile phones, coupled with community concern about public space. The authors insist on the dynamical influence of both technology development and law enforcement onto the type of market and its operation.

From an economic perspective, Caulkins and Reuter (2005) describe a model where dealers operate under limited rationality, providing one explanation for the fall in heroin and cocaine prices in the US despite increases in law enforcement intensity. They draw an important distinction between the initial decision to sell drugs, and the decision to continue selling drugs. Using prospect theory, they demonstrate the differences in risks and benefits. From a criminology perspective, Mazerolle et al. (2004) use cluster analytic techniques to identify types of drug-dealing places. The 6 different types were characterised by environmental features such as police calls for service, degree of commercial or residential activity, length of the street block, civil activity and civil disorder.

The main sources of information in relation to the structure of the heroin market(s) in Australia come from law enforcement reports that describe the high-levels of the market (importers and wholesalers), from ethnographic research that has largely focused on the retail levels of the market, and from routine epidemiological monitoring, such as the Illicit Drug Reporting Scheme (IDRS). Up until the mid-to-late 1990s, the transnational criminal organisations involved in drug smuggling to Australia had been seen as large international syndicates, with established infrastructure within Australia (ACPR 2003). It is now clear that this is no longer the case. The current predominate model is *transactional crime*. The drug trafficking is done by groups of relatively independent criminals who come together for specific transactions (but are not part of the one organisation). ACPR (2003:4) describe the current operations as fluid syndicates, with members that come and go, 'constantly forming, disbanding and reforming in response to the nature of the crime that they are committing'.

Ethnographic research is particularly valuable in understanding local drug markets. Maher (1996) reports findings in relation to the nature of the sample of users, the emergence of street-based injecting culture and the apparent resilience of the local drug market to law enforcement pressures. The author describes the street level market as characterised by freelance operations, with the heroin distributed by individuals or in units of small entrepreneurs, mostly user-dealers. Participation in the market by a user is 'easy to accomplish but often short-lived and sporadic' (Maher 1996: 20).

Complexity of illicit drug markets mirrors the complexity of illicit drug use itself. Unger and colleagues (2004) clearly summarise the challenges we face when trying to understand, describe, and eventually simulate users' behaviour:

Drug use is a result of a complex, dynamic interplay of posited risk and protective factors that operate at multiple levels of analysis. At the individual level, biological predispositions, personality traits, and cognitive mechanisms can increase or decrease the likelihood that adolescents will experiment with drugs, as well as the likelihood that they will become physically or psychologically dependent on drugs. At the interpersonal level, social influences from peers, family members, and other role models or networks can influence individuals' perceptions of the social norms surrounding drug use, which then can influence their own use of drugs.
(Unger et al. 2004: 1780)

According to Rhodes (2002), a harm reduction praxis founded on a risk environment framework encompasses social contexts that influence health and vulnerability in general as well as drug-related harms in particular. This inevitably leads to a consideration of non-drug and non-health specific factors in harm reduction, and in turn, points to the importance of what might be described as 'non health oriented interventions' in harm reduction.

The Australian heroin drought

In Australia, the advent of what is known as the heroin drought provides a paramount example of the complex interactions at stake and the conflicting analysis drawn by experts coming from different disciplines. According to Dietze and colleagues (Dietze et al. 2003), the supply of heroin in Victoria suffered a dramatic decline between late 2000 and early 2001, after a strong increase in heroin use and related harms in the late 1990s. This change in heroin supply was clearly reflected in reports of evolution of heroin overdoses. Relying on different sources of information, the authors argue that the drought was shadowed by a dramatic increase in amphetamine, benzodiazepine, and prescribed opioid use, resulting in a fairly constant number of injecting drug users in Victoria.

What happened in Australia from late 2000 was unique to that country. Despite a worldwide growth in the availability of stimulants—notably methamphetamine—no other country experienced a comparable shortage of heroin, or the extensive use of stimulants as an alternative to heroin. The Australian heroin drought is held up by the Australian Government as an example of law enforcement having a significant impact on the supply of drugs. The Australian Federal Police had seized 606 kg of heroin and dismantled a major drug trafficking syndicate, a few months before the drought. But Bush and colleagues (2004) argue that this explanation does not stand up to more detailed scrutiny as other factors were more influential. According to the authors, the most plausible explanation for both the heroin drought, and the increase in the availability of stimulants, is the strategic decisions and actions of the crime syndicates that supply the Australian market. Interestingly, Agar and Reisinger (2002) develop an equivalent rationale about the heroin drought that occurred in the USA during the mid-1970s. They counter-balance the impact of the 'French Connection' breaking up with the rise of Methadone-based replacement programs. The authors insist on the complex adaptive properties of the illicit drug use issues.

In order to have a dispassionate look at the question, one must first gather information coming from law enforcement, intelligence, treatment, and harm reduction sources. Then, this heterogeneous information must be critically analysed before being used to confront and explore the different plausible scenarios. In a broader context of substance misuse, Fuqua and colleagues (2004) rightfully claim that the whole process needs a transdisciplinary approach to describe such complex systems from more than one vantage point. This claim is particularly relevant in the case of illicit drug use.

Hence, it is not surprising that complexity theory has attracted an increasing number of scientists working in the domain of population health and epidemiology. For example, August and colleagues (2004) describe the complexity of prevention program implementation. The authors outline the challenges faced by developers of prevention programs in transporting scientifically proven or evidence-based programs (efficacy) into natural community practice systems (effectiveness). The intricacy of multiple interactions between individuals, the various time lines linked to different aspects of harm reduction, and contrasted social rationalities observed among field practitioners (prevention, law enforcement, harm reduction) contribute to the creation of a complex and unpredictable systems.

Complexity Theory and illicit drug use

In the broad context of health geography, Gatrell (2005) considers the primary characteristics of Complexity Theory, with particular emphasis given to networks, non-linearity and emergence. The author acknowledges its capacity to move away from reductionist accounts and to propose new perspectives on sociality and connectedness. Research on health inequalities, spatial diffusion, and resurgent infections, have much to learn from Complexity Theory provided that modelling results are inferred from 'good empirical work'. Gatrell rightfully underlines the fact that:

Metaphors and some of the methods used in complexity theory are essentially visual. Despite the disappearance of the graphical...from much of the research literature, the 'seeing eye' and the ability to detect and describe pattern remains at the forefront of many research methods, including health geography.
(Gatrell 2005: 2665)

Agar (2003) has developed over the years an empirical theory of illicit drug epidemic, called *Trend Theory*. They first look for a rapid increase in incidence. The assumption is that this rapidly increasing curve is an emergent property of systems that are themselves undergoing rapid change. They look at relevant segments of a society (clusters) where major ongoing changes may be linked to the drug. They also assume that changes are under way in the system of production for the illicit drug. Finally, they assume that change is also ongoing in the networks that link the production system with the population. Trends are dynamic and must be understood over time as they develop. Agar and Reisinger admit that:

The most difficult part of trend theory work is that each 'data point', if you will, represents a complicated research effort. A massive amount of different material must be gathered, where most of it does not directly ask or answer the questions that we have...With any luck, the effort to build a trend theory will help in some way as the drug field continues to struggle with that key epidemiological question: why these people in this place at that time?
(Agar and Reisinger 2003: 27)

In a previous paper, the same authors recognise that complexity underlies Trend Theory (Agar and Reisinger 2002). Complex adaptive systems consist of different actors in different sub-systems, all in continual change over time as they evolve with their environment. Complexity Theory also indicates that measures of the system as a whole—like epidemiological indicators of heroin addiction—are often emergent processes. Agar and Reisinger believe that explanation of a phenomenon of interest is not available in the location where that phenomenon takes place. Instead, events—most of them at remote social locations—unfold and interact over time, and the local phenomenon is only one of a number of factors involved. An explanation of a trend calls for a model of how that system works. A heroin trend increases when distant systems, by both chance and design, enter into interlocking feedback loops.

Epidemiologists have pioneered the use of Complexity Theory in Population Health. Outbreaks have been simulated through percolation processes into artificial networks or by means of emerging properties of artificial societies composed of interacting agents. For example, recent work form Meyers and colleagues

(2005) demonstrates how contact network epidemiology better explain the heterogeneity of SARS outbreaks worldwide, compared with traditional compartmental modelling. Likewise, Valente and colleagues (2004) use network level measures (centralisation, density, transitivity) to explore the impact of social networks onto drug use among adolescents. In this case, network analysis provides a technique to map specifically who has adopted evidence-based programs and where they are in the network. Hence, the network map provides important monitoring information indicating how well the practice is spreading.

Agar and Wilson (2002) provide a compelling example of Multi-Agent-Based Simulation in a context of youthful heroin experimenters in the Baltimore metropolitan area. The model is used to explore the impact of circulating stories of drug reputation on individual attitudes towards the drug. Based on ethnographic work, the model demonstrates a dampening effect of increased social connections, contrary to epidemiologic expectations. As described by the authors:

> To summarize...500 agents begin with normal, randomly distributed risk and a shared attitude set to some number with a parameter. The agents move around and, if they encounter heroin, they compare attitude to risk. If attitude is higher than T-risk, they try the heroin. And, if they try it, they have good or bad experiences, with some probability, and those experiences, should they occur, change their attitude by some amount. Agents also change their attitude, depending on the 'buzz' around the drug that they pick up as they move around the world. After a tick of the model, any adjacent agent might influence their attitude. The chances they do so, and the amount of the influence, will depend on the combined effect of both agents' experiences. Chances and amount also depend on whether the two agents are strangers or friends.
> (Agar and Wilson 2002: 49)

Mason and colleagues (2004) describe a slightly different approach to describe environmental impact on urban youth drug use. The approach incorporates individual, social, and geographical parameters to systematically understand the ecology of risk and protection for urban youth. Geographic Information Systems (GIS) derive spatial relationships and analyses between the specific locations where the teenagers are active, their subjective ratings of these locations, and objective environmental risk data. These social network and GIS data are merged to form a detailed description and analysis of the social ecology of urban adolescent substance use.

Even Chou and colleagues (2004), despite a strong empathy for statistical methods—a shared language between experts involved in drug use trans-disciplinary research—recognise that:

While the statistical models discussed later are based on assumptions of linear associations, nonlinear association can also be handled by some of these models. It should be noted that to understand and appreciate the dimensions of the process or phenomenon being studied, data-driven selection of either a linear model or a nonlinear model is critical. Using linear tools to study non-linear processes can yield misleading conclusions that impact the planning, implementation, and assessment of intervention programs.
(Chou et al. 2004: 1871)

Finally, Agar (2005) building on his previous work, recently argues for *emic* models, models grounded in what matters in the world of those being modelled. But most models are *etic*, in the sense that they are built on an outsider's view of the people and the world being modelled. In a pure positivist stand, etic models represent how the modeler thinks the world works; emic models represent how people who live in such worlds think things are. In a very inductive and post-normal move, Agar equips his artificial agents with rules of decisions coming from individual responses to ethnographic surveys. By doing so, he tries to explore and better understand the reasons for an early experimenter to become an addict, based on social ties. The author recognises that despite his commitment, some etic-based knowledge pervaded his model, but he emphasises the importance of a strong empirical experience to back up such complex modelling. This is the only current example of post-normal modelling in the domain of population health, unlike environmental studies where participatory modelling experiences are rapidly developing (Bousquet et al. 2003).

SimDrug—model description

Background

As part of the Drug Policy Modelling Project (DPMP),[1] our team was contracted to address a demand for new integrative approaches to support decision-makers and practitioners in implementing illicit drug policies. Our specific task was to present advantages and limitations of using a complexity-based approach for modelling illicit drug use and markets. Two key issues shaped the boundaries and content of the present project:

- finding a case study that would contain, a priori, as much complexity as possible, and would provide the information needed to build a consistent model; and

---

[1] http://www.turningpoint.org.au/research/dm_proj/res_dm_proj.htm http://www.turningpoint.org.au/-research/dm_proj/res_dm_proj.htm

- fitting into the actual structure of the DPMP in order to interact efficiently with relevant experts, and to avoid undesirable overlapping with other on-going research.

Advocating for a Multi-Agent System

Considering the 3 main scientific streams shaping complexity science, our team had to choose between Network Theory and Multi-Agent System for our initial approach, Dynamical Systems being already used by the other DPMP teams. Looking at the Australian illicit drug markets through a cross-scale approach, it seemed that urban districts constituting a drug scene involved most of the actors (with exception of importing syndicates and production cartels) while displaying a maximal complexity. This intermediate scale fits in between statistical accounts at the State or National levels, and ethnographic accounts of street-based interpersonal interactions and individual motivations. Social heterogeneity, spatial mobility, and abrupt changes characterise drug scenes. Global patterns and trends emerge from multiple interactions both distant and local.

The Melbourne heroin scene was rapidly perceived as the best option because of the following features:

- a well documented history of heroin use in Melbourne CBD and surrounding suburbs (hot spots) showing the cultural dimension of the local heroin scene;
- a trans-disciplinary team (sociology, epidemiology, and economics) already working on the case study and having developed a comprehensive database;
- a legitimate questioning of local authorities on balancing between law enforcement and harm reduction programs; and
- a retrospective view upon the conflicting explanations that arose after the so-called heroin drought.

Since most of the potential agents in the system were clearly identified but various aspects of their interdependent links were ill-defined, we decided to opt for a Multi-Agent System approach rather than a Network Theory one. Besides, trans-disciplinary communication advocated for a rather intuitive approach of modelling. The *building blocks* methodology attached to Multi-Agent Systems, and the visual paradigms (UML design) used to describe the modelling components, appeared to be highly relevant. The trans-disciplinary expert panel involved experts coming from ABM design and modelling, sociology, epidemiology, economics and criminology.

Model description

The model is created with the CORMAS platform (Bousquet et al. 1998), developed from the VisualWorks commercial software. CORMAS provides a SmallTalk-based environment to the developer where spatial visualisation, graphic results, and sensitivity analysis tools are already encapsulated. Hence, the modeller can concentrate on the development of the application only, without bothering with peripheral but time consuming tasks.

Time scale

We decided to work on a daily basis, meaning that one modelling time step is equivalent to a 24-hour day in reality. A first compromise among the group of experts was established around a fortnightly structure, but later developments showed that injecting behaviours needed more accurate time steps. Each simulation is run over a 5-year period. Even if the heroin drought period is our main target, we assume that different processes (with different response times) were at stake. Thus we take 1998–2002 as a period of reference. In terms of validation, this time bracket gives us the opportunity to test the robustness of the model by comparing a series of micro (agent level) and macro (system level) indicators with corresponding observed data. The model must be able to consistently reproduce pre-drought, crisis, and post-drought dynamics of the system.

Spatial environment

We decided to work on an archetypal representation of Melbourne's CBD based on a regular 50*50 square mesh. The size of the grid has been chosen accordingly to the number of users (3,000) and dealers (150) to be modelled and located in the environment. At this stage, there is no need to work on a real GIS-based representation since we mainly focus on social behaviours and interactions. Each cell, elementary spatial unit, corresponds to a *street block*. A *suburb* is defined as an aggregation of neighboring cells. 5 suburbs are created with different sizes and shapes, regardless of realistic features. In fact, the environmental mesh is a cellular automata able to process a large amount of information at the level of each cell. 2 special cells represent the *Police Station* and the *Treatment Centre*.

Street Block

The main characteristics (or attributes) of a *street block* are: the number of *overdoses*, *fatal overdoses* and *crimes* locally recorded. Each street block also has a *wealth* value, interpreted as a synthetic parameter indicating the social and material capital of the place. Initial values of wealth are randomly attributed and range between $100 and $500. Each time a crime is committed in a street block, its wealth value decreases by 5 per cent. Conversely, after a 10-day period without crime, the wealth value increases by 3 per cent. Wealth values are limited to a maximal value of $500. The initial wealth values come from ethnographic surveys of arrested offenders and correspond to the average money they can get from receivers. The increase and decrease rates are not calibrated yet. The concept of risk environment is encapsulated into the *risk* attribute. An empirical linear equation is used to calculate risk values at each time step:

risk = (10 * nb of crimes) + (10 * nb of overdoses) + nb of users on the street block

Risk values are used to calculate social dissatisfaction at the level of the suburb (see below), and to calculate the *conductivity* of a given street block to drug dealing. The following rules apply:

- One street block becomes *conducive* (a) if there is a dealer or (b) if risk > 20 or (c) if there are at least 4 conducive street blocks around.
- One street block becomes *non-conducive* if there is no dealer and (a) if the risk = 0 or (b) if there is at least 4 non-conducive street blocks around.

Suburb

Each suburb is able to calculate an average risk over its belonging street blocks. This overall risk is interpreted as a measurement of the social dissatisfaction of the local residents. When the corresponding value reaches a score of 5 or above the police station needs to intervene (see below).

Police Station

There is only one Police Station for the whole system. Constables without identified mission return to the Police Station. Likewise, arrested users, dealers, and wholesalers are transferred to the Police Station before being retrieved from the system. At each time step, the Police Station sends constables to suburbs with *suburb protest* values higher than 5.

Treatment Centre

For this initial prototype model, we created only one Treatment Centre that receives users who decide to undergo a treatment program. The overall capacity of the Centre corresponds to 1,000 patients. Three programs are available, differentiated by their duration and estimated success rates. Detoxification and therapuetic community programs are residential while users on methadone maintenance are still on the street and can still inject heroin.

Social entities

SimDrug includes different types of social agents: *users*, *dealers*, *wholesalers*, *constables*, and *outreach workers*. Obviously, these computer entities do not accurately mimic individual behaviours of their real life counterparts. In fact, each type represents a minimum set of characteristics and dynamics that allows the whole artificial population to display most of the properties observed in real societies. The trans-disciplinary work plays a paramount role in defining a consensual set of simplified rules for the corresponding agent to behave realistically.

Another issue deals with the creation of a closed or open system. In a closed system, the initial set of agents remains in the system during the whole simulation, with the exception of individuals having to die in the meantime. The only way to increase the population is to implement reproduction mechanisms at the level of the agents. This is a widely used solution among agent-based modellers as it helps to keep system dynamics partly under control. An open system allows the entry into and exit from the system of a given number of agents at any point in time. It becomes much more complicated to track back any single individual trajectory, but these systems suit bar-like problems much better (bar attendance, airport lounge flows, marketplace encounters).

We chose to implement an open system that sustains a constant number of *users*, *dealers*, and *wholesalers* (*constables* and *outreach workers* remain the same). At each time step, for a given number of users who die from overdose, escape addiction through treatment, or end up in jail, the equivalent number of new users will be created at the next time step. Likewise, a given number of arrested dealers or wholesalers will be automatically replaced. This strong assumption is based on the fact that no evidence supports the eventual change of users' or dealers' population sizes in Melbourne, beyond limited fluctuations.

User

Estimations for Melbourne give a range of 30,000 to 35,000 drug users considered as regular addicts (Dietze et al. 2003). In order to keep computing time into reasonable limits, we have decided to create a one-tenth model of the reality: 3,000 users are created in SimDrug. They are randomly located on the grid at the beginning of the simulation.

Welfare payments provide a regular fortnightly income of $200 to the users. This amount represents between 50 per cent and 80 per cent of real payments and takes into account withdrawal for other primary needs. Individual *cash* is increased with the profit made from crimes (burglary, shoplifting) or drug dealing.

The model does not include user agents changing their individual trajectory (or risk of developing greater dependency). Individual *drug need* is a constant value that indicates the agent's degree of addiction. We opted for the creation of three initial cohorts of users, based on ethnographic survey and clinical research:

- light addiction: 0.02 g/day for 30 per cent of users, equivalent to 1 fix/day;
- moderate addiction: 0.04-0.06 g/day for 54 per cent of users, equivalent to 2-3 fix/day; and
- severe addiction: 0.08 or 0.1 g/day for 16 per cent of users, equivalent to 4-5 fix/day.

At this stage, a user can buy and use only one type of drug at a time from his/her dealer. Each user is affiliated to one dealer's location and goes to the same hot spot as long as the dealer is selling drugs. As soon as the dealer disappears, all the affiliated users have to find another provider by walking around or contacting friends. Information regarding the drug bought is stored into the attribute *my drug*. In this prototype, we consider a street market with only two drugs available: *heroin* and *other* (being a generic term for amphetamines, cocaine, etc.).

A user will have a 0.5 per cent chance to declare an *overdose* when injecting heroin if one of the following conditions is fulfilled: (a) the previous drug injected was not heroin, increasing the risk of overdose, or (b) variation in quantity from previous injection > 0.02 g, or (c) variation in purity from previous injection > 15 per cent, or (d) exiting from an unsuccessful treatment period. A user declaring an overdose has a 90 per cent chance to be rescued if there is another user around to call for an ambulance. The two chance parameters are partially calibrated against global figures of fatal and non-fatal overdoses in Melbourne during the pre-drought period.

The attitude of users towards treatment programs is summarised within the attribute called *readiness for treatment*. The initial individual values are randomly picked between 10 and 50. A decrementing process—borrowed from literature on diffusion of innovation—slowly raises the motivation of the user each time he is targeted by an Outreach Worker (decrement: −1) or each time he is witnessing or experiencing an overdose (decrement: −1). The value of the attribute is reset at 20 each time a user comes out from an unsuccessful treatment period. The initial range of values is partially calibrated against the observed average chance for a real user to enter a treatment program over a one-year period. When the value of *readiness for treatment* has reached zero, the corresponding user has 20 per cent chance to enter a Detox program, 10 per cent to enter a Treatment Centre program, and 70 per cent to enter a methadone program. The actual implementation depends on the Treatment Centre's capacity to undertake the treatment. Detox and Treatment Centre are residential treatments while Methadone programs allow the user to continue to interact with others in the system. In the latter case, a user has a 7 per cent chance at each time step to consume illicit drugs as well. This percentage is coming from clinical research (2 days /month).

Dealer

The real number of dealers in Melbourne is a well kept secret. Hence, the expert panel decided to adapt estimated figures coming from the USA where population ratio between dealers and users range from 1/10 to 1/30. We decided for a 1/20 ratio, which partly corroborates corresponding ratios coming from Australian Courts. Thus, 150 *dealers* are initially created. At this stage, dealers can buy only one type of drug at a time from their wholesaler and then sell it to users. Initial *cash* amounts range randomly from $5,000 to $10,000. The question of the different mark-ups between wholesaler, dealer, user-dealer, and user embarrassed the expert panel for a while. Drawing from heterogeneous data and information, we agreed on the following:

- wholesaler's mark-up x 6.0;
- dealer's mark-up with user x 2.0;
- dealer's mark-up with user-dealer x 1.3; and
- user-dealer's mark-up with user x 2.0.

Initially, dealers are assumed to deal on the street market only. But they are able to assess the risk created by the presence of constables in their surroundings. As a consequence, they can choose to freeze temporarily their activities (ready to sell: no) or eventually to change their *deal type* from street market to hidden sale, according to a 20 per cent probability. This chance parameter has not been calibrated yet.

Wholesaler

Reliable figures from Australian courts indicate a ratio of 1/48 between defendants considered as wholesalers or importers, and small dealers. We decided to apply a very conservative ratio of 1/15 in SimDrug in order to take into account the eventual under-representation of *big fish* in the courts' figures. Hence, we created 10 *wholesalers* in the system.

Wholesalers are in charge of buying the 2 types of drug available on the market (heroin or other) and to supply the different dealers with one or the other. Initial *cash* amounts range from $50,000 to $100,000. They have to reset their stocks every 30 time steps while dealers come to buy more whenever they need. The availability of one drug or the other is given by the ratio between both. This ratio is considered as an externality of the model (depending on successful importation) and it is filled in from an external data file containing daily values of quantities, market prices, and purities. Wholesalers keep track of their usual clients. Hence, when police succeed in arresting a wholesaler, the corresponding dealers fall with him.

Constable

Initially, 10 *constables* were created and located at the *Police Station*. They can move randomly around the grid or target a specific street in response to a protest from the *suburbs*. In this case, they are tracking down dealers and user-dealers. They have 10 per cent chance to arrest a dealer, and 40 per cent to arrest user-dealers in the neighborhood. These figures were estimated from existing criminological studies.

The Police Station will send constables to a given location on a dealer chase if the average protest of the corresponding suburb reaches a value of 5. In reality, operations against wholesalers are often initiated by special units (drug squads) and rely on external intelligence or insider information. Hence, we decided that the Police Station has a 0.25 per cent chance of getting reliable information, of sending constables to the corresponding address and of operating a successful crackdown on a wholesaler. As mentioned above, the dealers linked to an arrested wholesaler are also retrieved from the system.

Outreach Worker

10 *outreach workers* were created and initially located at the *Treatment Centre*. Their aim is to convince users to undertake treatment programs. The Treatment Centre will send outreach workers to the street blocks displaying the highest overdose rates. As mentioned above, outreach workers have a purely mechanical effect: they decrease by 1 the value of the attribute *readiness for treatment* for all the users located on the same street block.

UML Structure

Several authors mentioned in the first chapters assert that research on illicit drug use needs a trans-disciplinary approach. Such an integrated approach requires a common language in order to first communicate, and then to build a consensual ontology. In the world of Complexity Theory—more specifically among the atomists—a common language is available. The Universal Modeling Language (UML) is developed around a series of visual paradigms (diagrams) that enable developers to share their knowledge with other experts and to encapsulate new knowledge into their project. Three main diagrams are usually used to describe the functionalities of a given model:

The class diagram describes the entities of the modeled system (classes) with their internal characteristics (attributes and methods) and external links with other classes. It corresponds to the casting of the model.

The sequence diagram describes the successive actions conducted independently by different classes or interactions between several classes. It corresponds to the storyboard of the model.

The activity diagram describes the intimate actions embedded into a given method. The exhaustive list of all the activity diagrams corresponds to the script of the model.

Class diagram

Figure 10.1 represents the Class Diagram designed with VisualParadigm system components are described through computer agents characterised by attributes and methods.

Figure 10.1. SimDrug Class Diagram (designed with VisualParadigm©)

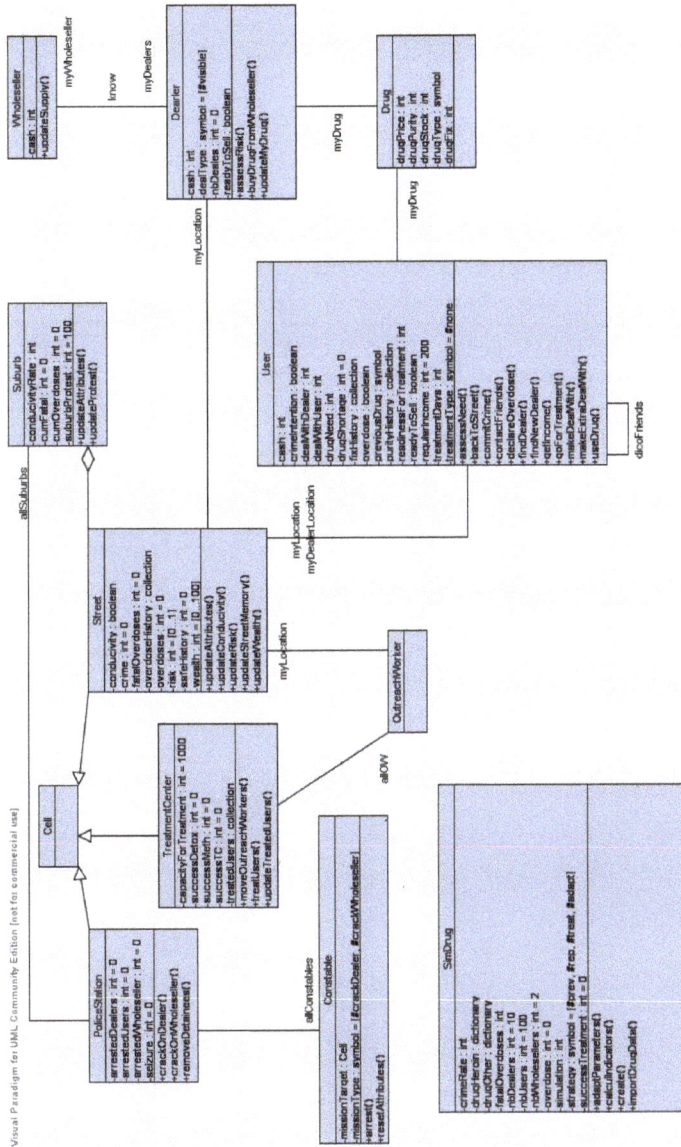

Modelling sequence

SimDrug is divided into 6 successive main stages:

1. resetting and updating population;

2. updating drug supply on the market;

3. activating users decision making process;

4. updating Treatment Centre performances;

5. updating street blocks status; and

6. activating police station and constables crackdowns.

Stage 1 aims at updating the population of agents, based on the changes triggered during the previous time step. All detainees are retrieved from the system and new users, dealers and wholesalers are created accordingly. Outreach workers are moved back to the Treatment Centre and dealers who were at their wholesaler's place go back to their street location. Stage 2 entails the methods for wholesalers and dealer's interactions towards drug supply. Wholesalers are given the opportunity to refill their supply once a month while dealers can visit their wholesaler as soon as their drug stock is sold out. Stage 3 focuses on the users' interactions with their environment and other agents. They start by assessing their need, looking at their available cash and drug, and decide whether they need to commit a crime. They then find their usual dealer (or, alternatively, a new dealer) and buy some drug. They use it at once and might declare an overdose. Stage 4 allows the Treatment Centre to manage new users entering treatments and on-going treated users reaching the end of their treatment duration. Stage 5 consists in updating the street blocks risk and conductivity status and calculating the new suburbs' protest values accordingly. Finally, Stage 6 allows the Police Station to adapt its strategy by reallocating constables on the grid and eventually performing successful crackdowns.

Activity diagrams

The following activity diagrams belong to Stage 3 and describe users' decision making process to assess their need (Figure 10.2), to use drug (Figure 10.3) and declare an overdose (Figure 10.4).

Figure 10.2. 'assessNeed' Activity Diagram (designed with VisualParadigm©)

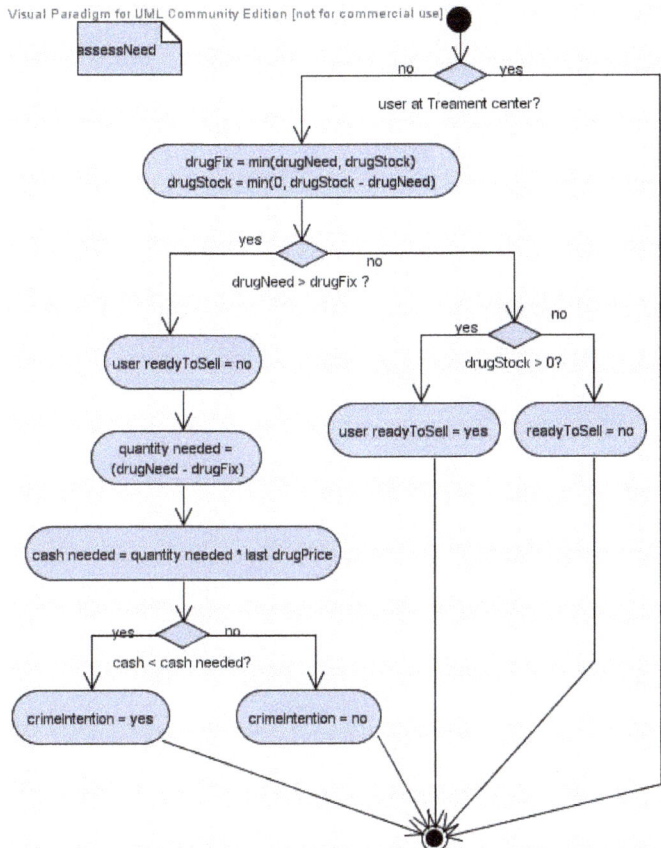

Non-treated users evaluate their needs by checking their available cash and drug stock. They are able to deduce the quantity needed and the corresponding cash required, based on the memory of the price they paid for their fix at the previous time step. This assessment will drive 2 context-dependent behaviours: users might need to commit a crime, or might be entitled to become user-dealer (if they have excessive cash).

Figure 10.3. 'useDrug' Activity Diagram (designed with VisualParadigm[c])

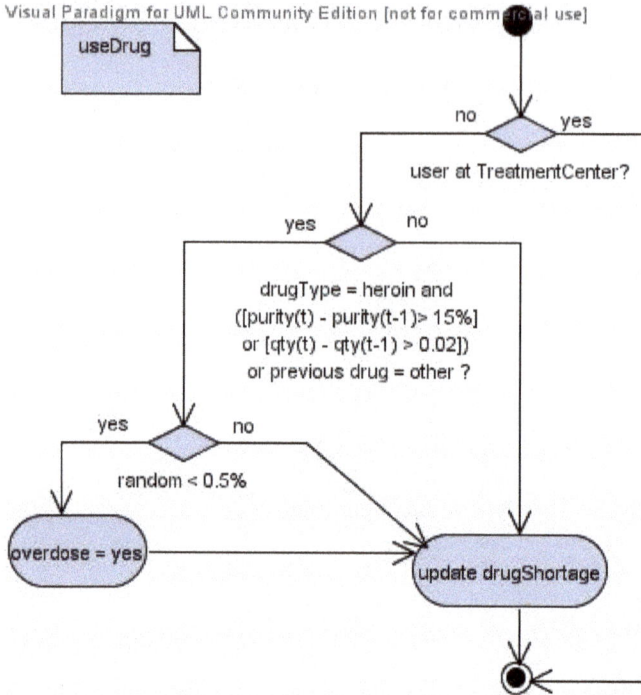

Users have a 0.5 per cent chance to declare an overdose when injecting heroin if they fulfil one of the following conditions:

- previous drug injected was not heroin;
- variation in quantity from previous injection > 0.02 g;
- variation in purity from previous injection > 15 per cent;
- exiting from unsuccessful treatment period.

Users declaring an overdose and close to another user able to call for an ambulance have a 90 per cent chance to be rescued.

Figure 10.4. 'declareOverdose' Activity Diagram (designed wtih VisualParadigm⁰)

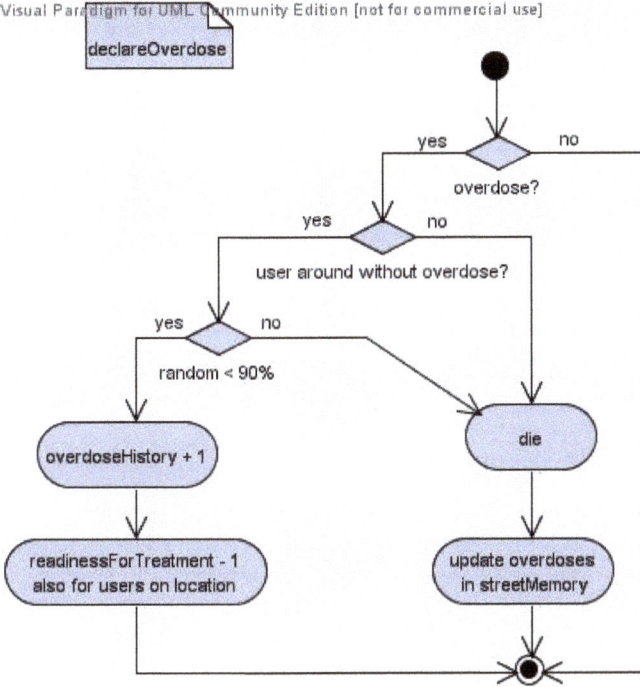

SimDrug—preliminary results

The CORMAS platform encapsulates sensitivity analysis tools and provides output data directly into Excel format files. Each scenario is run 10 times and CORMAS allows for the recording of output variables such as: the total number of crimes per time step; the cumulative number of non-fatal and fatal overdoses; the number of users injecting heroin or another drug per time step; the number of users under each type of treatment; the number of successfully treated users; the police station efficiency as regards crackdowns; the proportion of users-dealers, and more.

The base scenario has been set up with the parameters and values described in the previous section. It contains 3,000 Users, 150 Dealers and 10 Wholesalers. These figures are not subject to sensitivity analysis so far and remain unchanged for all the scenarios. The base scenario is used as a reference to derive sensitivity analysis on a chosen set of parameters summarised below:

• number of Constables: 10
• number of Outreach Workers: 10
• chance for a user to declare an overdose: 0.5 per cent

- chance for a user declaring an overdose to be rescued: 90 per cent
- wealth decreased by 5 per cent when a crime is committed on a given street block
- wealth increased by 3 per cent after a 10-day period with no crime
- crackdown on dealers occurs for suburbs with a protest value > 5
- chance for a user-dealer to be arrested during a *crackdown on dealer* mission: 40 per cent
- chance for a dealer to be arrested during a *crackdown on dealer* mission: 10 per cent
- chance for the police station to arrest a wholesaler at each time step: 0.25 per cent
- treatment capacity at the Treatment Centre: 1000

As for the input data featuring drugs' characteristics, we have agreed on a very simplified set of values. Both drugs, *heroin* and *other*, are equally available on the market. Hence, wholesalers spend half of their money on *heroin* and the other half on *other*. Both drugs have the same purity (30 per cent) which remains constant through the simulation. Wholesalers buy heroin for \$150/g and *other* for \$125/g. Hence, for the base scenario, we have decided to discard the impact of drug availability, quality and price in order to calibrate and analyse the remaining parameters.

Results

Overdoses

The proposed rules to declare an overdose are consistent with real data regarding overdoses and fatal overdoses. On an average, 1100 overdoses occur over a 4-year period, amongst which 150 are fatal (Figure 10.5). On an average, these figures correspond to a 9.2 per cent p.a. rate of overdose, and a 1.2 per cent p.a. rate of fatal overdose over the entire population of users. Statistics for Victoria in 1998-1999 provide an estimated 10 per cent and 1 per cent for the observed values.

Figure 10.5. Simulated Total and Fatal Overdoses over the 4-year period

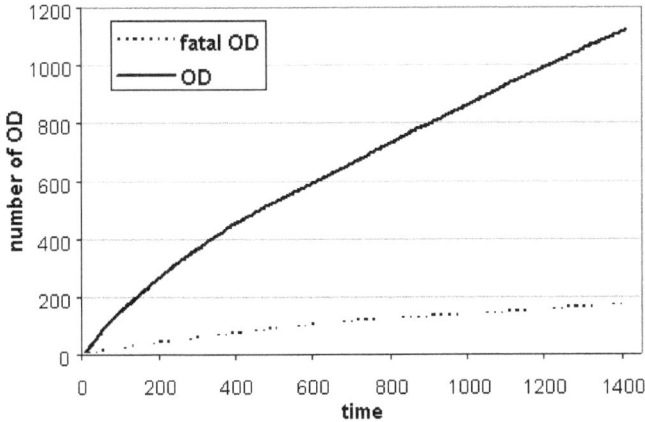

Treatment

After 3 years, on average, 800 users are under treatment at any given time step: 750 are on methadone (among which 50 still inject heroin), 40 are in a Treatment Centre, and only 10 are in detox. Statistics from treatment programs in Victoria indicate that 70 per cent of real users are involved in a treatment program over a period of 12 months. Our 26 per cent rate is much more conservative but corresponds to a proportion of users who take on a program until its date of completion, unlike official statistics. At the end of the simulation, 1,000 users have been successfully treated (800 thanks to methadone, 170 thanks to Treatment Centre and 30 thanks to detox). According to the simulations, methadone treatment happens to be, by far, the most efficient way to withdraw from heroin addiction.

Crime and Hot Spots

Crime rate follows a 15-day periodic pattern driven by the CentreLink-like payment periodicity. Crime rate increases as users' available cash decreases over the fortnight period and falls again when users receive their next payment. On average, 800 crimes are committed per time step. This outcome needs to be discussed and validated against real data.

In terms of spatial changes, the locations of *hot spots* on the grid evolve over time as a result of the constables patrolling the grid in response to suburban protests. This spatial mobility of hot spots can be viewed as an emerging property of the system, as no rules have been set up at the local level (street blocks) to define hot spots' patterns. Figure 10.6 shows the position of hot spots at the beginning of the simulation (left) and the extension and displacement and of hot spots at the end of the simulation (right).

Figure 10.6. Evolution of hot spot locations over time from initial condition (left) to final state (right)

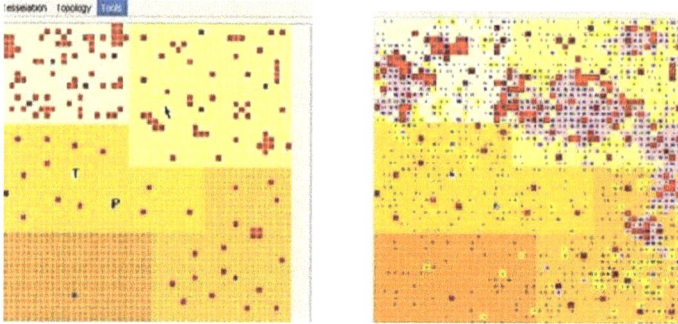

Dealer cash

At the end of the simulation, dealer cash ranges between $40,000 and $800,000. On an average, a dealer earns $2,400/week. These figures are close enough to the ones coming from police records ($3,500 to $4,000/week) if we take into account that a significant number of new dealers in the model fail to establish a profitable business.

User-dealer

At the end of the simulation, 300 users are also user-dealers, corresponding to 10 per cent of the population of users. This result is consistent with current estimates provided by the expert panel.

Sensitivity Analysis

For each parameter, several scenarios were tested in order to evaluate the sensitivity of the model to different values of the parameter. Thus, each scenario corresponds to a change in only one parameter in order to avoid overlapping effects. Table 1 summarises the effect, measured in degree of correlation ('+++' means high correlation, but not necessarily positive), of tested parameters on the output variables.

Table 10.1. Influence of tested values on selected variables

Tested Values / Variables	crime rate	dealer cash max	OD	fatal OD	pop heroin	pop other	pop no drug	pop shortage	pop treatment	arrest dealer	arrest user	arrest wholesalers	success treatment	fix with dealer	fix with user	user-dealers	seizure
outreach workers [1 – 10 – 20 – 50 – 100]	–	–	+++	++	++	+	+++	+	+++	+	++	–	+++	+	+	+	–
constables [1 – 10 – 50 – 100]	+	+++	–	–	–	++	–	++	–	+++	+++	–	–	++	–	+++	–
update wealth [0 – 3% – 5%]	++	+	–	–	+	++	+++	++	+++	–	+++	–	+++	++	++	+++	–
arrest User-Dealer [10% – 25% – 40%]	–	–	–	–	–	–	–	–	–	–	++	–	–	++	++	++	–
arrest Dealer [5% – 10% – 20%]	+	+	+	+	–	–	–	–	+	++	–	–	–	–	–	–	–
arrest wholesaler [0.1% – 0.25% – 0.5%]	–	–	+	++	–	–	–	–	–	–	+	+	–	–	–	–	++
suburb protest [3 – 5 – 7]	+	++	–	–	–	++	–	–	–	+++	+++	–	+	–	+	++	–
treatment capacity [600 – 800 – 1000]	–	–	–	–	–	–	++	++	+++	–	–	–	++	–	–	–	–

Outreach workers

The number of outreach workers influences strongly the overdose rates and the number of users undergoing treatment programs. This influence seems to take off beyond 20 agents located on the grid (Figure 10.7). This impact on the amount of treated users is a direct consequence of the ability of the outreach worker agents to modify individual readiness for treatment. The clear impact on overdose rates is more interesting as any user quitting an unsuccessful treatment increases his chances of overdose due to the withdrawal period. Clearly, non-linearity between tested values and variables opens a window of opportunity to run cost-efficiency analysis amongst mixed strategies.

Figure 10.7. Influence of outreach workers on overdose rates (left) and treated user rates (right)

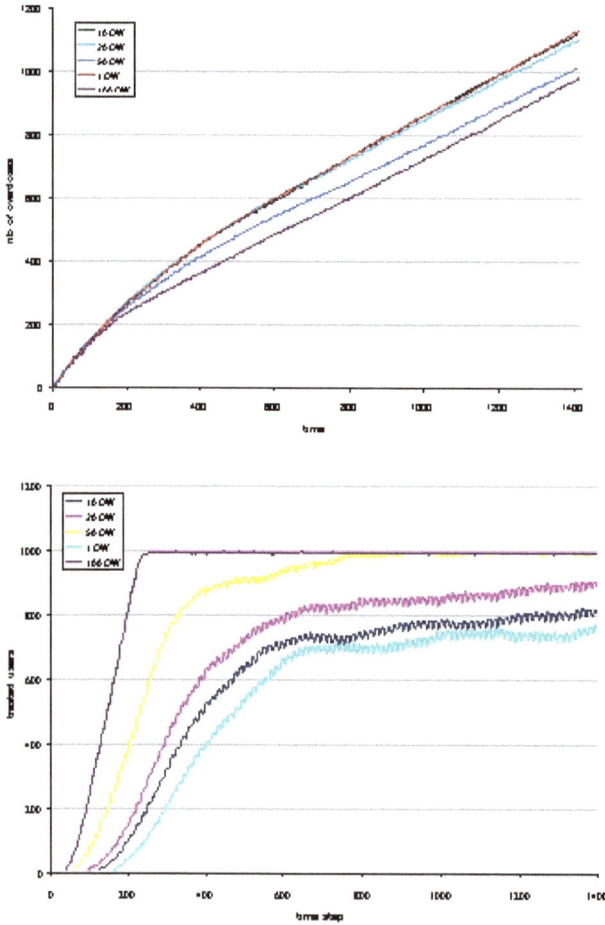

Constables

Increasing the number of constables from 10 to 100 has a clear and expected positive influence on the number of arrested dealers (Figure 10.8). However, looking at dealers' maximum income, it is more surprising to notice the lack of major impact when comparing the scenarios with 50 and 100 constables. A plausible explanation lies in the ratio between constables and dealers that does not generate great difference for ratios above 1 constable for 3 dealers. But again, cost-efficiency needs to be assessed for such large ratios that would probably stretch law enforcement capacities beyond limits.

Figure 10.8. Influence of constables on number of arrested dealers (left) and maximum dealer's cash (right)

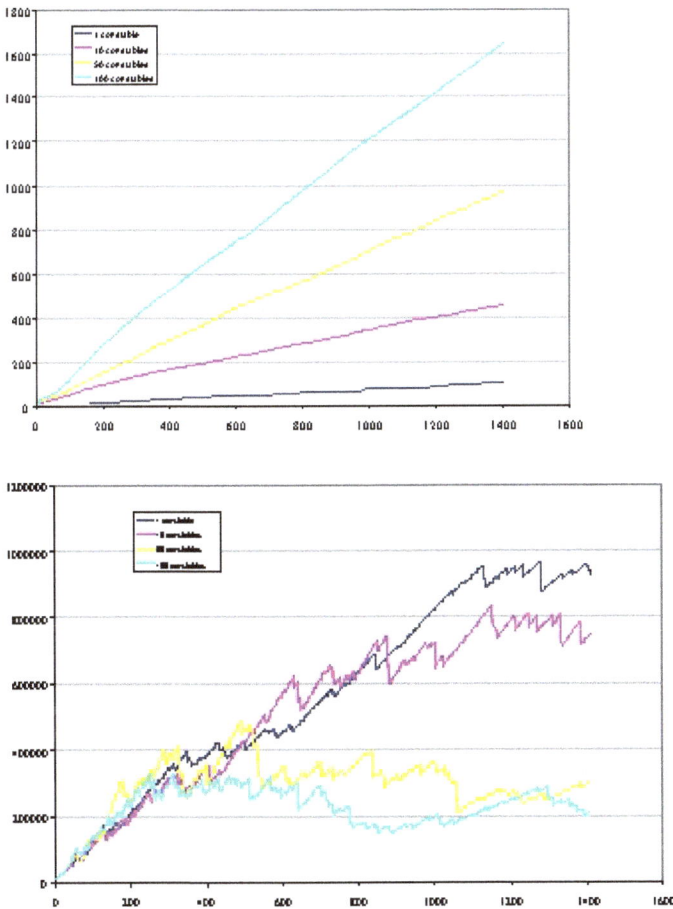

Non drug-related variables

Interestingly, the way *wealth* attributes of the *street blocks* are updated significantly influences most of the output variables. Obviously, the amount of wealth available on *street blocks* drives users' revenues from crimes. It constrains the possibility to fulfll one's drug needs, and it impacts on the number of user-dealers. Consequently, the number of arrested users is also affected. Beside, it affects the number of treated users by reducing the chance for users to reach the required stage of readiness without being caught by the constables beforehand (Figure 10.9). This outcome seems to validate some experts' claims about the necessity to better take into account *non-drug-related* environmental factors in order to understand these markets.

Figure 10.9. Number of users under treatment according to increasing values for wealth updating rate

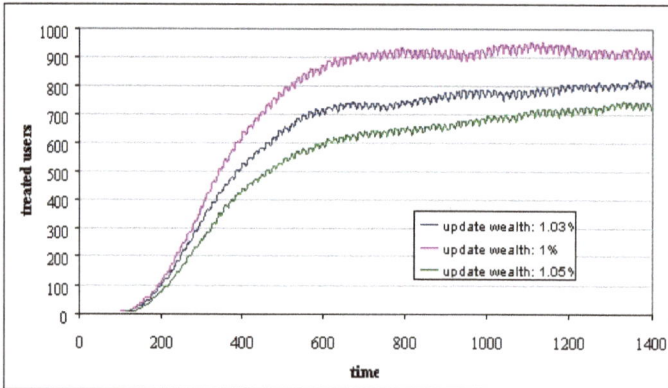

Simulating the heroin drought

The striking figures linked with the heroin drought concern the number of fatal and non-fatal overdoses reported in Victoria at that time. Within a few months, fatalities fell from an average 300 per annum to an equivalent of 40 per annum during the drought peak, resulting in a 52 per cent permanent decrease in the number of casualties from the drought onset (Dietze et al. 2003). Despite all our efforts, it was impossible to set up a scenario for SimDrug to display such a dramatic response without pushing some parameters to highly unrealistic values. Hence, the expert panel analysed our initial assumptions again. It was decided to successively modify 2 essential features:

- transforming SimDrug into a closed system rather than an open one, thus, removed agents are not replaced in the system; and
- modifying the input data files in order to take into account the observed availability of heroin during the simulated period.

A closed system design succeeded in creating a sharp fall in the number of overdoses, due to the simple fact that a decreasing number of users populated the system. But the system never recovers after the simulated drought, the market simply collapses. Besides, there is no evidence so far that the overall population of injecting users in Victoria significantly changed between 1998 and 2002. Nevertheless, it is probable that pre-drought conditions influencing individual decision to inject heroin had some effect. Hence, SimDrug's degree of openness should be reviewed according to some pioneering work by Agar (2005).

Given the fact that modelling an illicit drug market based on 2 equally available drugs does not depict the reality of the heroin trade in Melbourne, we have decided to use heroin's purity, quantity, and price data derived from Dietze et al. (2003). The *other* drug's availability was calculated in order to secure a constant overall availability of drugs on the market. The authors acknowledge that this first-pass assumption needs to be validated against further evidence. Figure 10.10 compares simulated fatal overdoses from this new scenario with the ones coming from the base scenario. While the base scenario—assuming that heroin covers 50 per cent of the market at any time—provides a nearly steady rate of 35 casualties per annum, the new scenario shows a sharp decrease—around time step 800—which corresponds to the heroin drought period, from 60 casualties per annum before the drought onset, to a mere 30 casualties afterwards. Though this 50 per cent decrease is consistent with findings from Dietze and colleagues (2003), it has to be noticed that if our 1/10 scale were correct, the pre-drought simulated figures double the ones reported in reality. The same analysis and conclusions can be derived from results on total overdoses.

Figure 10.10. Number of fatal overdoses derived from the base-scenario and from real data

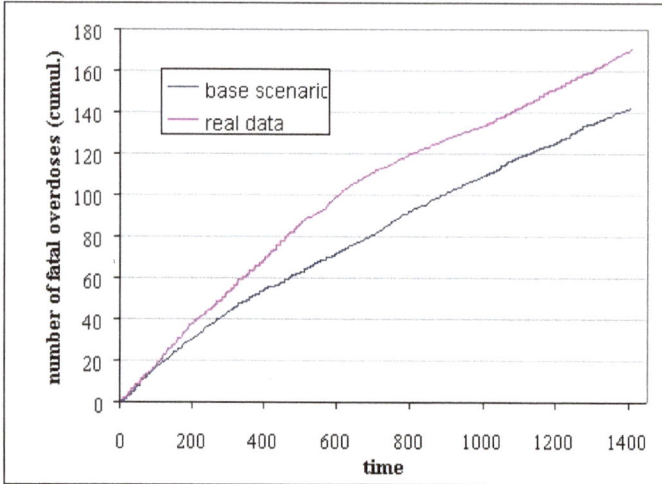

Conclusion

This chapter presents our attempt to build a first agent-based model dedicated to study the illegal drug market in Melbourne during the heroin drought period. As described by Gorman and colleagues (Gorman et al. 2004), drug use related problems are heterogeneously distributed with respect to population and geography and they need to be considered as essentially based on local interactions. SimDrug has been conceptualised and implemented in order to capture the primary community structures and relationships that support drug use and related outcomes. Geography and local interactions are embedded within the structure of the spatial grid divided into 5 archetypal suburbs. Using the propriety of the cellular-automata, SimDrug allows for diffusion processes, such as hot spot displacement, to occur. Interactions amongst agents could be increased by creating converging sites where massive connections arise, such as shopping malls or central stations. Moreover, as argued by Gorman et al. (2004:1725):

> Models that capture the behavior of…complicated community systems and control strategies that modify them must, therefore, combine available data, statistics, and spatiotemporal dynamics.

One of the main advantages of SimDrug is its ability to gather and blend, within the same tool, data (mainly second-hand) coming from very diverse sources. The structure is already flexible enough to integrate more information, as the prototype will evolve. The next stage will focus on transforming this data-collecting oriented platform into a discussion-oriented tool by improving the economic components. Integrating cost-efficiency analysis will help explore combined strategies by adjusting the allocated resources between treatment (outreach workers and Treatment Centre) and repression (Police Station and constables). We can also incorporate harm reduction programs (through new rules for outreach workers regarding needle syringe distribution). SimDrug encompasses great expectations to be used as a tool to confront and generate discussion among stakeholders and policy-makers. However, as pointed out by Gorman et al. (2004), such an approach will never provide an optimal solution, but rather numerous possible context-specific solutions with potential outcomes being highly uncertain. In fact, the best global solution may be a collection of local solutions tailored to local circumstances and needs. Obviously, such an approach needs to be carefully explained and tools need to be genuinely tailored in order to appeal to policymakers, who would normally favor large scale standardised interventions that promise to deliver assured, definite, and extensive outcomes.

Acknowledgments

The authors wish to thank the Colonial Foundation Trust for funding the Drug Policy Modelling Project (DPMP). They also need to thank Gabriele Bammer, Jonathan Caulkins, and Peter Reuter for their helpful comments.

References

Agar, M. (2003) The story of crack: Towards a theory of illicit drugs trends. *Addiction Research and Theory* 11(1): 3–29.

Agar, M. (2005) Agents in living color: towards emic Agent-Based Models. *Journal of Artificial Societies and Social Simulation* 8(1). Available at http://jasss.soc.surrey.ac.uk/8/1/4.html

Agar, M. and H.S. Reisinger (2002) A heroin epidemic at the intersection of histories: the 1960s epidemic among African Americans in Baltimore. *Medical Anthropology* 21(2): 115–56.

Agar, M. and D. Wilson (2002) DrugMart: heroin epidemics as Complex Adaptive Systems. *Complexity* 7(5): 44–52.

August, G.J., K.C. Winters, G.M. Realmuto, R. Tarter, C. Perry and J.M. Hektner (2004) Moving evidence-based drug abuse prevention programs from basic science to practice: 'Bridging the efficacy-effectiveness interface'. *Substance Use And Misuse* 39(10–12) : 2017–53.

Australasian Centre for Policing Research (ACPR) (2003) Enhancing the links between drug misuse prevention programs and crime prevention programs. Adelaide, Australian Centre for Policing Research.

Australasian Centre for Policing Research (2003) The impact of general law enforcement on the illicit drug market. ACPR Discussion paper, March.

Available at http://www.acpr.gov.au/pdf/drugs/Impact%20of%20general.pdf

Bousquet F., I. Bakam, H. Proton and C. Le Page (1998) 'CORMAS : Common-Pool Resources and Multi-Agent Systems'. Lecture Notes in Artificial Intelligence 1416: 826-37.

Bousquet, F., O. Barreteau, P. d'Aquino, M. Etienne, S. Boissau, S. Auber, C. le Page, D. Babin and J. C. Castella (2003) Multi-agent systems and role games: an approach for ecosystem comanagement. In M. A. Janssen (ed.) *Multi-Agent Approaches for Ecosystem Management*. Cheltenham: Edward Elgar.

Bradbury, R.H., D.G. Green and N. Snoad (2000) Are ecosystems complex systems? In T. Bossomaier and D. Green (eds), *Complex Systems*, pp. 339-65. Cambridge: Cambridge University Press.

Bush, W., M. Roberts and M. Trace (2004) Upheavals in the Australian drug market: heroin drought, stimulant flood. The Beckley Foundation Drug Policy Programme, Drugscope Briefing Paper No. 4.

Caulkins, J. and P. Reuter (2005) Analyzing illicit drug markets when dealers act with limited rationality. In F. Parisi and V.L. Smith (eds), *The Law and Economics of Irrational Behaviour*. CA: Standford Uni Press

Chou, C.P., D. Spruijt-Metz and S.P. Azen (2004) How can statistical approaches enhance transdisciplinary study of drug misuse prevention? *Substance Use and Misuse* 39(10–12): 1867–1906.

Dietze, P., P. Miller, S. Clemens, S. Matthews, S. Gilmour and L. Collins (2003) The course and consequences of the heroin shortage in Victoria. NDLERF Monograph no. 6. Adelaide: Australasian Centre for Policing Research.

Dorn, N., K. Murji and N. South (1992) *Traffickers: Drug Markets and Law Enforcement* . London: Routledge.

Fuqua, J., D. Stokols, J. Gress, K. Phillips and R. Harvey (2004) Transdisciplinary collaboration as a basis for enhancing the science and prevention of substance use and abuse. *Substance Use and Misuse* 39(10–12): 1457–1514.

Gatrell, A.C. (2005) Complexity theory and geographies of health: a critical assessment. *Social Science and Medicine* 60(12):2661–71.

Gorman, D.M., P.J. Gruenewald, P.J. Hanlon, I. Mezic, L.A. Waller, C. Castillo-Chavez, E. Bradley and J. Mezic (2004) Implications of systems dynamic models and control theory for environmental approaches to the prevention of alcohol and other drug use-related problems. *Substance Use and Misuse* 39(10–12): 1713–50.

Maher, L. (1996) Illicit drug reporting system trial: ethnographic monitoring component. NDARC Technical Report No. 36. Sydney: NDARC.

Mason, M., I. Cheung and L. Walker (2004) Substance use, social networks, and the geography of urban adolescents. *Substance Use and Misuse* 39(10–12): 1751–77.

May, T. and M. Hough (2004) Drug markets and distribution systems. *Addiction Research and Theory* 12(6): 549–63.

Mazerolle, L., C. Kadleck and J. Roehl (2004) Differential police control at drug-dealing places. *Security Journal* 17(1): 61–9.

Rhodes, T. (2002) The 'risk environment': a framework for understanding and reducing drug-related harm. *International Journal of Drug Policy* 13: 85–94.

Ritter, A. (2005) *A Review of Approaches to Studying Illicit Drug Markets.* DPMP Monograph Series. Fitzroy: Turning Point Alcohol and Drug Centre.

South, N. (2004). Managing work, hedonism and the borderline between the legal and illegal markets: two case studies of recreational drug users. *Addiction Research and Theory* 12(6): 525–38.

Unger, J., L. Baezconde-Garbanati, P.H. Palmer, E. Nezami and J. Mora (2004) What are the implications of structural/cultural theory for drug abuse prevention? *Substance Use and Misuse* 39(10–12): 1779-1820.

Valente T.W., P. Gallaher and M. Mouttapa (2004) Using social networks to understand and prevent substance use: a transdisciplinary perspective. *Substance Use and Misuse* 39(10–12): 1685–1712.

View from an Energy Expert

The simulation of power system operation has always been necessary in order to ascertain the costs of production, especially fuel consumption and the expected reliability of power systems. With the coming of digital computers in the late 1960s, the techniques were improved significantly over the hand/graphical calculations that were used earlier. They were typically based on general averages or the simulation of sample days of operation and, initially, were considered accurate enough.

Later in the same decade, the problem of poor availability of large generating units forced a rethink of the approaches, aided by the availability of improved computing power. Monte-Carlo simulations of hourly operation over all days of a year, embodying detailed chronological effects, were developed. Probabilistic simulations, involving the manipulation of probability distributions to give expected values of major variables, were developed later, and proved suitable for longer term studies.

These techniques were aimed at modeling the main sources of variability affecting the utilities at the time-generating unit forced and planned outage rates, load forecasts and, for some systems, hydroelectric inflow variability. The techniques were generally quite accurate and achieved widespread use in the utility industry during the 1970s. The author of this introduction was heavily involved in this phase of development.

But with the coming of competitive markets and greatly improved generation plant performance in the 1990s, the major variables shifted to those associated with the independent bidding decisions made by multiple independent generation owners, load forecasts, now affected by individual decisions on demand management, new patterns of flow over transmission networks and the variable output of some renewable resources, especially wind farms. A new approach to power system simulation was required, but has proven difficult to achieve.

The multi-agent based simulation model representing Australia's National Electricity Market (NEM) developed by the CSIRO shows great promise in overcoming the problems of the conventional approach. Its structure is such that it is capable of representing the various decisions made by the 'agents' participating in the NEM and of the combined effect of these decisions on supply reliability, electricity costs and prices, greenhouse gas emissions, and other secondary outputs. The Australian market allows participants a greater degree of freedom compared with that of other markets and is prone to manipulation by the exercise of market power by the relatively small number of players. It is quite difficult to model correctly.

The continued development of the CSIRO model is to be supported and the progress thus far is impressive.

Dr Robert R Booth
Managing Director
Bardak Ventures Pty Ltd
Doonan, Queensland

11. NEMSIM: Finding Ways to Reduce Greenhouse Gas Emissions Using Multi-Agent Electricity Modelling

David Batten and George Grozev

Abstract

This chapter outlines the development of an agent-based simulation model that represents Australia's National Electricity Market (NEM) as an evolving system of complex interactions between human behaviour in markets, technical infrastructures and the natural environment. Known as NEMSIM, this simulator is the first of its kind in Australia. Users will be able to explore various evolutionary pathways of the NEM under different assumptions about trading and investment opportunities, institutional changes and technological futures, including alternative learning patterns as simulated agents grow and change. The simulated outcomes help the user to identify futures that are eco-efficient, such as maximising profits in a carbon-constrained future. Questions about sustainable development, market stability, infrastructure security, price volatility and greenhouse gas emissions can be explored with the help of the simulation system.

Introduction

For more than a decade, electricity industries have been undergoing worldwide regulatory reform, with the general aim of improving economic efficiency. In many places, these changes have culminated in the appearance of a wholesale power market. There are various, sometimes contradictory, conclusions about the performance of such restructured electricity markets. In this new context, the operation of the generating units no longer depends on centralised state or utility-based procedures, but rather on the decentralised decisions of various firms—profit-maximisers in different states of economic health with different long-term plans. Market performance depends largely on how each market participant responds to market design—including the rules, market observations, operational procedures and information revelation.

Recently, interest has grown in the use of market-based policy instruments for environmental purposes—to introduce climate change regulation into the electricity sector and reduce greenhouse gas emissions. Australia is an early adopter of both electricity industry restructuring and market-based environmental instruments. However, the mixed performance of these schemes to date illustrates the need for considerable care in the design of market-based approaches (MacGill et al. 2004). For example, the opportunities for market participants to exercise market power are real and observable (Hu et al. 2005).

Firms engaged in these new electricity markets are exposed to higher risks, so their need for suitable decision-support models has increased. Regulatory agencies also require analysis support models so as to monitor and supervise market behaviour. The problem is that traditional electrical operational models are a rather poor fit to these new circumstances, since the new driving force for operational design—market behaviour—was excluded. Furthermore, purely economic or financial models used in other economic contexts do a relatively poor job of explaining behaviour in electricity markets. Nowadays, convincing electricity market models must consider at least 3 interrelated, dynamic processes: market participants' behaviour; the physical limitations of production and transmission assets; and the environmental outcomes associated with electricity generation, transmission and consumption.

This chapter outlines the development of an agent-based simulation model that represents Australia's NEM as an evolving system of complex interactions between human behaviour in markets, technical infrastructures and the natural environment. This simulator is the first of its kind in Australia. Users will be able to explore various evolutionary pathways of the NEM under different assumptions about trading and investment opportunities, institutional changes and technological futures, including alternative learning patterns as participants grow and change. The simulated outcomes will help the user to identify futures that are eco-efficient, for example, maximising profits in a carbon-constrained future. Questions about sustainable development, market stability, infrastructure security, price volatility and greenhouse gas emissions can be explored with the help of the simulation system.

Some physical and behavioural characteristics of Australia's NEM are outlined. This serves as a background for the following section, in which it is argued that Australia's wholesale and contract market system should be treated together as a complex adaptive system. Then we review 3 approaches to electricity market modelling: optimisation, equilibrium and simulation models. The next section argues the case for agent-based simulation models, on the grounds that markets and their participants co-evolve by adaptive learning, though all are pursuing profit-maximising goals. Our empirical work reveals that they adapt their behaviour in response to several dynamic factors, including their own costs, experiences and observed market outcomes. We then describe NEMSIM, the National Electricity Market Simulator currently under development. The greenhouse gas emissions calculator incorporated in NEMSIM is discussed. The final part describes the simulated outcomes, tables and reports that are produced by NEMSIM followed by a summary of future work.

Australia's national electricity market

The NEM in Australia was launched on 13 December 1998 and incorporates 5 States and Territories: Queensland, New South Wales, Australian Capital Territory, Victoria, and South Australia. In 2005, the Tasmanian electricity grid will be linked with the mainland if the Basslink project is completed. The NEM is not a truly national market, since Western Australia and the Northern Territory are not physically connected. It is a gross pool-type market and it operates under the administration of the National Electricity Market Management Company (NEMMCO). As such, the NEM displays many features in common with other pool-type markets after restructuring, for example, the original England/Wales pool prior to New Electricity Trading Agreements (NETA).

There were 89 registered market participants at the end of 2003. They include 41 generator companies, 20 network service (transmission and distribution) providers, 29 market customers and 9 traders. Several participants are registered in more than one category and some are registered as intending to participate. There are more than 350 physical generation units, of which about 190 can dispatch electricity. The total NEM installed capacity is around 37 gigawatts (GW). In the NEM, a settlement day starts at 04:00 and ends at 04:00 the next day. Each settlement interval is a half-hour period starting on the hour or half-hour. For example, settlement interval 6 denotes the period from 06:30 to 07:00. Each dispatch interval is a 5-minute interval.

Operating the NEM

The NEM is operated as follows (Figure 11.1):

*   Scheduled generators submit offers in 10 price bands, stacked in an increasing order in 10 incremental quantity bands over a settlement day. These quantity offers correspond to the 10 price bands for each of the 48 settlement intervals, and must be received by NEMMCO before noon of the previous day (i.e., a day before the real dispatch).
*   Regional network operators prepare demand forecasts, whereupon NEMMCO runs a linear program to dispatch generation and meet demand every 5 minutes. This program aims to maximise the value of trade based on dispatch bids or offers from market participants along with other ancillary services, subject to constraints on the physical network and the generating units. Thus less expensive generating units are dispatched first, but the price offered by the most expensive generating unit dispatched determines the price for that dispatch interval.
*   The settlement price for each settlement interval is the average of all prices over the six dispatch intervals in the settlement interval.

Figure 11.1. Inter-connectors in Australia's National Electricity Market

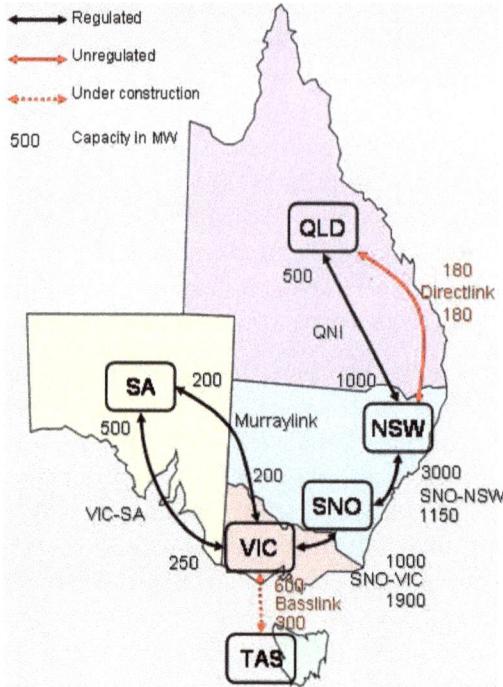

The price cap (or VOLL—value of the lost load) in the NEM is currently $10000/MWh, an increase from $5000/MWh from April 2002 onward.[1] In practical terms, generators can bid their price bands up to this figure, knowing that the market price is capped by this limit.

Market information plays an important role in participant's decision making and in ensuring reliable dispatch operations. NEMMCO provides market participants with information such as load forecasts, pre-dispatch (and price sensitivity analysis) data, dispatch data, as well as medium- (7 days) and long-term (2 years) forecasting of supply scenarios or system adequacy. Market participants are permitted to adjust their bids in response to the latest information. For example, they can rebid the quantity for a previously bid settlement interval up to 5 minutes before dispatch, although the price bands are fixed for the entire settlement day.

On its website,[2] NEMMCO publishes the bidding and dispatch data for all scheduled generating units shortly after the end of a settlement day. Yesterday's files contain bids/offers and rebids made by generating units and dispatchable

[1] Prices are in Australian dollars.

[2] www.nemmco.com.au

loads for each day's settlement intervals. They also record rebid explanations, since early 2002. The daily dispatch files include regional information such as regional reference prices, total demand and dispatchable generation. Although Barmack (2003) has expressed concerns about the revelation of bidding data to the public, in some earlier work we have demonstrated that there are several positive aspects of such data revelation (Hu et al. 2005).

Several other researchers have studied the market's performance during this period of restructuring. Wolak (1999) examined the first 3 months of the NEM's operation and observed that wholesale prices in Victoria and NSW dropped initially compared with those prevailing prior to the introduction of the NEM. Outhred (2000) concluded that the extent of competition is high and that the market performs well. By way of contrast, Short and Swan (2002) reported a mixed performance based upon their calculation of Lerner indices—defined as the ratios of the difference of bid price and marginal price to the bid price—for the regions in the NEM.

Some of our recent research identified various strategies used by generators in the market. The data we examined showed that two-thirds of the generating units are relatively inactive in terms of changing bidding strategies. However, the minority of generators that do respond actively to changes in market conditions are base-load units that tend to dominate the supply side. These 30-or-so active generating units are the *giants* in their regions. Their greater size affords them more opportunities to try different combinations of capacity offers in different price bands. Also, their sheer size renders them a more secure and confident position. They can more easily afford to try different strategic bids since part of the bulk of their capacity can be committed to very low price bands, safe in the knowledge that it is almost certain to be dispatched. This releases the remainder of their capacity for experimental bidding in strategic price bands.

Also, our research shows that larger generators are more likely to use quantity offers rather than price offers to improve their market positions. This is especially true in periods of peak demand. In terms of traditional game theory, such a result suggests that NEM generators may behave more like players in Cournot competition (by using quantity as a strategic variable). We shall return to this issue when we discuss several modelling and simulation approaches in sections 4 and 5.

Market power issues in the NEM are also of interest. Whether deliberate or otherwise, the strategy of capacity withholding by generators has the effect of raising prices. Such outcomes have been observed in several deregulated electricity markets, like California (Borenstein et al. 2002), Britain (Green and Newbury 1992; Wolfram 1999) and Australia (Short and Swan 2002). In the NEM, capacity withholding can be accomplished by offering capacity only in very high price bands, or making one or more generating units unavailable. Our research shows that capacity withholding is used by several larger firms who own the larger generating units (Hu et al. 2005). Since these generating units have the capability to influence regional prices, this may partly explain why they are engaged in strategic bidding.

Generators are allowed to rebid their capacity commitment in response to load changes and other factors. Via rebidding, they can take advantage of the information provided by NEMMCO in the pre-dispatch phases to improve revenue streams. There is concern over rebidding, partly because it may help certain generators to exercise market power. Some peaking units use rebidding to adapt to changes in market conditions, instead of using the daily offer/bid opportunities. They offer identical quantity bands through all settlement intervals and then rebid at short notice. Moreover, these rebids can occur several times within a single settlement interval. This action highlights the volatility of demand and the rapidity of changes in market prices, which are not easy to predict and are further compounded by the technical advantages of those generating units that can startup rapidly.

Electricity markets are complex adaptive systems

Nowadays, the NEM faces a number of additional challenges. It is not only the exercise of market power that can cause dramatic price fluctuations from day to day. Temperature variations and network congestion play a similar role. Since retail prices are pegged, some retailers and market customers find their profit margins squeezed. Each generating unit (coal-fired, gas-fired, hydroelectric or renewable) has different start-up costs, start-up times and performance variables. In a climate of uncertainty, choosing a generator unit and figuring out a profitable price at which to service newly posted demand is a challenging and risky business.

Clearly, the market for electricity has become more complex. In an evolving world of power generation, transmission and distribution, individual firms make use of increasingly specific and timely feedback of bid prices, costs, realised demand and operating information to enhance managerial decisions. Just possessing information holds little competitive advantage. Private firms and government organisations invest in various analytical and decision-making tools, as well as other sophisticated devices to gather pertinent information. Faster feedback requires faster adaptive reaction. Those who receive and respond to feedback more quickly gain competitive advantage (for example, revising their bids or choosing not to bid at a time of volatile price movements). The challenge is to respond to information quickly yet profitably.

Economic volatility is only one aspect of the problem. A growing need to mitigate ecological impacts—especially greenhouse gas emissions—has highlighted a need to develop methods capable of addressing economic and ecological uncertainties consistently within an integrated framework. It is now believed that greenhouse gases contribute significantly to global warming. In Australia, the stationary energy sector accounts for almost 50 per cent of these emissions. Electricity generation dominates this sector, with 64 per cent of the emissions, and is thus the major culprit in terms of total emissions.

An additional complexity is that day-to-day bidding strategies in the NEM are affected by hedged positions in other contract markets, such as the Over the Counter (OTC) Electricity Market (mainly Swaps and Caps, with some Swaptions and other Options) and the use of Renewable Energy Certificates (RECs). Other jurisdictional schemes designed to reduce greenhouse gas emissions have been introduced by the States (MacGillct al. 2004). In order to understand planning and decision making over different time horizons, these new schemes must be considered. Interdependencies between the spot (NEM) and contract markets are an important factor in determining hedging decisions of generators, retailers and other market participants. Such hedged positions are important because they influence incremental investment in the medium term and thus the closure decisions of agents.

Nowadays, electricity markets are an evolving system of complex interactions between nature, physical structures, market rules and participants (Figure 11.2). Participants face risk and volatility as they pursue their goals, and make decisions based on limited information and their mental models of how they believe the system operates. There is a wide diversity of agents; they use different strategies; they have different capacities; they use different generation technologies; they have different ownership; they are physically located at different locations, and they face different grid constraints. In summary, their objectives, beliefs and decision processes vary markedly. Such a diversity of inputs may be expected to lead to a rich diversity of market outcomes.

Figure 11.2. The NEM as a Complex Adaptive System

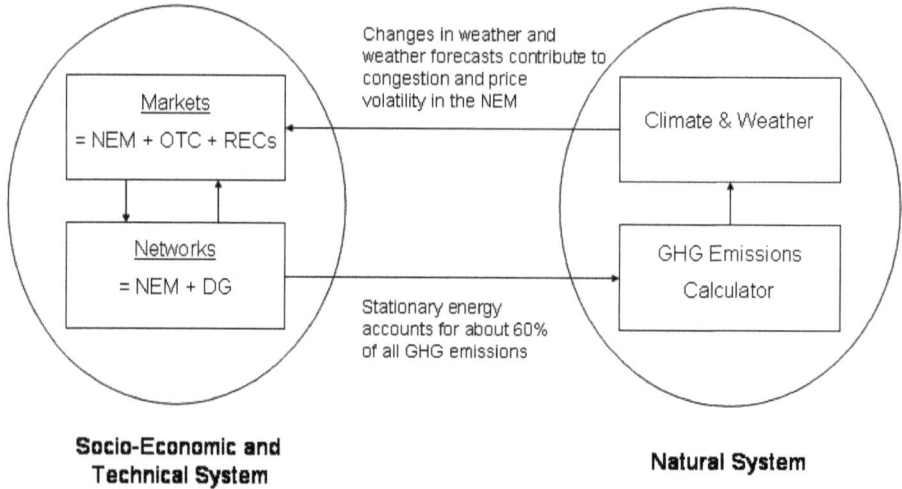

The NEM possesses all the intrinsic features of a complex adaptive system: a largish number of intelligent and reactive agents, interacting on the basis of limited information and reacting to changes in demand (due to weather and consumer needs). As no single agent is in control, some (for example, generator agents) may profit more than others. The result may be a considerable price degree of price volatility, inadequate reserves, demand uncertainty and higher levels of greenhouse gas emissions.

A key question then arises: what kinds of energy-economy models (if any) are capable of handling all these complexities? In the next section, we shall review several approaches in an attempt to answer this question.

Approaches to electricity market modelling

A large number of papers have been devoted to modelling the operation of deregulated power systems. In this section, we draw upon some earlier reviews to compare and contrast the main approaches based upon model attributes. Such attributes can help us to understand the advantages and limitations of each modelling approach.

From a structural viewpoint, the approaches to electricity market modelling reported in the technical literature can be classified according to the scheme shown in Figure 11.3. They fall into three main classes: optimisation, equilibrium, and simulation models. Optimisation models focus on the profit maximisation problem for a single firm competing in the market, while equilibrium models represent the overall market behavior—taking into consideration competition among all participants. Simulation models are regarded increasingly as an alternative to equilibrium models when the problem under consideration is too complex (for example, too nonlinear or dynamic) to be addressed within a traditional equilibrium framework.

Figure 11.3. Some trends in electricity market modelling

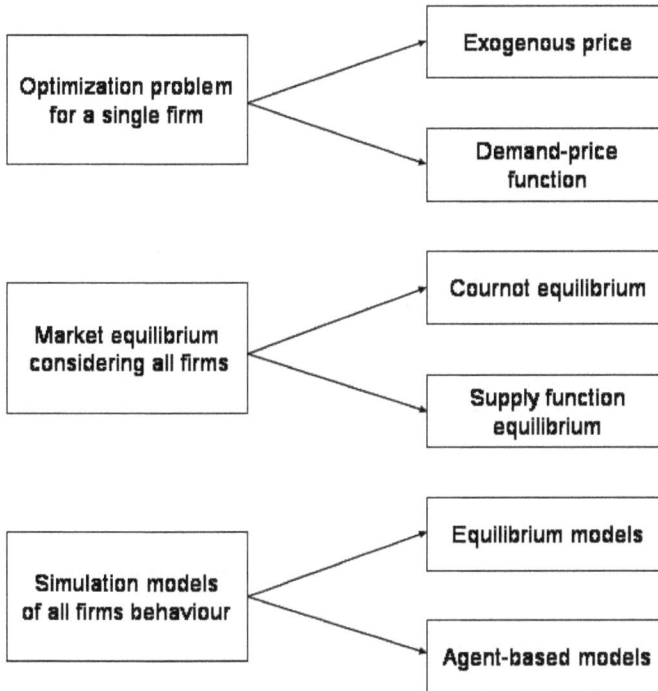

Source: Ventosa et al. 2005

Various assumptions are often made on the objectives, strategies, beliefs and capabilities of market participants. In game theory models, for example, participants are assumed to be rational in the sense that they can obtain and explore all the relevant information in order to deduce the best outcome. As we shall see shortly, some of these rigid assumptions can be relaxed with the help of agent-based simulation, since participants may employ different strategies and be subject to different sets of rules to guide their behaviour. They may have access to different information and possess different computational capabilities. The challenge then is how to assign a particular agent the appropriate set of behavioural rules and computational capabilities.

Previous review articles (Kahn 1998; Day et al. 2002; Ventosa et al. 2005) have focused mostly on the equilibrium models found in game theory. Kahn's survey was limited to 2 types of equilibrium resulting from firms in oligopolistic competition: *Cournot equilibrium* , where firms compete on a quantity basis; and

Supply Function Equilibrium (SFE), where they compete on both quantity and price. Although both models are based on the Nash equilibrium concept, the Cournot approach is usually regarded as being more flexible and tractable.

Day et al. (2002) conducted a more detailed survey of the modelling literature, listing the strategic interactions that have been, or could be, included in power market models as:

- Pure Competition;
- Generalized Bertrand Strategy (Game in prices);
- Cournot Strategy (Game in quantities);
- Collusion;
- Stackelberg (Leader-follower games);
- Supply Function Equilibria;
- General Conjectural Variations; and
- Conjectured Supply Function Equilibria.

Their conclusion was that the CSF approach to modelling oligopolistic competition is more flexible than the Cournot assumption, and more computationally feasible for larger systems than the standard supply function equilibrium models. We shall discuss the CSF approach further in the next section.

Trends reported in Ventosa et al. (2005) followed similar lines, noting that most models used to evaluate the interaction of agents in wholesale electricity markets have persistently stemmed from game theory's concept of the Nash equilibrium. For the first time, however, some simulation models were included (albeit briefly). We shall discuss simulation models further in following sections.

Cournot equilibrium models

Previously, we revealed that larger generators in the NEM use quantity offers rather than price offers to improve their market positions, especially in peak periods. In other words, they seem to act like players in Cournot competition. The assumptions underlying a Cournot solution correspond to the Nash equilibrium in game theory. At the solution point, the outputs (quantities dispatched) fall into an intermediate zone between fully competitive and collusive solutions. In effect, a second firm becomes a monopolist over the demand not satisfied by the first firm, a third over the demand not satisfied by the second, and so on.

Since the Cournot solution represents a kind of imperfect collusion technique, we must ask if this equilibrium concept approximates the reality of Australia's NEM? Although generators are forbidden from changing bid prices in the rebidding process, by shifting quantity commitments up or down between different price bands they can achieve a similar effect to changing prices directly. In reality, therefore, both quantity (directly) and price (indirectly) serve as decision variables. Thus the Cournot assumption may not be appropriate for NEM generators. Furthermore, by expressing generators' offers in terms of quantities only (instead of offer curves), equilibrium prices are determined by the demand function. This shortcoming tends to reinforce the idea that Supply Function Equilibrium (SFE) approaches may be a better alternative to represent competition in the NEM (Rudkevich et al. 1998).

In an electricity market context, a Cournot solution posits rather shortsighted behaviour on the part of generating agents. It implies that each of them modifies its bids in response to the bids and dispatches of others, without allowing for the fact that others may react in a similar manner. There is no evidence that most NEM generating agents behave in this manner, although small groups of them may do so.

Supply Function Equilibrium Models

In the absence of uncertainty and knowing competitors' strategic variables, Klemperer and Meyer (1989) showed that each firm has no preference between expressing its decisions in terms of a quantity or a price, because it faces a unique residual demand. When a firm faces a range of possible residual demand curves, however, in general it expects a greater profit in return for exposing its decision tool in the form of a supply function (or offer curve) indicating those prices at which it is willing to offer various quantities to the market. This SFE approach, originally developed by Klemperer and Meyer (1989), has proven to be an attractive line of research for the analysis of equilibrium in wholesale electricity markets.

To calculate an SFE requires solving a set of *differential equations* , instead of the typical set of algebraic equations that arises in traditional equilibrium models, where strategic variables take the form of quantities or prices. Thus SFE models have considerable limitations concerning their numerical tractability. In particular, they rarely include a detailed representation of the generation system under consideration. Originally developed to address situations in which supplier response to random or highly variable demand conditions is considered, perhaps their attraction nowadays is the possibility of obtaining reasonable medium-term price estimations with the SFE methodology.

Conjectural variations approaches

A recent strategy has been to employ the Conjectural Variations (CV) approach described in traditional microeconomic theory. The CV approach can introduce some variation into Cournot-based models by changing the conjectures that generators may be expected to assume about their competitors' strategic decisions, in terms of the possibility of future reactions (CV). Day et al. (2002) suggest taking this approach in order to improve Cournot pricing in electricity markets. For example, one could assume that firms make conjectures about their residual demand elasticities or about their rivals' supply functions (Day et al. 2002). In the context of electricity markets, the latter is called the Conjectured Supply Function (CSF) approach. The CV approach can be viewed as generalising Stackelberg models (in that the conjectured response may not equal the true response). Also, superficially it resembles the SFE method described above.

Discussion

It has been argued that Cournot and Bertrand assumptions may be inappropriate for pool-type auction markets like the NEM, in which every firm bids a supply function for each generating unit. In this case, the decision variable is the bid function's parameters. For this reason, SFE has been chosen as the basis of many power market models. The resulting equilibria represent an intermediate level of competition, lying between Bertrand and Cournot solutions. However, a few additional problems remain: some equilibria are not unique; a large range of outcomes is possible; and equilibria for SFE models are difficult to calculate—indeed, none may exist.

Equilibria for SFE models have proven particularly difficult to calculate for large systems with transmission networks and a significant number of generators with limited capacity. The reasons are that the generating firm's optimisation problem on a network is inherently non-convex (and hence a challenge to solve) and an equilibrium solution may not exist. Unless strong restrictions are placed on the form of the bid functions (such as linear with only the slope or intercept being a variable), modellers have been forced to make unrealistic assumptions such as all firms having identical marginal cost functions.

Although the asserted realism of the SFE conjecture makes it attractive for markets without significant transmission constraints, it is not a practical modelling method if realistic details on demand, generation, and transmission characteristics are desired. This rules out the SFE approach for any serious representation of the NEM. Needless to say, most SFE studies have been designed for very simple systems (for example, 1–4 nodes). Alternatively, when larger networks are considered, SFE models search over only a handful of strategies to find the optimal strategy for each of 2 firms, or bids are restricted to a linear function with either fixed slope or intercept.

Simulation models

As discussed above, equilibrium models impose limitations on the representation of competition between agents in an electricity market. In addition, the resulting set of equations is frequently difficult or impossible to solve. The fact that power systems are based on the operation of generation units with complex physical constraints further complicates the situation. Simulation models are an alternative to equilibrium models when the problem under consideration is too complex to be addressed within a formal equilibrium framework. This is certainly the case in the medium to long-term, when investment decisions, hedging strategies and learning processes become important endogenous variables.

Simulation models typically represent each agent's strategic decision dynamics by a set of sequential rules that can range from scheduling generation units to constructing offer curves that include a reaction to previous offers submitted by competitors. The great advantage of a simulation approach lies in the flexibility it provides to implement almost any kind of strategic behaviour, including feedback loops between the auction market and forward contract markets. However, this freedom also requires that the assumptions embedded in the simulation be more carefully (and empirically) justified.

Some simulation models are closely related to equilibrium models. For example, Day and Bunn (2001) propose a simulation model that constructs optimal supply functions, to analyse the potential for market power in the England and Wales Pool. This approach is similar to the SFE scheme, but it provides a more flexible framework that enables us to consider actual marginal cost data and the asymmetric behaviour of firms. In this simulation model, each generation company assumes that its competitors will keep the same supply functions that they submitted the previous day. Uncertainty about the residual demand curve is due to demand variation throughout the day. The optimisation process to construct nearly optimal supply functions is based on an exhaustive search, rather than on the solution of a formal mathematical programming problem. The authors compare the results of their model for a symmetric case with linear marginal costs to those obtained under the SFE framework, which turns out to be extraordinarily similar. In the next section, we shall explore a subfield within the realm of simulation models that is attracting increased attention for the modelling of electricity markets: agent-based simulation.

Agent-based models of electricity markets

The interactions within an electricity market constitute a repeated game, whereby a process of experimentation and learning changes the behaviour of the firms in the market (Roth and Erev 1995). A computational technique that can reflect these learning processes and model the structure and market clearing mechanism (with a high level of detail) would appear to be necessary. The most promising technique at this point in time is agent-based simulation (Batten, 2000).

Agent-based simulation provides a more flexible framework to explore the influence that the repetitive interaction of participants exerts on the evolution of wholesale electricity markets like the NEM. Static models neglect the fact that agents base their decisions on the historic information accumulated due to the daily operation of market mechanisms. In other words, they have good memories and learn from past experiences (and mistakes) to improve their decision making and adapt to changes in several environments (economic, physical institutional and natural). This suggests that adaptive agent-based simulation techniques can shed light on features of electricity markets that static equilibrium models ignore.

A brief review of earlier agent-based models

Bower and Bunn (2000) present an agent-based simulation model in which generation companies are represented as autonomous adaptive agents that participate in a repetitive daily market and search for strategies that maximise their profit based on the results obtained in the previous session. Each company expresses its strategic decisions by means of the prices at which it offers the output of its plants. Every day, companies are assumed to pursue 2 main objectives: a minimum rate of utilisation for their generation portfolio and a higher profit than that of the previous day. The only information available to each generation company consists of its own profits and the hourly output of its generating units. As usual in these models, the demand side is simply represented by a linear demand curve.

Such a setting allowed the authors to test a number of potential market designs relevant for the changes that have recently occurred in England and Wales wholesale electricity market. In particular, they compared the market outcome that results under the pay-as-bid rule to that obtained when uniform pricing is assumed. Additionally, they evaluated the influence of allowing companies to submit different offers for each hour, instead of keeping them unchanged for the whole day. The conclusion is that daily bidding together with uniform pricing yields the lowest prices, whereas hourly bidding under the pay-as-bid rule leads to the highest prices.

The introduction of NETA in the UK represented a good opportunity to test the usefulness of large-scale agent-based simulation to provide some insights on market design. The agent-based platform enables a detailed description of the market, taking into account discrete supply functions, different marginal costs for each technology, and the interactions between different generators. Bunn and Oliveira (2001) followed that work by developing a simulation platform that represents, with much more detail, the way that market clearing in NETA was designed to function. This platform models the interactions between the Power Exchange and Balancing Mechanism; considers that generators may own different types of technologies; considers an active demand side, including suppliers; and takes into account the learning dynamics underlying these markets as a process by which a player selects the policy to use in the game by interacting with its opponents. In later work, they adapt and extend this simulation platform to analyse if the 2 particular generators in the Competition Commission Inquiry had gained enough market power to operate against the public interest (Bunn and Oliveira 2003).

Researchers at Argonne National Laboratory in Chicago have developed the Electricity Market Complex Adaptive System (EMCAS) model (North et al. 2002; Veselka et al. 2002). Like the above-mentioned simulation models developed at the London Business School, the EMCAS model is an electronic laboratory that probes the possible effects of market rules by simulating the strategic behaviour of participants. EMCAS agents learn from their previous experiences and modify their behaviour based on the success or failure of their previous strategies. Genetic algorithms are used to drive the adaptive learning of some agents, and pool, bilateral contract and ancillary services markets are included. The EMCAS model is arguably the most sophisticated agent-based electricity model to date, embodying more development hours than other simulation models of its type.

At Iowa State University, Leigh Tesfatsion and her colleagues have examined market power experimentally in an agent-based computational wholesale electricity market operating under different concentration and capacity conditions (Nicolaisen et al. 2001). Pricing is determined by a double auction with discriminatory midpoint pricing. Buyers and sellers use a modified Roth-Erev individual reinforcement learning algorithm to determine their price and quantity offers in each auction round. High market efficiency is generally attained, but the aggregate measures used are too crude to reflect the opportunities for exercising market power that buyers and sellers face. Their results suggest that the precise form of learning behaviour assumed may be largely irrelevant in a double auction system.

Taylor et al. (2003) developed an agent-based model to simulate the complexity of the large-scale Victorian gas market in south-eastern Australia. The model can be used to elicit possible emergent behaviour that could not be elicited otherwise under an uncertain future of deregulation and restructuring. Like an electricity market, the complexity in the gas market derives from the uncertain effects of a multiplicity of possible participant interactions in numerous segments, such as production, storage, transmission, distribution, retailing, service differentiation, wholesale trading, power generation and risk management. The agent-oriented programming platform devised for this work has the potential to overcome the limitations of traditional approaches (discussed earlier) when attempting to operate in changing environments. A similar platform has been adopted for our National Electricity Market Simulator.

Agent behaviour in Australia's national electricity market

Among the 90 registered participants in the NEM, most fall into categories based on the role they perform in the market. [3] These categories are generators,

---

[3] Some participants play more than a single role in the NEM and therefore belong to more than one category.

Transmission Network Service Providers (TNSPs), distribution network service providers (DNSPs), market network service providers (MNSPs), customers (retailers and end-users), and traders. NEMMCO, key regulators (such as NECA and the Australian Competition and Consumer Commission), and a number of other organisations (like the Council of Australian Governments and the Australian Greenhouse Office) are among the additional actors. For convenience, in the remainder of this chapter we shall refer to all of these relatively autonomous actors as *agents*.

Agents are intelligent and adaptive, meaning that they make operational and strategic decisions on the basis of the information available to them and the market's rules, and they modify their own strategies on the basis of new information that comes their way. In many decision situations, however, some agents can do no better than exhibit purposive but contingent behaviour. Although they have specific goals of their own (for example, to maximise profits, market share, utilisation factors, etc.), their ability to attain these goals is largely beyond their own control. Furthermore, earlier research on the NEM has shown that demand side participants have very little ability to influence outcomes compared with those on the supply side.

In an adaptive market such as the NEM, no single agent can control what *all* the other agents are doing. Many of the collective outcomes are not obvious, because simple summation to linear aggregates is impossible. The outcomes are governed by and dependent on a system of nonlinear interactions between individual agents and the market environment, i.e., between agents and groups of other agents or between agents and the whole market. In such limited-information situations, some agents can and do exert more influence than others. Groups of agents may benefit from informal partnerships or tacit collusion. Bidding and rebidding is unlikely to be competitive under these circumstances, since individual agents or coalitions may find it more profitable to search for opportunities to exert market power.

In electricity markets, we often find examples of locally interacting agents producing unexpected, large-scale outcomes. Problems experienced by the Californian electricity market provide a vivid illustration. During the summer of 2000, wholesale electricity prices in California were nearly 500 per cent higher than they were during the same months in 1998 or 1999 (see Figure 11.4). This explosion of prices was not only unexpected but sustained. Unlike previous price spikes observed in the US wholesale markets, the California experience proved to be more than a transient phenomenon of a few days' duration. It persisted until roughly mid-June 2001. Increases in gas prices and consumer demand, reduced availability of power imports, and higher prices for emissions permits are known to have contributed to price rises. But these market fundamentals are insufficient to explain the extraordinary gap between realised prices and competitive benchmark prices. Evidence of capacity withholding by suppliers (generators or traders) has been found, and this, together with other factors, is thought to have led to such remarkable price increases and market manipulations during 2000/2001.

Figure 11.4. Electricity prices in California, 1998-2000

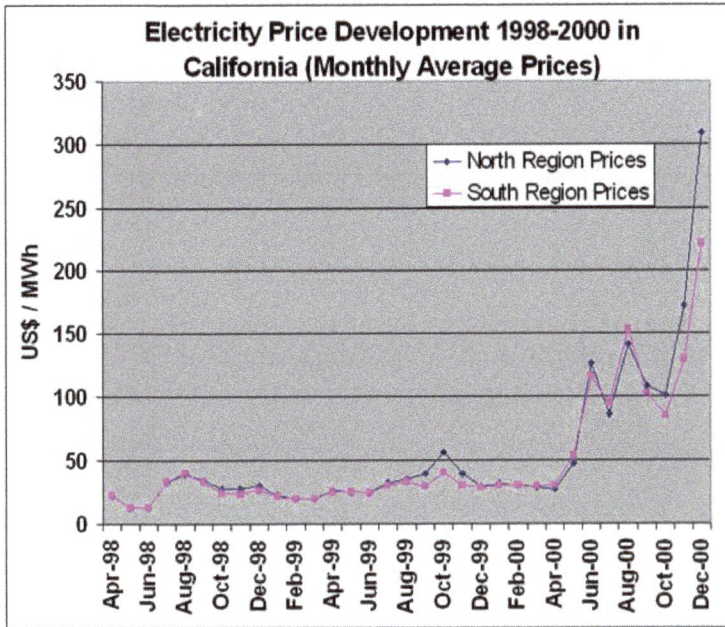

Electricity Price Development 1998-2000 in California (Monthly Average Prices)

Source: California Electricity Wholesale Price Review, 2001

Although some observers have argued that the NEM spot market is more robust to gaming than its Californian equivalent, the NEM is still vulnerable to market power.[4] For example, generator companies can profit more than customers by exploiting the bidding rules set by the National Electricity Code Administrator (NECA). Because no single body has control over market outcomes, agents exploiting anomalies can reap rich rewards. The rest of the market suffers accordingly. Recent spot prices have been observed to fluctuate from a low of around $5/MWh to a high of $10,000/MWh. Some of this volatility is caused by the weather, industrial action or diurnal/seasonal peaks and troughs in demand, but a significant proportion is not. Rather, it is an intrinsic by-product of the way the trading system and its rules operate.

---

[4] See Outhred (2000), pp.19-20.

NEMSIM: the National Electricity Market simulator

NEMSIM is an agent-based simulation model that represents Australia's NEM as an evolving system of complex interactions between human behaviour in markets, technical infrastructures and the natural environment. This simulator is the first of its kind in Australia. Users of NEMSIM will be able to explore various evolutionary pathways of the NEM under different assumptions about trading and investment opportunities, institutional changes and technological futures—including alternative learning patterns as participants grow and change. The simulated outcomes will help the user to identify futures that are eco-efficient, for example, maximising profits in a carbon-constrained future. Questions about sustainable development, market stability, infrastructure security, price volatility and greenhouse gas emissions could be explored with the help of the simulation system.

An overview of NEMSIM is shown in Figure 11.5. The NEMSIM project is part of CSIRO's Energy Transformed Flagship research program, which aims to provide innovative solutions for Australia's pressing energy needs. Motivation for the project is the Flagship's mission to develop low emission energy systems and technologies. Electricity generation is a substantial source of greenhouse gas emissions contributing 35 per cent of Australia's total emissions.

Figure 11.5. An overview of NEMSIM

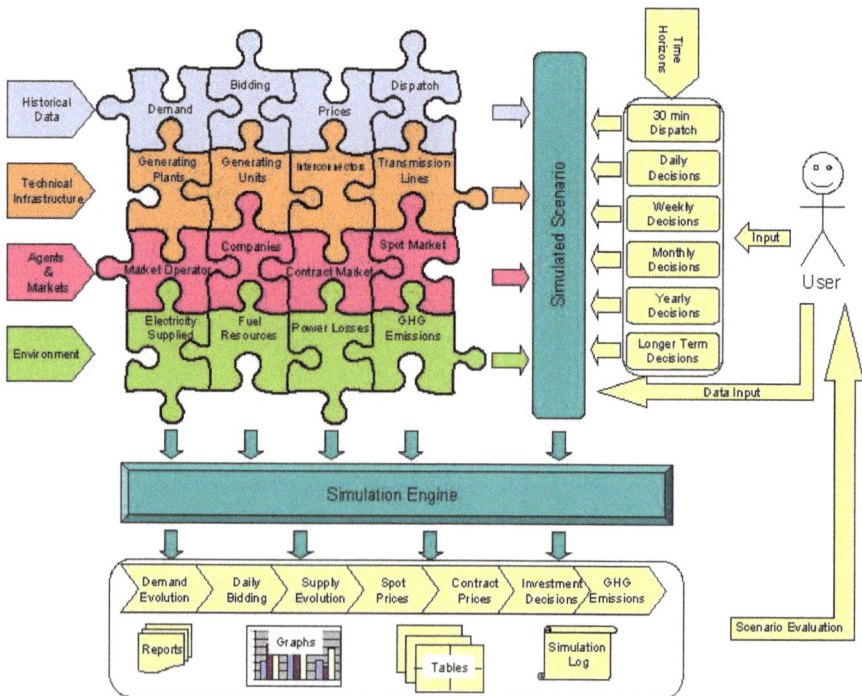

Learning from experience

The development of NEMSIM is facilitated by 6 years of historical market data on demand, pricing and power dispatch for the NEM. Being a dynamic, repetitive and information-rich system, the NEM offers a huge amount of data. NEMSIM uses this data to extract representative patterns of regional demand (on a daily, weekly and seasonal basis), regional prices, supply and demand growth, etc. The historical database includes an extensive time series of bidding data, an essential source of information about market participants and their trading strategies. In NEMSIM, silicon agents (for example, representing generator firms, network service providers, retail companies, a market operator, etc.) buy and sell electricity in a simulated trading environment. Agents have different goals. As well as maximising profits, some may wish to increase market share, diversify generation sources or work more closely with end-users. Short-term strategies (for example, bidding tactics in the NEM) are affected by medium-term strategies (for example, hedged positions in the OTC markets). In turn, both are affected by investment decisions and other changes over the longer term. NEMSIM treats agents as being uniquely intelligent, making operational and strategic decisions using the individual information available to them. Also, they are adaptive, learning to modify their behaviour in order better to realise their goals. Learning algorithms can allow agents to *look back* (learn from their historical performance), *look sideways* (learn from other participants' strategies) and *look ahead* (take future plans and forecasts into account).

In an adaptive market such as the NEM, no single agent has control over what all the other agents are doing. Because of the way the market is structured, however, the marginal bidder can exert more influence on market outcomes. The overall outcome is not always obvious, because it depends on many factors. The aim of NEMSIM is to provide a platform where a population of simulated agents interact, constrained only by realistic rules and the physical grid system. Agents' individual and collective behaviours co-evolve from the bottom up, producing both expected and unexpected emergent outcomes at the system level.

NEMSIM environments

Simulated agent life in NEMSIM unfolds in three *environments*: firstly, a *trading environment*, in which transactions can occur in interlinked spot and forward contract markets; secondly, a *physical grid* of sites, generation units, lines and interconnectors across which electricity flows, and; thirdly, a *natural environment*, which provides energy resources and accumulates greenhouse gas emissions. Each environment is separate from the agents, on which the agents operate and with which they interact.

The physical environment of transmission and distribution infrastructure imposes several constraints on electricity market operations. For example, the quantity of power sold from one region to another is constrained by the transmission capacity between the 2 regions. NEMSIM represents the diversity of objects and attributes associated with generation and transmission infrastructure in a simplified form. Some key objects (with corresponding attributes in parentheses) are: generating plants (location, unit composition), generating units (maximum capacity, generation technology, ramp rates, efficiency and emission factors) and interconnectors (transmission technology, adjacent regions, losses).

NEMSIM as a greenhouse gas emissions calculator

The effects of anthropogenic and natural Greenhouse Gas (GHG) emissions on the absorption of terrestrial radiation and global warming have been extensively studied in the last 2 decades. It is scientifically proven that changes in the concentrations of GHG can alter the balance of energy transfers between the atmosphere, land, oceans and space and that the increase concentrations of GHG will increase energy absorption by Earth, producing global warming. With a share of approximately 55 per cent, carbon dioxide (CO_2) is the main component of GHG emissions. During the last 150 years, atmospheric CO_2 has increased from about 280 parts per million by volume (ppmv) to about 372 ppmv (US Environmental Protection Agency 2004). Without intervention, it will exceed 550 ppmv by the end of this century.

There have been many efforts to mitigate and/or reduce GHG emissions, including international initiatives such as the Climate Convention and the Kyoto Protocol. However, the total anthropogenic GHG emissions are projected to continue to grow despite current initiatives, with one of the main causes being increasing demand for electricity generation.

Electricity generation is a substantial source of greenhouse gas emissions. According to the Australian Greenhouse Office (2004) it contributed 182 Mt of carbon dioxide equivalent emissions (Mt CO_2-e) in 2002, or 33 per cent of net national emissions (550 Mt CO_2-e) in Australia. The net emissions are calculated across all sectors under the accounting provisions of Kyoto Protocol for Australia.

One method of calculating the potential GHG emissions due to the future electricity generation in Australia is given in (Graham et al. 2003). This study aims to evaluate the impact of future development options of the electricity market in terms of GHG emissions and costs. It is based on a portfolio simulation framework and CSIRO's Electricity Market Model (Graham and Williams 2003) that is similar to the bottom-up type of dynamic optimisation models such as MARKAL. [5] In this chapter, 5 GHG targets (business as usual, 2 moderate and 2 extreme emissions targets) were studied in conjunction with seven key data assumptions (for example, more abundant gas available, CO_2 capture and sequestration feasible/infeasible, high/low demand growth).

NEMSIM, as an agent-based simulation tool of the NEM, is capable of calculating the GHG emissions associated with electricity generation of a given simulation scenario. The method is also bottom-up type aggregation. The advantages of NEMSIM are that its simulation framework allows quite precise modelling of the emissions up to the level of each generating unit, accommodating a variety of changes in the operational, technological, company, market and regulatory values and parameters. We believe that the main advantage in this sense is the agent-based framework that allows slow changes to accumulate over long periods and sudden changes to be introduced due to the emerging events based on the decision making and interactions of the participating agents.

Calculating the GHG emissions due to electricity generation is implemented in NEMSIM on a fossil fuel consumption basis using fuel (generation technology) specific emission factors. A set of generation technologies are modelled, for example, conventional black coal (pulverised fuel), conventional brown coal (pulverised fuel), and natural gas simple cycle. The set of generation technologies is specified in an XML input file and one example is given in Table 11.1. The XML input file format allows changes to generation technologies and their parameters to be done easily. The main attributes of each generation technology are the emission GHG factor (f) and the net energy efficiency (e), how much of the embodied energy of the fossil fuel is transformed into electricity energy.

Table 11.1. XML fragment with Generation Technologies description

```
<GenerationTechnologies>
    <GenerationTechnology id="BLACK_PF" name="Conventional black coal PF"
        emissionFactor="89.92" netEfficiency="0.376"/>
    <GenerationTechnology id="BROWN_PF" name="Conventional brown coal PF"
        emissionFactor="93.53" netEfficiency="0.28"/>
    <GenerationTechnology id="NG_SC" name="Natural Gas Simple Cycle"
        emissionFactor="51.59" netEfficiency="0.38"/>
    <GenerationTechnology id="NG_CC" name="Natural Gas Combined Cycle"
        emissionFactor="51.59" netEfficiency="0.534"/>
    <GenerationTechnology id="HYDRO" name="Hydro"
```

[5] MARKAL, developed by the International Energy Agency over a period of more than 20 years, is a generic model aiming to represent evolution of a specific energy system over a period of up to 50 years at the national, state or regional level.

```
        emissionFactor="0" netEfficiency="1"/>
    <GenerationTechnology id="WIND" name="Wind"
        emissionFactor="0" netEfficiency="1"/>
  </GenerationTechnologies>
```

Each generating unit is assigned one of the defined generation technologies. The net energy efficiency can also be defined for a selected generating unit (in that case it overrides the same attributes of the assigned generation technology) to allow flexibility in long term, when due to the different reasons the efficiency may change. GHG emissions for a given simulation period or simulation scenario are estimated by using the following formula:

$$q = \frac{3.6\,g\,f}{10^3\,e}$$

where: q is the amount of CO_2 equivalent emissions expressed in t;

g is the amount of the electricity generation expressed in MWh;

f is the emission factor for a given generation technology expressed in kt CO_2-e/PJ;

e is the net energy efficiency of the generation technology/generating unit (dimensionless);

and $3.6\ 10^{-3}$ is a conversion factor from kt/PJ to t/MWh.

The direct values of emission factors are used, which means that emissions that are associated with extraction and production of the fossil fuels used are not considered. This approach allows easier comparison between different plants, companies and regions, however, its accuracy is inferior as it does not include indirect emissions that are usually several per cent from direct and they show moderate variability by region, company and technology.

Some examples of NEMSIM output windows (in tabular and graph form) for GHG emissions are shown in Figure 11.6. They display company data and the values are illustrative only. A regional summary of GHG emissions is shown in Figure 11.7. Again the numbers are only illustrative. Other options for output windows and reports for GHG emissions are available within NEMSIM.

Figure 11.6. NEMSIM GHG emissions example windows for a generator plant: tabular form (top); graphical form (bottom)

GeneratorPlant Bayswater - CO2 tonnes

Show Format

Interval	BW01 (tonnes)	BW02 (tonnes)	BW03 (tonnes)	BW04 (tonnes)
Mon 26 Apr: 0430	210.93	210.93	210.93	210.93
Mon 26 Apr: 0500	210.93	210.93	210.93	210.93
Mon 26 Apr: 0530	210.93	210.93	210.93	210.93
Mon 26 Apr: 0600	210.93	210.93	210.93	210.93
Mon 26 Apr: 0630	198.02	198.02	210.93	210.93
Mon 26 Apr: 0700	210.93	210.93	210.93	210.93
Mon 26 Apr: 0730	198.02	198.02	198.02	210.93
Mon 26 Apr: 0800	210.93	210.93	210.93	210.93
Mon 26 Apr: 0830	210.93	210.93	210.93	210.93
Mon 26 Apr: 0900	198.02	198.02	198.02	210.93
Mon 26 Apr: 0930	210.93	210.93	210.93	210.93
Mon 26 Apr: 1000	198.02	198.02	198.02	210.93
Mon 26 Apr: 1030	210.93	210.93	210.93	210.93
Mon 26 Apr: 1100	198.02	198.02	198.02	198.02
Mon 26 Apr: 1130	198.02	198.02	198.02	198.02
Mon 26 Apr: 1200	198.02	198.02	198.02	198.02
Mon 26 Apr: 1230	198.02	198.02	198.02	198.02
Mon 26 Apr: 1300	198.02	198.02	198.02	198.02

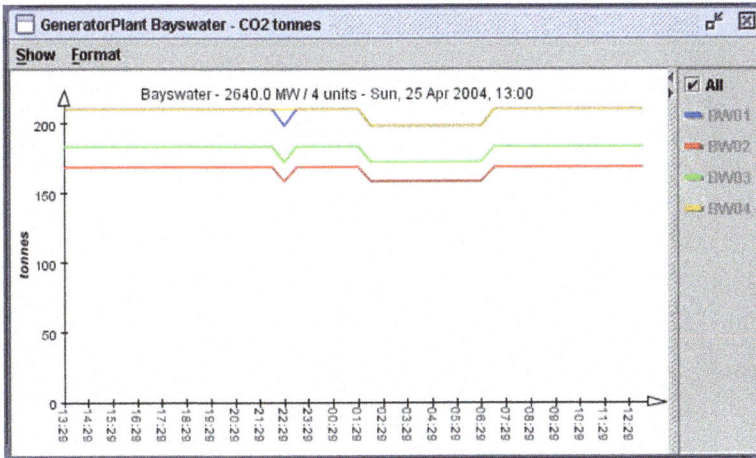

GeneratorPlant Bayswater - CO2 tonnes

Show Format

Baywater - 2640.0 MW / 4 units - Sun, 25 Apr 2004, 13:00

tonnes

All
BW01
BW02
BW03
BW04

Figure 11.7. NEMSIM regional summary window for GHG emissions

Time	New South Wales	Queensland	Snowy	South Australia	Tasmania	Victoria
Sun 25 Apr. 1730	5365.13	4546.14	0.0	902.17	0.0	4054.66
Sun 25 Apr. 1800	5365.13	4534.05	0.0	893.62	0.0	4385.35
Sun 25 Apr. 1830	6075.4	5303.93	0.0	867.84	0.0	4716.05
Sun 25 Apr. 1900	6075.4	5098.78	0.0	915.1	0.0	4716.05
Sun 25 Apr. 1930	6075.4	5009.61	0.0	959.85	0.0	4716.05
Sun 25 Apr. 2000	5838.64	4407.6	0.0	964.35	0.0	4716.05
Sun 25 Apr. 2030	5838.64	4569.81	0.0	949.56	0.0	4716.05
Sun 25 Apr. 2100	5838.64	4350.17	0.0	977.92	0.0	4716.05
Sun 25 Apr. 2130	6075.4	5455.05	0.0	949.2	0.0	4716.05
Sun 25 Apr. 2200	7700.98	3707.74	0.0	1210.59	0.0	4054.66
Sun 25 Apr. 2230	6803.78	3707.74	0.0	1330.8	0.0	3721.03
Sun 25 Apr. 2300	7117.16	3707.74	0.0	1328.76	0.0	3625.42
Sun 25 Apr. 2330	6797.49	3707.74	0.0	1328.01	0.0	3553.27
Mon 26 Apr. 0000	7557.34	4197.5	0.0	1330.91	0.0	3481.11
Mon 26 Apr. 0030	6490.4	3707.74	0.0	1317.54	0.0	3517.19
Mon 26 Apr. 0100	7269.58	3707.74	0.0	1200.74	0.0	3711.25
Mon 26 Apr. 0130	7355.75	3707.74	0.0	1331.51	0.0	3481.11
Mon 26 Apr. 0200	6785.4	3707.74	0.0	1200.21	0.0	3707.83

Region Summary - CO2 Emissions tonnes — Show

Simulated outcomes

The purpose of NEMSIM is not to predict the future, but to generate and explore the various alternative futures that could develop under different conditions. NEMSIM generates various *What if?* scenarios under different input definitions. The simulator can also be used to show the possible evolutionary trajectories of a given scenario under given conditions. For example, the introduction of more distributed generation into the marketplace involves a transition from the current paradigm of the centrally-dispatched electricity grid to new, more decentralised ones. This may involve new markets, new brokers, new technology and new grid structures. NEMSIM is a generative tool that can identify the transition states needed to reach specific final states such as these.

References

Australian Greenhouse Office (2004) National Greenhouse Gas Inventory–2002. Factsheet 1. Energy: Stationary sources and fugitive emissions. Canberra: Commonwealth of Australia.

Barmack, M.A. (2003) What do the ISO's public bid data reveal about the California market? *The Electricity Journal* 24: 63–73.

Batten, D.F. (2000) *Discovering Artificial Economics: How Agents Learn and Economies Evolve*. New York: Westview Press.

Borenstein, S., J. Bushnell and F. Wolak (2002) Measuring market inefficiencies in California's restructured wholesale electricity market. *American Economic Review* 92(5): 1376–405.

Bower, J. and D.W. Bunn (2000) A model-based comparison of pool and bilateral market mechanisms for electricity trading. *The Energy Journal* 21(3): 1–29.

Bunn, D. W. and F. S. Oliveira (2001) Agent-based simulation: an application to new electricity trading arrangements of England and Wales. *IEEE Transactions on Evolutionary Computation* 5(5): 493–503.

Bunn, D.W. and F. S. Oliveira (2003) Evaluating individual market power in electricity markets via agent-based simulation. *Annals of Operations Research* 121: 57–77.

Day, C. and D. Bunn (2001) Divestiture of generation assets in the electricity pool of England and Wales: a computational approach to analyzing market power. *Journal of Regulatory Economics* 19 (2): 123–41.

Day, C.J., B.F. Hobbs and J.S. Pang (2002) Oligopolistic competition in power networks: a conjectured supply function approach. *IEEE Transactions on Power Systems* 17(3): 597–607.

Graham, P., P. Coombes, N. Dave, D. Vincent, G. Duffy (2003) Options for electricity generation in Australia. Technology Assessment Report 31, Cooperative Research Centre for Coal in Sustainable Development.

Graham, P.W. and D.J. Williams (2003) Optimal technological choices in meeting Australian energy policy goals. *Energy Economics* 25(6): 691–712.

Green, R.J. and D.M. Newbery (1992) Competition in the British electricity spot market. *Journal of Political Economy* 100 (5): 929–53.

Hu, X., G. Grozev and D. Batten (2005) Empirical observations of bidding patterns in Australia's national electricity market. *Energy Policy* 33(16): 2075–86.

Kahn, E. (1998) Numerical techniques for analyzing market power in electricity. *Electricity Journal* 11(6): 34–43.

Klemperer, P. and M. Meyer (1989) Supply function equilibria in oligopoly under uncertainty. *Econometrica* 57(6): 1243–77.

MacGill, I., H. Outhred and K. Nolles (2006) Some design lessons from market-based greenhouse gas regulation in the restructured Australian electricity industry. *Energy Policy* 34(1): 11–25.

Nicolaisen, J, V. Petrov and L. Tesfatsion (2001) Market power and efficiency in a computational electricity market with discriminatory double auction pricing. ISU Economic Report No. 52, August.

North, M., G. Conzelmann, V. Koritarov, C. Macal, P. Thimmapuram and T. Vaselka (2002) E-laboratories: agent-based modelling of electricity markets. American Power Conference, Chicago, April 15–17.

Outhred, H. (2000) The competitive market for electricity in Australia: why it works so well. *Proceedings of the 33rd Hawaii International Conference on System Sciences*, Hawaii.

Roth, A.E. and I. Erev (1995) Learning in extensive form games: experimental data and simple dynamic models in the intermediate term. *Games and Economic Behaviour* 8:164–212.

Rudkevich, A., M. Duckworth and R. Rosen (1998) Modelling electricity pricing in a deregulated generation industry: the potential for oligopoly pricing in a Poolco. *The Energy Journal* 19(3): 19–48.

Short, C. and A. Swan (2002) Competition in the Australian national electricity market. *ABARE Current Issues*, January, 1–12.

Taylor, N.F., M.P. Harding, G.S. Lewis and M.G. Nicholls (2003) Agent-based simulation of strategic competition in the deregulating Victorian gas market. Paper presented at the 2003 WDSI Conference.

US Environmental Protection Agency (2004) US Emissions Inventory 2004: Inventory of U.S. Greenhouse Gas Emissions and Sinks: 1990–2002.

Ventosa, M., A. Baillo, A Ramos and M. Rivier (2005) Electricity market modeling trends. *Energy Policy* 33: 897–913.

Veselka, T., G. Boyd, G. Conzelmann, V. Koritarov, C. Macal, M. North, B. Schoepfle and P. Thimmapuram (2002) Simulating the behaviour of electricity markets with an agent-based methodology: the electricity market complex adaptive system (EMCAS) model. International Association for Energy Economists (IAEE) North American Conference, 6–8 October, Vancouver, B.C.

Wolak, F.A. (1999) *An empirical analysis of the impact of hedge contracts on bidding behaviour in a competitive electricity market*. Stanford, California: Stanford University.

Wolfram, C.D. (1999) Measuring duopoly power in the British electricity market. *American Economic Review*, 89(4): 805–26.

Viewpoint from a Regional Adviser

Within Pacific island countries, fresh water is essential to human existence and a major requirement in agricultural and other commercial production systems. The water resources of small island countries are fragile due to their small size, lack of natural storage and competing land use. They are extremely vulnerable to natural and anthropogenic hazards, including droughts, cyclones and urban pollution.

Kiribati is one of the small island developing states in the Pacific, consisting of 33 islands spread over a territorial area of over 3 million square kilometres. The total land area amounts to only 726 square kilometres and all but one of the 33 islands are low-lying coral atolls. South Tarawa supports the highest proportion of the Kiribati population with 43 per cent, approximately 45,000 people, living on a stretch of land of around 18 square kilometres.

South-Tarawa's water supply is sourced from groundwater lenses pumped from galleries on water reserves allocated by the government on adjacent islands. The potable water supply from the existing system, which is based on only 30 L/capita/day, is insufficient. Most of the freshwater lenses on South Tarawa have been polluted and augmentation by rainwater collection at the household level is not widespread enough. The occurrence of water-born diseases such as diarrhoea can be attributed to people still using shallow, open hand-dug wells which are contaminated by leaking sewage systems, soak pits or pig pens.

Given the rising demand for a sustainable urban water supply, the development of additional water resources is a government priority. Land issues compounded by the reality of land shortage and complex family land ownership has meant that the water reserves set aside for public water supply have been under increasing pressure from squatters and agricultural uses, leading to conflicts and vandalism of public assets.

Groundwater studies are currently executed to study the feasibility of creating additional water reserves on two other islands. This provided an excellent opportunity for the team of CIRAD and ANU to study the social impact of this development and assist the Government of Kiribati and other stakeholders to avoid the problems encountered on the existing reserves.

The authors have shown that a useful dialogue can be generated through innovative approaches such as multi-agent-based simulations coupled with role-playing games. They demonstrated that the complex situation of the management of Tarawa's water resources can be captured in a model that allows the collection, understanding and merging of viewpoints from different stakeholders.

Small island hydrology may be considered as relatively simple, using straightforward water balance models to determine the sustainable yield from a freshwater lens that forms naturally under an island according to the Ghyben-Herzberg principle. In contrast, the management of water resources on a densely populated island such as South-Tarawa is quite complex due to the specific socio-political and cultural structures relating to traditional community practices, rights and interests, which are interwoven with colonial and 'modern' practices and instruments.

The surveys carried out in the first stage of the project resulted in a better understanding of people's perception of water management and sanitation issues. This provided a good basis for the agent-based model developed in stage 2 and the role-playing game of stage 3 that managed to engage government as well as local community stakeholders in a difficult though meaningful dialogue. I will be eager to learn of the outcomes of stages 4 and 5, which will see the formalisation of different scenarios and final discussions with the different stakeholders.

Marc Overmars BSc PhD
Water Resource Adviser

South Pacific Appied Geoscience Commission (SOPAC)
Suva, Fiji

12. AtollGame: A Companion Modelling Experience in the Pacific

Anne Dray, Pascal Perez, Christophe LePage, Patrick D'Aquino and Ian White

Abstract

Multi-Agent Systems (MAS) have been developed to study the interaction between societies and the environment. Here we use MAS in conjunction with a Companion Modelling (ComMod) approach to develop a Negotiation Support System for groundwater management in Tarawa, Republic of Kiribati. In agreement with the complex and dynamic nature of the processes under study, the ComMod approach requires a permanent and iterative confrontation between theories and field circumstances. Therefore, it is based on repetitive back and forth steps between the model and the field situation. The methodology applied in Tarawa relies on 3 successive stages. First, a Global Targeted Appraisal focus on social group leaders in order to collect different standpoints and their articulated mental models. These collective models are partly validated through Individual Activities Surveys focusing behavioural patterns of individual islanders. Then, these models are merged into a single conceptual model that is further simplified in order to create a role-playing game. This game is played during iterative sessions, generating innovative rules and scenarios. Finally, when the rules become too complex, a computer based version of the game replaces the board version. Stakeholders can explore the possible futures of freshwater management in Tarawa and eventually agree on an equitable collective solution.

Introduction

Low coral islands are heavily dependent on groundwater for freshwater supplies. The availability, quality, and management of groundwater are central to sustainable development and poverty alleviation in many developing small island nations. Increasing populations, growing per capita demand and restricted land areas limit water availability and generate conflicts (Falkland and Brunel 1993).

This study is carried out in the Republic of Kiribati, on the low-lying atoll of Tarawa (Figure 12.1). The water resources are predominantly located in freshwater lenses on the largest islands of the atoll. The water table is typically 0.8 m to 1.6 m below ground surface. Groundwater is supplemented by rainwater on most of these islands. South Tarawa is the capital and main population centre of the Republic. The water supply for the urban area of South Tarawa is pumped from horizontal infiltration galleries in groundwater protection zones called *water reserves* on Bonriki and Buota islands. These currently supply about 1300 m³/day, equivalent to about 30 l/capita/day of freshwater, representing 60 per cent of the needs of South Tarawa's communities. Rainwater tanks and local private wells supply the rest (White et al. 2002).

Figure 12.1. Tarawa Atoll (Bonriki and Buota islands are on the lower right of the atoll)

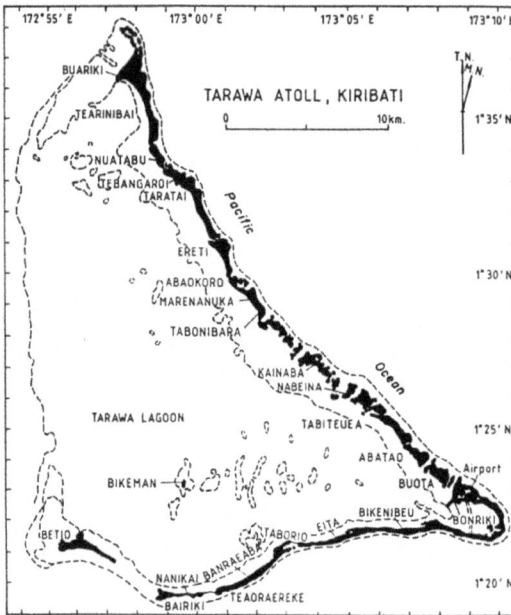

The government's declaration of water reserves over privately owned land has lead to conflicts, illegal settlements and vandalism of public assets. Water consumption per capita tends to increase towards western-like levels, threatening the sustainability of the system. As well, pollution generated by the 45,000 inhabitants of South Tarawa has already contaminated all the freshwater lenses, with the exception of Buota and Bonriki reserves (White et al. 1999).

The government is now conducting intensive groundwater investigations on the islands of Abatao and Tabiteuea in order to delineate the freshwater lenses and to provide more accurate estimates of the sustainable yields from these islands. Depending on the results of the investigations and community discussions, the groundwater resources could be partly used to supplement the water supplies from Bonriki and Buota to South Tarawa. However, already available information underlines the necessity to take into account the social impact of such implementation, in order to avoid the problems encountered on Bonriki and Buota.

Our project aims at providing the relevant information to the local actors, including institutional and local community representatives, in order to facilitate dialogue and to help devise together sustainable and equitable water management practices. Multi Agent-Based Simulations (MABS) coupled with a role-playing game have been implemented to fulfill this aim. They provide powerful tools for studying interactions between societies and their environment (Bousquet et al. 2002). They have the potential to greatly reduce conflict over natural resource management and resource allocation. In order to collect, understand and merge viewpoints coming from different stakeholders, the following 5-stage methodology is applied: collecting local and expert knowledge; blending the different viewpoints into a game-based model; playing the game with the different stakeholders; formalising the different scenarios investigated in computer simulations; and exploring the simulated outcomes with the different stakeholders. The first 3 stages are described successively in this chapter.

Collective knowledge

Theoretical assumptions

We acknowledge the constructivist theory in socio-psychology and believe that the nature of individual representations is socially constructed through people's interactions with their physical and social environment (Descola 1996). We agree on the fact that these adaptive mental models can be partly elicited through Knowledge Engineering-based techniques and translated into conceptual models (Becu et al. 2003). We assume that social groups carry collective representations of their environment and that these mental models can be partly elicited from wisely selected representatives. We argue that individuals belonging to the same group share the same representation, but their behaviour is driven by personal motivations and tacit knowledge. Thus, they can temporarily dismiss part of the shared representation.

Our methodology includes 2 sets of interview. The first , called Global Targeted Appraisal (GTA), is focusing on groups' representatives, or *Group Voices*, and its objective isto elicit representations rather then individual behaviours. The second, called Individual Activities Survey (IAS), is conducted with individuals belonging to the same groups.

Global targeted appraisal

Prior to the interviews, a short survey helped selecting relevant spatial and social groups, and understanding their hierarchical links in order to identify 30 *Group Voices* belonging to different religious, cultural, administrative, educational and gender groups. They were then interviewed individually at their place through semi-structured interviews, in order to highlight their understanding of the main interactions between local people and water resources. The interview was divided into three exercises:

- exercise 1: photo interpretation dealing with Tarawa overall features and issues;
- exercise 2: cognitive mapping focusing on the interviewee's home island; and
- exercise 3: card game focusing on water cycle and human use interactions.

For the first exercise, 4 successive groups of photos referring to different aspects of Tarawa's environment and activities were given to the interviewee. For each group, the interviewer first gave the general topic of discussion and asked the interviewee to describe important elements in the photos related with the topic (Figure 12.2). The topics to be discussed with the interviewee were: population, landuse and landownership; social and economic activities; climate and environment; water resources and water use; and environmental pollution and water quality.

Figure 12.2. Elder man in Abatao commenting on photos of economic activities

For the second exercise, the interviewee was provided with a sheet of paper and asked to draw the location or the spatial distribution of the island's key features. Key features were grouped according to topics largely overlapping the ones described for South Tarawa. The interviewer interacted directly on the map with the interviewee. When one group of key features had been displayed on the map, prompting questions used during the first step were reused in order to crosscheck information consistency and to outline the island's specificities.

The third exercise focused on the way the interviewee understood and represented water management processes. Water management included natural water cycles along with human activities (consumption, pollution, protection). It was based on a card game between the interviewer and the interviewee (Figure 12.3). The interviewer had 60 cards in his hands, representing elements that seemed important for water management processes. At first, the cards *Coconut*, *Rainfall* and *Groundwater* were displayed on a board, and the interviewee was asked to provide other elements in order to complete the natural water cycle. The interviewee was also asked to draw links between cards (action, relation, impact) and to describe them in a few words. The second set of cards displayed was: *Household*, *Pig* and *Vegetable Crop*. The topic was about direct water management from local settlers. The last set of cards included *Government*, *PUB* (the local water agency) and *South Tarawa Residents*. This part focused more on water management at the institutional level.

Figure 12.3. Example of Card Game's flowchart

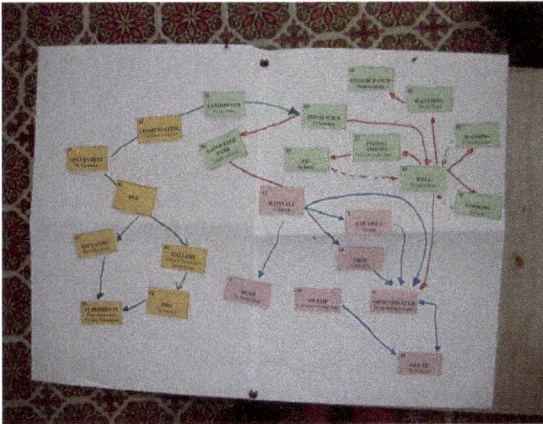

natural water cycle elements (pink), human activity elements (green), and institutional elements (orange)

Individual activities survey

The objectives of this second field survey were to partly validate the models elicited during the Global Targeted Appraisal (GTA) and to quantify some relationships already described by the *Group Voices*. In order to achieve these objectives, 24 people were selected from among the 4 islands and submitted to specific individual interviews. This time, the interview relied upon a more structured questionnaire form, focusing on local facts and personal activities and behaviour. Interviews were generally held in the central house or shelter with viewing distance to the well and other domestic facilities. Questions were grouped within 5 topics, which were very similar to the ones used in the GTA: demography and landownership; activities and landuse; water resource and water use; improving water-use and sanitation; and water reserve.

Processing elicited knowledge

Results from the GTA

First exercise

The first exercise of the GTA intended to establish confidence with the interviewees and to let them browse general topics without focusing immediately on sensitive local issues. Thus, we started from demographic evolution and constraints and ended with global problems of pollution on South Tarawa. From a psychological viewpoint, we wanted to confirm the ability of the interviewees to develop dynamic rationales from static material (photos). A second objective was to test our ability to elicit their mental constructs.

Some interviewees made extensive use of the photos during the interview, others focused mainly on one of them to develop their rationale. On several occasions, interviewees developed their ideas without referring to the photos at all, in some sort of story telling mode. Pre-selected prompting questions were instrumental in keeping the discussion alive during most of the interviews. The *Transcript Analysis* technique was adapted and applied to each recorded interview (Newell 1982; Shadbolt and Milton 1999). Three lists of elements (words or group of words) were completed: social/institutional elements, spatial/geographical elements, and passive/biological elements. Semantic links between these elements were defined according to the grammatical structure of the transcript (Figure 12.4).

Figure 12.4. Partial view over an associative network

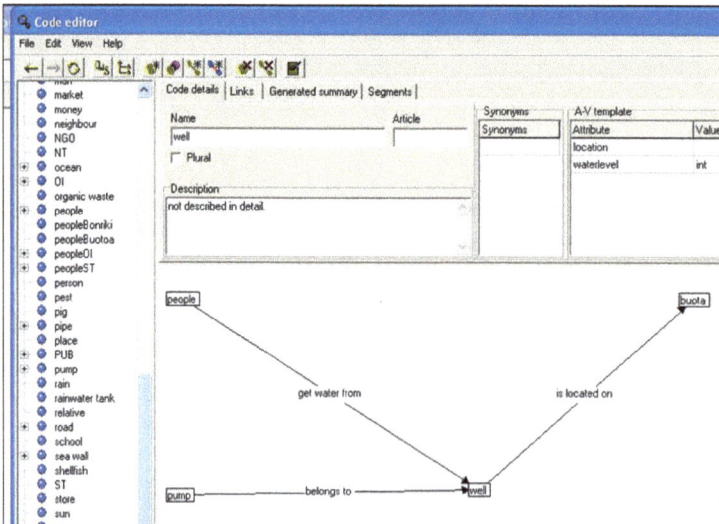

Codes and links are individually labelled

All interviewees were fully aware of the population increase on South Tarawa and its impact on society and environment. Increased pollution, land pressure and alcoholism linked with unemployment were the main issues described. Most interviewees stressed the fact that climate pattern and weather features had changed over time, but concerns about coastal erosion and sea level rise issues were not yet shared in the population. The concepts of soil water infiltration and groundwater recharge were well admitted by a large majority of the interviewees. However, the concept of evaporation was still fairly vague.

Even if the traditional view *my land/my well/my water* was still recognised, the science-driven concept of *one island/one water* was widely accepted. The impact of tidal wave and flood/drought on the water lens had often been mentioned and discussed. Illegal tapping into the PUB pipes, and illegal settling on the water reserves, were issues openly discussed by several interviewees. With regard to pollution, most interviewees mentioned solid waste in the first place, especially cans and tins. They also acknowledged the impact of pigs on the environment, as they were causing bad smells and spreading disease. Only half of the interviewees perceived the direct link with water lens contamination. Human wastes were often not mentioned at all.

Second exercise

The second exercise of the GTA intended to focus on specific issues related to the interviewee's island. Most of the time, it was mainly an update of more general considerations expressed during the first exercise, and the discussion often concentrated on the specific features of the island's society and environment. From a psychological viewpoint, our objective was to confirm the interviewees' ability to represent spatial entities, to manipulate these entities on the map and to describe dynamic processes directly on the map. These were essential elements to validate the use of a role-playing game later on. A striking outcome from the exercise was that the 4 islands not only constituted a spatial continuum between South and North Tarawa, they represented a temporal bridge between present and past time as well. Social, environmental and economic evolution on Bonriki was largely replicated on the 3 other islands. It was especially true regarding land tenure issues.

Regarding water issues—remembering that we didn't ask the interviewees about their own feelings but more generally about the thoughts of local people—the situation was contrasted. Where water reserves already existed, angry and happy people coexisted. The opposition mainly focused on the level of financial compensation (land lease), but one has to recognise that this claim was strongly backed by a barely sustainable pressure on land tenure. This pressure had already boosted the land market to unexpected levels. Access to their water through the PUB reticulated system was also described as a fair claim. On Abatao and

Tabiteua, the fears crystallised around the landownership issues. Beyond the financial bargaining game, traditional livelihood and environmental harmony were genuine arguments. Bonriki and Buota cases were often taken as examples of the way things shouldn't be done.

Third exercise

The last exercise was meant to directly elicit individual knowledge of water management processes. The card game was designed to disaggregate these representations into unit elements and causal links. As mentioned previously, some of the elements had been discussed during the first exercise (in a broader context), and it was interesting to observe whether these elements might reappear during the card game. From a psychological viewpoint, beyond the ability of the interviewees to disaggregate their mental constructs into basic elements, we were interested in verifying their acceptance to play the game by the rule and their capability of justifying a given choice. Again, these hints were valuable in the perspective of a role-playing game implementation.

The story was divided into 3 stages of evolution (pristine island, remote settlers and interconnected islands), and cards were distributed at each stage. A first analysis consisted in counting the number of cards and noting the elements cited. Overall, the interviewees took an average of 28 cards out of 60, and most of them were much more comfortable with the second theme. The reference to their daily life obviously helped to build useful analogies, whereas the level of abstraction required with themes 1 and 3 was more challenging. An interesting outcome was the fact that a large majority of people were able to describe an infiltration-like process linking the rainfall input with the groundwater recharge. The freshwater lens itself was seldom perceived as a specific entity, but rather an attribute of the soil itself (*water in the soil*). The sanitation elements were often skipped from the representation as if they were not part of the water cycle. This was relatively surprising as most of these people referred, when prompted, to pollution mechanisms during the photo interpretation exercise. Finally, the interaction with South Tarawa was described through engineering elements and links. The institutional and legal aspects were merely represented through negotiation between landowners and government (*Compensation* card).

Different table breakdowns gave more details about specific socio-cultural subgroups (Figure 12.5). First, higher educated people had a better understanding of complex processes (*Evaporation, Taxation*), and a better awareness of modern technologies (*Pump, Gallery*). This group hardly overlaps with a precise age group. If the young/middle age group mentioned *Evaporation* more often then their elders, this group was mainly characterised by a strong focus on financial negotiations with the government about the water reserves (*Landowner and Compensation* elements).

Figure 12.5. Overall table ranking most quoted elements during the card game

Overall	Stage 1		Stage 2		Stage 3	
	mean	top elements	mean	top elements	mean	top elements
overall (26)	9	Tree 85% Soil 69% Ocean 69% Infiltrating 50% Evaporating 46%	12	Watering 100% Well 96% Cooking 96% Washing 88% Feeding 85%	8	Pipe 96% PUB Tap 65% Gallery 65% Compens. 65% Reserve 62%

We were interested also in checking whether the government expert and council member groups would display specific features. So far, the latter group didn't show any specificity compared with the overall results, but the first group clearly demonstrated its (partial) belonging to the high education group (*Evaporation* and *Taxation* elements). Beyond this characteristic, the experts were almost the only ones to mention the pollution and sanitation elements in the water cycle (*Pollution*).

In conclusion, the GTA allowed us to collect valuable first-hand information. This information was used to design a bottom-line model of freshwater management on the atoll. A Unified Modelling Language (UML) based version of this model was used as a template to build a role-playing game, called *AtollGame* (Figure 12.6). A critical question was whether people shared equivalent representations of the hydrogeological processes. The GTA successfully demonstrated that almost everybody was able to describe an infiltration process leading to the recharge of an island-wide freshwater lens. This was of prime importance in order to step into the next phase of the negotiations.

Figure 12.6. UML-based Class Diagram representation of the common ontology

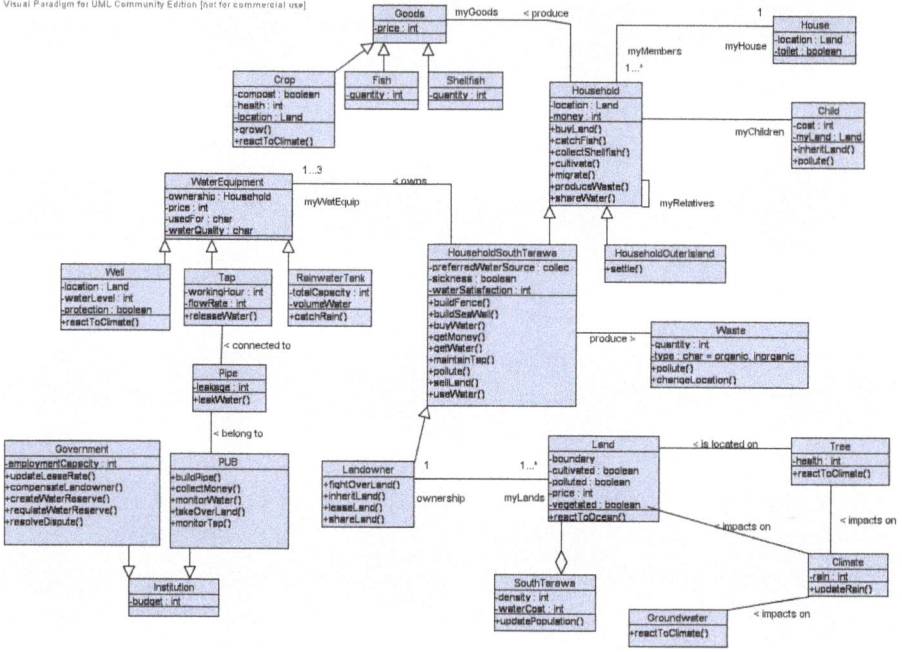

Finally, the ability of the interviewees to explain their mental constructs, to manipulate spatial entities on the map, to break down complex systems into simple elements and to accept to play games by the rules, confirmed the potential of success of a role-playing game during the next phase.

Results from the IAS

As mentioned in the methodology, the IAS was designed to partly validate the models elicited during the GTA, and to quantify some relationships described during the GTA. 24 people, all but 3 of whom were landowners, were interviewed.

Family members ranged from 3 to 18 on Tabiteua and Abatao, and from 3 to 16 on Buota and Bonriki. On an average, each family was composed of 8 members. We considered family to be the close family living in the same compound and sharing some facilities. Most of the time it included 3 generations from direct descend. The average size of a block of land was 2.5 acres. Most of the time, the blocks were collectively owned by different members of the family (between 2 and 10). The maximum size of a block was approximately 6 acres. Very often, other related families, according to the *extended family* concept, also occupied these blocks. On average, 3 other families were living on the same block as the interviewee, although the average value does not express the diversity (between 0 and 10).

In terms of employment, there was a huge difference between Tabiteua/Abatao where only one interviewee relied on a regular job and Buota/Bonriki, where nearly 75 per cent of the families had at least one regular income. On average, they generated AU$425 of monthly income. Subsistence economy was based on fishing (fish, shellfish), local product manufacturing (thatching, string), or local food marketing (coconut, papaya, toddy). On average, they generated a mere AU$140/week, but it was fairly hazardous to extrapolate this figure into a monthly income as most of these activities are irregular and subject to fundraising initiatives. Special attention was given to vegetable cropping activities. They mainly concerned Abatao (100 per cent of interviewees) and Bonriki (50 per cent of interviewees), where the average income jumped towards AU$250/week with a more regular pattern controlled by market niches (schools, hospital, and restaurants).

Each interviewee, with one exception, had access to a personal well. Again, there was a large difference in the use of alternative water sources between Tabiteua/Abatao (only one person collecting rainwater in drums) and Buota/Bonriki, where water collection was more diversified (38 per cent use of drums, 19 per cent use of PUB water, and 11 per cent use of rainwater tanks). On an average, families used 460 l of water daily. Given an average family size of 8 people, the consequent 58 l/person/day represented a rather high estimate of the usual figures quoted in the literature (between 30 and 50 l/person/day). One has to remember that most of the interviewees didn't face any problem of water availability, and that most of them enjoyed very good water quality. Hence, a resource with free access didn't limit consumption.

Furthermore, most interviewees confessed that everyone in the family, including children, was getting water from the well on an instant demand basis (90 per cent of cases). Consumption from families regularly watering their vegetable plots jumped to an average 100 l/person/day. Nearly 95 per cent of the interviewees considered their water as safe and didn't recall any health incident linked with water quality. Nobody on Tabiteua/Abatao considered that their island faced pollution problems, whereas 45 per cent on Buota/Bonriki complained about an increasing threat. It was an interesting that:

- 95 per cent recognised that grey water was just thrown away around the house;
- 25 per cent recognised that they were discarding solid waste on the beach; and
- 30 per cent recognised that they were discarding solid wastes in the bush (pit, hole).

Only a few interviewees mentioned watering the garden with grey water or composting the domestic waste. Nevertheless, 29 per cent admitted to regularly burning solid wastes. Regarding the Water Reserve issues, responses were slightly different for Tabiteua /Abatao (forecasted) and Buota/Bonriki (implemented). 44 per cent of the interviewees on Buota/Bonriki remembered that the government provided information about pumping, 31 per cent were not present at that time and only 25 per cent denied receiving any information. Nevertheless, nearly 50 per cent were not satisfied with the implementation by the government, 19 per cent had no comment, and only 31 per cent were satisfied with the implementation. In order to agree on the actual pumping, 31 per cent mentioned higher financial compensation, 19 per cent access to PUB water and facilities, 19 per cent requested the recognition of their rights. 44 per cent expressed strong doubts about the government's ability to manage properly the water pumping and distribution, whereas 25 per cent felt fairly confident about it, and 31 per cent had no opinion. On Tabiteua/Abatao the situation was clearer as we were discussing an ongoing process. All of the interviewees recognised that the government informed them beforehand, but 63 per cent disagreed with the implementation as they felt forced into it. While 50 per cent of the interviewees would like to enter financial negotiations with the government, 38 per cent remained strongly opposed to any move towards a water reserve on their island.

In conclusion, we would like to underline the fact that the IAS confirmed the fact that sanitation issues were largely disconnected from the water management consideration for most of the interviewees. Hence, we confirmed that the gap in the mental constructs elicited during the GTA largely overlapped with the behavioural models displayed during the IAS.

AtollGame: the model

Elements of design

As previously stated, the agent-based model and the corresponding role-playing game were designed according to the different viewpoints, converging or conflicting, recorded during the interviews. Freshwater lenses were perceived as global and undivided resources on each island, but few interviewees described the lens entity and its properties. Therefore, we decided to implement a very simple reservoir-like entity in the model. Through different descriptions, water infiltration into the soil is acknowledged by a majority of interviewees. Water uptake by vegetation and evapotranspiration processes were perceived far less and so we decided to use a very simple water balance model linking the groundwater with the atmosphere. Seasonality of the climate was differently perceived, and we decided to create a quasi-random rainfall allocation rule. Sea level rise and global change influences were kept as scenarios to explore.

Daily water use was generally described in same terms. Only a few interviewees linked solid wastes and grey water production to the infiltration process. We decided to focus mainly on daily water demand, and to let the sanitation issues arise from the game playing sessions later. Landowners, traditional or new buyers, are the essential actors in the negotiations with the government and it was decided that each active agent in the model, or player in the game, had to become a local landowner. The connection between land tenure issues and water management was an essential element. It drives the land-use restrictions and land leases discussions. For this reason it was decided to design the model around conflicting land and water allocation rules.

The population increase, mainly through immigration, was perceived as a threat in terms of water consumption, pollution generation and pressure on the land, and it was decided to submit the agents/players to increasing numbers of new settlers on their land. Financial issues linked with water management mainly dealt with land leases, equipment investment and, seldom, with water pricing. It was decided to allocate different types of income to the agents/players in order to activate these mechanisms.

The virtual landscape

AtollGame was created with VisualWorks, using the CORMAS platform developed by Bousquet et al. (1998). AtollGame includes:

- Spatial active entities: *AtollCell;*
- Social entities: *Household* and *AgentPUB;* and
- Spatial passive entities: *Landuse, Wateruse* and *WaterBalance.*

Two 80 acre virtual islands were created, each one on a 25x45 regular spatial grid with hexagonal cells (Figure 12.7). Each unit cell corresponds to a 490m² land area. We decided to work on environments representing virtual islands in order to prevent stakeholders from feeling personally tackled. Island 1 corresponds to a low population island (50 families) where the government is already pumping freshwater. Island 2 corresponds to an overcrowded island (200 families), already polluted and depending on a freshwater adduction pipe for drinking water. Each hexagonal cell provides isotropy properties used to model freshwater lenses by generating isopiezometric circles. Each cell holds one *Landuse* passive entity, which can be tree, crop or bare soil type, and one *WaterBalance* passive entity.

Figure 12.7. AtollGame environment

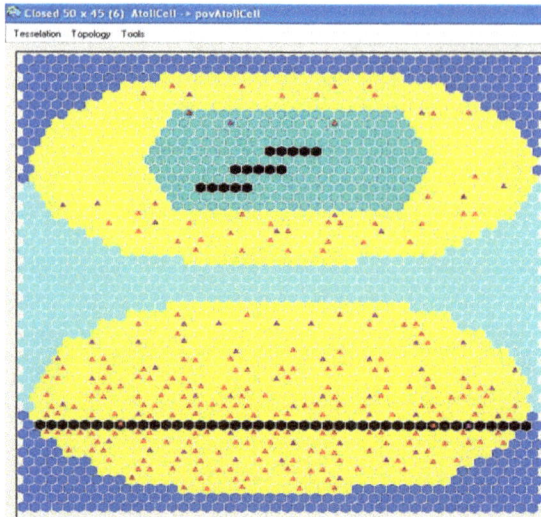

AtollGame environment with top island featuring Water Reserve and pumping stations, and bottom island featuring a water distribution pipe. Triangles represent the landowners (in purple) and their relatives (in red).

The modelling time step corresponds to a 10 day period and the simulations are limited to 1 year. Several modelling viewpoints were created, directly accessible during the simulations, in order to represent key features of the system. The *Landuse viewpoint* visualises the eventual changes between the 3 land-use types. The *LensDepth viewpoint* allows the depth evolution of the Lens to be followed during the simulations. The *WellWater viewpoint* represents the water salinity evolution, varying between fresh, brackish and salty.

Water balance and hydro-geological models

The recharge of the freshwater lenses is directly controlled by the infiltration rate through the unsaturated soil layers. The water balance is simulated within AtollGame using a slightly modified version of the mass-conservation driven model proposed by Falkland (1992) for South Tarawa. This 3-reservoirs-based model, called WATBAL, uses rainfall and Potential Evapotranspiration (PET) as input data. Runoff is not taken into account because of the high permeability of the coral sand soils. The first reservoir intercepts the rainfall at the vegetation level. The second reservoir corresponds to the soil water storage. The water entering the third reservoir corresponds to the recharge of the freshwater lens. Recharge of the lens may occur only after plants have satisfied their water requirements. Tree crops (mainly coconut trees in this case) are able to extract water directly from the lens. This model was adapted to AtollGame by bringing down the hydrological calculations at the level of each cell through its *Water-*

Balance and *Landuse* entities. Each instance of these two classes can operate its specific part of the water balance. Hence, AtollGame takes the spatial heterogeneity of the processes and their time dependence into account.

The shape and the depth of the freshwater lenses are calculated according to the model proposed by Volker et al. (1985). This model predicts the depth of the freshwater lens and the thickness of the transition zone from the recharge and uptake values, according to the maximum length of the lens. Two strong assumptions limit the use of this simple 2D-model: the recharge is constant, and the lens is in a steady-state condition. Hence, the model is often used for long-term predictions based on ten years averaged data. This vertical, 2D representation had to be adapted to the AtollGame distributed grid (Perez et al. 2003).

Some cells have been selected and designated as lens centres or nuggets. Using the isotropy property of the grid, each nugget is surrounded by concentric circles of isopiezometric cells. The orthogonal distance between the lagoon and ocean shores, crossing the nugget gives the value of the radius (L). The distance between 2 nuggets may be smaller than their respective radius; in this case, a common cell is given the deepest value calculated at each time step. The global shape of the lens corresponds to overlapping bowls (Figure 12.8).

Figure 12.8. Representation of the freshwater lens in AtollGame

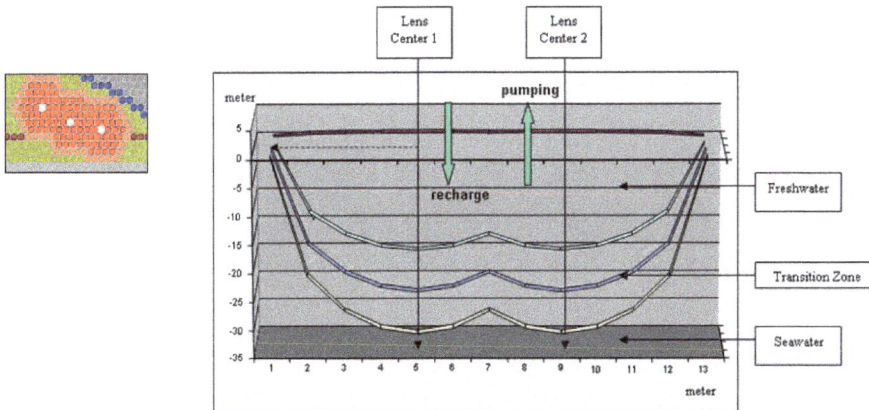

Nuclei and izopiezometric areas (left), and corresponding volume (right)

The hydrological model, using outputs from the water balance model, provides an update of the cell's attribute *depth* at each time step. This attribute is then used to specify the water quality of the lens by updating the cell's attribute *wellWaterQuality* according to a simple rule: if the depth is lower than 1.6 m, the water is considered as salty, if the depth is higher than 3.1 m the water is said to be fresh, and in between the water is brackish.

Social entities

Two classes of social entities were defined. On Island 1, 50 *Household* agents have been created. They all represent a family but only 11 are *landowners*, the rest are *relatives*. On Island 2, there are 200 *Household* agents, 42 of whom are *landowners*. The initial locations of the households are saved in the environment. The main attributes of each *Household* are: the size, the drinking and domestic water requirements, a list of *Wateruse* equipment, and a consumption satisfaction index. Theoretical demands are set up to estimated levels of 20 l/day/person for drinking water on Island 1, and 40 l/day/person on Island 2. For both islands, the domestic water consumption has been fixed at 40 l/day/person. Households are all provided with a well, some are given a rainwater tank and some, only on Island 2, have a connection to the water pipe. The decision process dealing with water consumption follows a simple rule: households satisfy their drinking water needs from their rainwater or PUB connection, if any, and supply their domestic water demand from their well. If they only have a well, they use it for both purposes, taking the risk of drinking brackish or salty water according to their well water quality. The water availability is updated at each time step according to the type of equipment. Groundwater extraction from individual wells is limited by a maximum depletion rate in the vicinity of the well.

One process accounts for population growth through the introduction of new relatives at the beginning of each time step. The *AgentPUB* class contains only one instance representing the Public Authority Board, in charge of the water distribution among the population. At this stage of the model, social interactions focus only on the competition for water between households and the management decisions from PUB. The *AgentPUB* is characterised by the volume of water pumped from Island 1 and given to Island 2. The pumping rate is initialised at 150 m^3/time step. The distribution among the different households on Island 2 is driven by their ranking distance along the main pipe. The *AgentPUB* can modify its pumping rate according to the average recharge rates.

AtollGame: the role-playing game

Conceptual framework

A role-playing game has been designed as a medium of communication based on the existing conceptual model. It is meant to open or develop the communication between stakeholders. In a well-designed role-playing game, players are aware of the issues at stake, but allow themselves to express their views and behave accordingly to their beliefs. Another fundamental characteristic is the ability of the role-playing game to generate collective scenarios that will explore new management avenues. In order to achieve these tasks, the role-playing game must:

- *represent* simplified features and processes encountered in reality (in particular, biophysical processes, social interactions and spatial descriptions should be understood and accepted by all players as plausible assumptions);
- *secure*, at every stage of the game, the neutrality of the selected rules and of the Game Master decisions (this *fair game* is instrumental in helping players build self-confidence and advocate their viewpoints); and
- *create* opportunities for players to comment, modify and improve the rules (the game is intentionally designed with a rudimentary set of rules that needs improvement).

Thus, players come progressively from playing against each other towards a situation where they appropriate the game collectively. Finally, when players have realised the collective benefit of the game, they tend to explore more complex situations and to implement more rules. Most of the time, the increasing complexity tends to bring the artificial game and the real environment together. What is at stake with this approach is to give the local stakeholders the capacity to build their future together. Instead of asking foreign expertise to provide an hypothetical best solution, this expertise is used to help local people adopt equitable management of the resource. Of course, manipulation, lobbying, struggle for power are inherently part of the process, but in an armless and controlled environment.

The game

Two game boards, identical to the computer-based visual maps, are used. Board 1 represents a scarcely populated island where the government is already pumping freshwater. Board 2 represents an overcrowded island where the government needs to provide drinking water (Figure 12.9). On each island, 8 players are allocated numbered locations. On the first board, the water reserve boundaries, and pumping galleries are physically represented. On the second board, the main pipe is delineated by a stripe of adhesive. Each player takes the role of a landowner who needs to provide enough water to his/her family during the game. In order to do so, players have to invest wisely in different types of equipment.

Figure 12.9. Island 1 (left) and Island 2 (right)

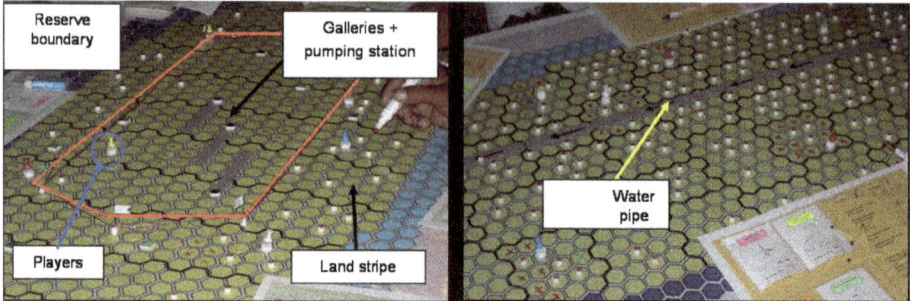

The players

At the beginning of the game, each player randomly draws a card that defines his/her personal profile in the game. The players' profile includes: job type, family size, water needs and initial equipment. On Island 1, amongst the 8 players, 1 is a public servant, 3 are seamen and 4 have no job. On Island 2 the distribution is slightly different: 3 public servants, 3 seamen and 2 have no job. They earn virtual money from their jobs (2–6 tokens) in the form of matches. The relatives living or arriving on their land at the beginning of each round can cost them money (–2 or –1 tokens), or provide some additional income (0–2 tokens).

The duration

The game spreads over 2 daily sessions. Within each daily session, 4 rounds (each equivalent to a 3-month season) are to be played. The first round corresponds to a good rainy season (550 mm), the second one to a very bad season (190 mm), the third and fourth rounds only replicate the data from the first round. On Island 1, the pumping rate from the government is steady and corresponds to a 150 m^3/day. On Island 2, the government is providing the local residents 150 m^3/day through the pipe.

The objective

The individual objective of the players is to minimise the number of angry or sick people in their household. People may become *angry* because they didn't have enough water to drink during the round. People may become *sick* if they drank unhealthy water during the round. Unhealthy water is when the water is polluted or salty. *Pollution* depends on the number of people living on the island, and contaminating the freshwater lens. *Salty water* depends on the recharge rate of the freshwater lens and the location of the players on the island. Rainfall and pumping rates affect the level of recharge. Rainwater or water from the adduction pipe are safe to drink.

In order to provide drinking water to their family, players are given *buckets* at the beginning of the game. One bucket can store 20 litres each day. One person is supposed to need 20 litres of drinking water each day. The initial number of buckets is lower then the family's needs by 2 or 3 buckets. In order to provide enough water to their family, the players have to *buy* equipment that will increase their storing capacity, counted in equivalent-buckets. A manual *pump* costs 2 tokens and provides 2 more buckets of storing capacity. A *rainwater tank* costs 3 tokens and provides 5 more buckets of storing capacity. A *PUB tank* costs 3 tokens and provides 5 more buckets of storing capacity.

Normally, the rainwater tank and the *PUB tank* automatically refill at the end of each round. But, if the rainfall during the round was not sufficient, the rainwater tanks remain empty. The same happens with the PUB tanks if the adduction pipe cannot provide enough water during the round.

Players can decide to farm vegetable gardens in order to increase their income. Each *crop* card costs 1 token and needs 4 extra buckets of water for irrigation at each round. The profit from the crop depends on the climate. If the round was rainy, then the crop provides 2 extra tokens to the player. If the round was dry, the crop failed and there is no extra income.

Playing rounds

First session

During the first session, dedicated to individual strategies, players have to accommodate new relatives on their land and adjust their income correspondingly. According to their available cash, they may choose to invest on new equipment and/or decide to crop. The rainfall conditions influence groundwater salinity, storage in rainwater tanks and crop yields. This session allows players to perceive the impact of the location on their well's water quality: the closer to the lagoon or ocean, the saltier groundwater becomes during drought periods.

At the end of the first session, one player on Island 1 is given the opportunity to sell part of his land to a new (virtual) settler, he can accept or not. At the same time, on Island 2, one player is given the opportunity to leave his land and to relocate on Island 1 if he can make a deal with another player on that island. This is meant to introduce the first interaction between the 2 tables.

Second session

The second session is dedicated to collective decision-making. First, the game master introduces one collective event card on each island. On Island 1, the card mentions that the government has decided to get rid of all the settlers and crops located on the reserve. Players have to relocate their relatives and they loose the removed crops. On Island 2, the card mentions that the government has decided to raise a connection fee from each dwelling connected to the pipe. One player, selected randomly, is given the task of fulfilling the government decisions. This player leaves his *landowner* role and becomes a *Water Agency* player. As such, he freely negotiates with players on both islands. His task is amplified with the introduction, at the second round, of one collective event on Island 2: due to financial issues, the government cannot maintain the distribution pipe properly. As a consequence, the discharge falls to 75 m^3/day. Another event is introduced on Island 1 during the third round: due to water shortage on Island 2, the government decides to increase the pumping rate from 150 m^3/day to 250 m^3/day.

Within the computer-based simulator, the *AgentPUB* entity represents the water agency in charge of the water distribution among the population. The *AgentPUB* controls the volume of water pumped from Island 1 and transferred to Island 2. The initial pumping rate on Island 1 is adjusted according to the demand. The distribution on Island 2 is driven by the Household entity's position along the pipe.

Outcomes from the role-playing game

At first, it was encouraging to see that representatives from the different islands displayed different viewpoints about the water reserves. The group meetings organised in the villages prior the workshop allowed for a really open debate. On the institutional side, the position of the different officers attending the workshop demonstrated a clear commitment to the project. All the participants showed the same level of motivation either to express their views on the issue or to genuinely try to listen to other viewpoints. Participants also accepted to follow the rules proposed by the Game Master, especially the necessity to look at the problem from a broader perspective.

During the first session, the players quickly handled the game and entered into interpersonal discussions and comparisons. The atmosphere was good and the game seemed playful enough to maintain the participants' interest. Most players experienced the fact that individual strategies were strongly dependent on environmental uncertainties. Interaction between the 2 tables started when one player from the Island 2 was given the opportunity to move to Island 1. The bargaining process that emerged from this new situation illustrated the actual tensions existing around the land tenure market in the real situation. The connection between water management and land allocation issues was clearly demonstrated by the players' behaviour.

The second day, the introduction of a water agency and the selection of its director created a little tension among the participants. But, after a while, the players accepted the new situation as a gaming scenario and started to interact with the newly created institution. On Island 1, the decision to remove crops and settlers from the water reserve immediately generated animated discussions among players. On Island 2, the recovery of service fees from the players connected to the distribution pipe had the same effect. At that stage, players started to merge arguments based on the game with other ones coming directly from reality. On Island 1, players entered direct negotiations with the Director of the water agency. On Island 2, opposing discussions occurred between players willing or not to pay the fee.

Finally, the Game Master introduced the fact that the water agency was no longer able to maintain the reticulated system due to a poor recovery of the service fees. It had the immediate consequence of a sharp decrease of the quantity of water offered on Island 2. At this stage, all players gathered around Island 1 and entered a collective debate on the water reserve management. Consequently, players were asked to go back to their table and to list solutions to improve the situation on their island. When the 2 lists were completed, the Game Master helped the participants to build a flowchart of financial, technical and social solutions (Figure 12.10), taking into account issues from both islands.

Figure 12.10. Flowchart of financial, technical and social solutions

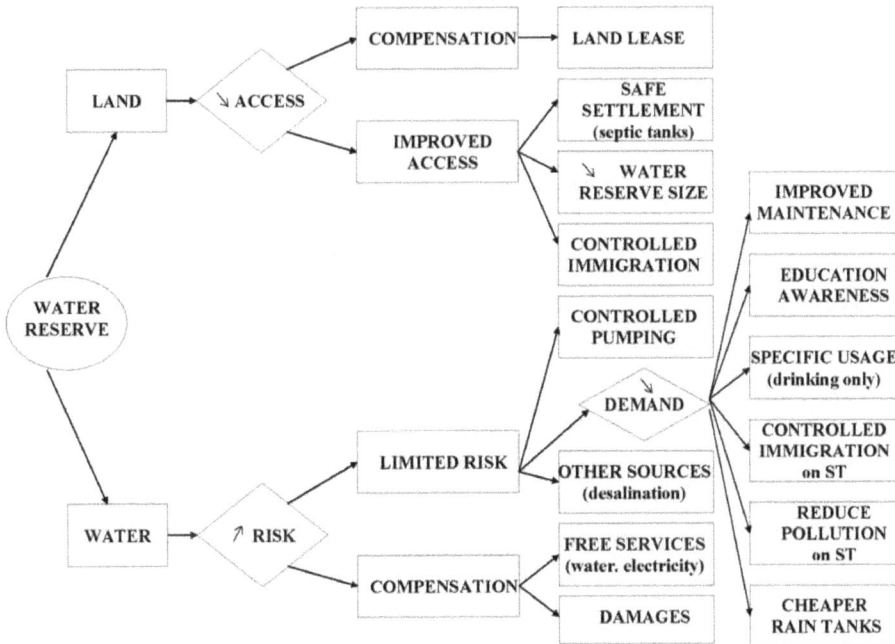

Exploring scenarios

A collective analysis of the flowchart concluded that the actual situation was largely unsustainable either from a financial or social viewpoint. The government relies on the land leases (top part of the flowchart in Figure 12.10) in order to secure social acceptance of the water reserves. The land market already pushes land prices to levels that can't be matched by government leases. Beside, other technical solutions (desalination plants or improved distribution) are not yet directly linked with the water exploitation issues on the islands. Some local residents claim that the environmental risks created by the pumping in the water reserves should be compensated as well. These *Damages* (bottom part of the flowchart) should be given to all the permanent residents of the island, besides the land leases granted to the concerned landowners. But fewer claims were made for negotiated and regulated use of the water reserves, somehow weakening the environmental risk claim.

The flowchart (Figure 12.10) provides a set of interdependent solutions that should be explored in order to gradually unlock understanding of the present situation. Whether we look at already existing or forecasted water reserves, the following guidelines are highly relevant:

- the financial solutions could be mitigated with technical solutions including regulated access to the water reserves or participatory management of pumping;
- the water exploitation issues could be more strongly linked with the water distribution ones, and eventually with the sanitation ones (keeping in mind that the latter are generally disconnected from the other issues for most people).
- exploring the middle part of the flowchart would enable more *consensual* stakeholders to participate in negotiations that are presently dominated by more extremist views; and
- management issues on the existing water reserves (Bonriki and Buota) and implementation issues on the forecasted ones (Abatao and Tabiteuea) are inherently interrelated. On one side, creating new water reserves without confronting the actual problems on the existing ones is not viable. On the other side, the introduction of new actors in the debate helps reduce the actual bipolar confrontation between landowners and the government on the existing water reserves.

Discussion and perspectives

Instant debriefing

At the end of the 2-day workshop, the project team held what we called an *instant debriefing*. Beyond our satisfaction of having conducted a playful and fruitful exercise, our analysis focused on the ways to transform the scenario-flowchart into a viable road map for the government. Careful study of the memos (videos, notes, and game spreadsheets) revealed the existence of 2 types of strategic behaviour among players that would strongly influence the outcomes. Final discussions around the flowchart were highjacked by a minority of *pseudo-players*. These were local stakeholders who came with a strong agenda in mind and tried to enforce their views throughout the game. As requested, they were indeed outspoken representatives but were not prepared to compromise. The role-playing game failed to modify their position and, for example, they locked the discussions into endless arguing about financial compensations. However, the game helped more consensual players to move some distance from these extremist views and to advocate a more flexible approach to future negotiations.

In contrast, some representatives of government agencies appeared to be *virtual-players*, without any mandate for taking a decision to their home institution. As previously mentioned, only one player was part of the SAPHE Steering Committee, all others were not directly involved in the decision making process regarding the water reserves issues. These players played a fair game but hardly attempted to defend their agency's policy. Hence, the role-playing game was only perceived as a mere exercise in communication.

It was highly recommended that the next steps involve a sequential process for interactions. First, experts from the relevant government agencies should be confronted with a new version of the computer-based simulator, including most of the options present in the existing flowchart. The experts would help select affordable scenarios for the government and tune the parameters. Then, these scenarios should be presented in the different villages through collective meetings where people would have the opportunity to interact with the computer simulations. Evaluation of the government criteria and scenarios would lead to the creation of newly modified ones. At last, the government experts and the island's representatives should meet again to assess the remaining options and hopefully agree on an equitable management scenario.

Distant debriefing

At the end of May 2004, the project team returned to Tarawa to implement the strategy outlined above. Unfortunately, soon after our arrival, we were informed that the SAPHE Steering Committee had decided to organise meetings with the local communities on Tabiteuea and Abatao. The official objective of these meetings was to present the design of the pumping galleries to the local residents and to seek their agreement in principle. During separate meetings with members of the Steering Committee, the project team tried to underline the inherent risk linked to this hasty decision:

- The SAPHE Steering Committee relied on a design extracted from a technical report (Falkland 2003) that was not meant to be a final blueprint for implementation. The report carefully investigated the hydro-geological conditions prevailing on Abatao and Tabiteuea and provided some guidance in terms of positioning (distance to actual settlements). There was no mention of land ownership issues, demographic growth or pollution control.

- The SAPHE Steering Committee was not yet able to provide complete information to the local communities about financial arrangements, land use constraints or other compensation claims eventually raised by some local residents. Beyond, the local negotiation issue, these elements are instrumental in evaluating the economic viability of such a technological option. Falkland (2003) was very cautious regarding this specific aspect in his recommendations.

- The SAPHE Steering Committee didn't take into account the outcomes of the role-playing game workshop, which provided a tentative road map for further negotiations, including financial and technical aspects. It was obvious that an upfront confrontation would re-ignite the water lease issue. Besides, the present state of mind of a majority of residents on Abatao needed to be dealt with cautiously in order to avoid any more damage.

Despite the concerns above, the SAPHE meetings were confirmed but the Steering Committee accepted that our project team attend the meetings as observers. As expected, the meetings held in Tabiteuea and Abatao would have benefited from a careful analysis and better understanding of the outcomes of the role-playing game workshop. Whether they agreed with the water reserve, local residents constantly referred to our final flowchart. These meetings resulted in back-stepping and a sterile process in which a majority of local residents focused again on the financial aspects of the question.

The reason for this apparent backlash from the Steering Committee is partly to be found in the absence of crucial *meta-players* during the role-playing game sessions. As shown on Figure 12.7, the Asian Development Bank (ADB), the project contractor, and the government cabinet control the financial, technical, and political agendas of the SAPHE project. One year before the end of the loan-based project, technical delivery from the contractor became essential in order to secure the last payments and to turn on the loan reimbursement countdown. At the political level, the Cabinet was entangled in multiple discussions with the ADB and other funding agencies. The Steering Committee, instead of being a driving force, had become an arena for external and conflicting pressures.

Despite our genuine claim that the results from the role-playing game were far from being detrimental to the SAPHE project, and that further negotiations with local communities would need far less than 6 months, the financial and technical agendas prevailed. Beyond the timing issue, it is also assumed that acknowledging the results of our project would have been considered by some members of the Steering Committee as a recognition of their incapacity to tackle the problem in the first place.

The incapacity at the political level to handle the uncertainty of the situation was demonstrated by a first decision to freeze the SAPHE project (October 2004), and then to accept its implementation (January 2005) without modification. The frustrating consequences were that more than 6 months were lost in the process and that the local communities ended up as frustrated as usual. An interesting final twist is the actual willingness of PUB to resume collaboration with our research team in order to develop a more participatory approach in the future. This struggle between centralised and decentralised management is epitomised in the concept of polycentric institutions developed by Ostrom in her most recent work (2005). Technical agencies such as PUB often rely on deductive scientific approaches to reach outcomes that would need more inductive and flexible solutions. But flexibility means that one must assume some uncertainty during the implementation and give up hope of deterministic and predictable solutions (Bradshaw and Borchers 2000). What is true at the technical level becomes paramount at the political one.

There is a need for integrating not only the participation but, more importantly, the engagement of local communities in projects that concern their future. Following Aslin and Brown (2004), we argue that local communities need to be involved not only in the analysis of the results (consultation) or the choice of the possible scenarios (participation), but in the knowledge creation itself (engagement). This is the *post-normal* way chosen, for example, by colleagues working on Companion Modelling approaches (Bousquet et al. 2002).

References

Aslin, H.J. and V.A. Brown (2004) *Towards Whole of Community Engagement: A Practical Toolkit*. Canberra: Murray-Darling Basin Commission.

Becu, N., F. Bousquet, O. Barreteau, P. Perez and A. Walker (2003) A methodology for eliciting and modelling stakeholders' representations with Agent Based Modelling. *Lecture Notes in Artificial Intelligence* 2927: 131–49.

Bousquet, F., I. Bakam, H. Proton and C. Le Page. (1998) Common-Pool Resources and Multi-Agent Systems (Cormas). *Lecture Notes in Artificial Intelligence* 1416: 826–37.

Bousquet, F., O. Barreteau, P. d'Aquino, M. Etienne, S. Boissau, S. Aubert, C. Le Page, D. Babin and J.-C. Castella (2002) Multi-agent systems and role games: collective learning processes for ecosystem management. In M.A. Janssen (ed.) *Complexity and Ecosystem Management: The Theory and Practice of Multi-agent Systems*, pp. 248–85. Cheltenham: Edward Elgar.

Bradshaw, G.A. and J.G. Borchers (2000) Uncertainty as information: narrowing the science-policy gap. *Conservation Ecology* 4(1): 7. Available at http://www.consecol.org/vol4/iss1/art7/

Descola, P. (1996) Constructing natures: symbolic ecology and social practice. In P. Descola and G. Palsson (eds), *Nature and Society: Anthropological Perspectives*, pp. 82–102. London: Routledge.

Falkland, A.C. (1992) Review of Tarawa freshwater lenses, Republic of Kiribati. Hydrology and Water Resources Branch. ACT Electricity and Water, Rep 92/682, Canberra, Australia (unpublished report).

Falkland, A.C. and J.P. Brunel (1993) Review of hydrology and water resources of the humid tropical islands. In M. Bonell, M.M. Hufschmidt and J. Gladwell (eds.), *Hydrology and Water Management in the Humid Tropics*, pp 135-63. Cambridge, England: Cambridge University Press–IAHS.

Falkland, T. (2003) Review of Groundwater Resources Management for Tarawa. Kiribat SAPHE Project: Mid-Term Review, Loan No 1648-KIR (SF). Ecowise Environmental report No EHYD 2003/10, prepared on behalf of Asian Development Bank.

Newell, A. (1982) The knowledge level. *Artificial Intelligence* 18:87–127.

Ostrom, E. (2005) *Understanding Institutional Diversity*. Princeton, N. J.: Princeton University Press.

Perez, P., A. Dray, I. White, C. Le Page and T Falkland (2003) AtollScape: Simulating Freshwater Management in Pacific atolls, Spatial processes and time dependence issues. In D. Post (ed.) *Proceedings of the International Congress on Modelling and Simulation* , 14-17 July 2003, Townsville, Australia. MODSIM 2003, vol. 4, pp. 514–8.

Shadbolt, N. and N. Milton (1999) From knowledge engineering to knowledge management. *British Journal of Management* 10 (4): 309-22.

Volker, R.E., M.A. Mariño, and D.E. Rolston, (1985) Transition zone width in groundwater on ocean atolls. *Journal of Hydraulic Engineering* 111(4): 659–76.

White, I., A. Falkland, L. Crennan, P. Jones, B. Etuati, E. Metai and T. Metutera (1999) Issues, Traditions and Conflicts in Groundwater Use and Management. UNESCO-International Hydrological Programme, Humid Tropics Programme. IHP-V Theme 6. Technical Documents in Hydrology No. 25. UNESCO, Paris.

White, I., T. Falkland, P. Perez, A. Dray, P. Jones, T. Metutera and E. Metai (2002) An integrated approach to groundwater management and conflict reduction in low coral islands. In *Proceedings of the International Symposium on low-lying coastal areas. Hydrology and Integrated Coastal Zone Management, 9–12 Sept 2002, Bremerhaven, Germany*. UNESCO HIP, pp. 249–56.

View from a Principal Scientist

Western Australia's North West Shelf (NWS) contributes $6 billion to the national economy and is an economically significant land and sea region in Australia. It produces the majority of Australia's domestic and exported oil and gas, and supports commercial fisheries, aquaculture, salt production, iron ore processing, shipping, and a rapidly expanding tourism industry.

The rapid growth of marine industries in and around the NWS has led to complex and somewhat fragmented, management and regulatory structures. The Western Australian Government recognised that a collaborative approach to integrated environmental management was essential to balancing and managing multiple-uses of the NWS ecosystem.

The North West Shelf Joint Environmental Management Study (NWSJEMS) was established in 2000 to develop and demonstrate practical and science-based methods that support integrated regional planning and environmental management of the NWS. A range of agent-based computer tools were developed to model the various complex biological and physical processes for predicting ecosystem and human impacts. However, the novel aspect of the Study was the application of The Management Strategy Evaluation (MSE) approach which linked these tools to provide, for the first time, a way of integrating the key environmental, social and economic processes. All of these modelled processes are complex and poorly understood, and the models used in MSE are designed to represent the uncertainty in understanding and predictions. In this way MSE was used to evaluate the effectiveness of prospective management strategies in achieving defined objectives for 4 major sectors operating on the NWS: petroleum, conservation, commercial fisheries and coastal development. This is the first time an attempt has been made to apply a complex systems science approach in the form of MSE to multiple-use management of a whole regional ecosystem.

The MSE approach provides decision-makers with the information and predictions of the range of consequences from prospective management actions from which to base a decision, given their own objectives, preferences and attitudes towards risk. It deals explicitly with multiple and potentially conflicting objectives, and with scientific uncertainty.

Applications of MSE are often to find a strategy that achieves the stated objectives robustly across all the scenarios that are represented. Alternatively MSE can be used to identify the particular scenarios under which an otherwise desirable strategy fails and to design an adaptive strategy to detect and correct such failure if it occurs in the real world. So the models in MSE have a fundamentally different function to models used solely for prediction. One of the key attractions of MSE is that it can proceed even if the models used contain many uncertainties and give highly uncertain predictions of the future because the emphasis is on finding strategies that can succeed despite these uncertainties.

The MSE models and information inputs are challengingly complex, but the NWSJEMS is demonstrating that the application of MSE to a regional ecosystem is achievable.

Chris Fandry BSc, PhD
Senior Principal Research Scientist
CSIRO Marine and Atmospheric Research

13. Multiple-Use Management Strategy Evaluation for Coastal Marine Ecosystems Using *InVitro*

A.D. McDonald, E. Fulton, L.R. Little, R. Gray, K.J. Sainsbury and V.D. Lyne

Abstract

The Management Strategy Evaluation (MSE) framework has been applied in a multiple-use setting to demonstrate practical science-based methods that support integrated regional planning and management of coastal and marine ecosystems. Multiple-use MSE has, so far, focused on 4 sectors: oil and gas, conservation, fisheries and coastal development. For each sector a selection of development scenarios, provided by the relevant interest groups, is represented. These scenarios include prospective future sectoral activities and their impacts, and the sectoral response to management policy and strategies. The agent-based modelling software InVitro is well placed for analysing prospective social and ecological impacts of multiple-use management strategies in a risk-assessment framework such as MSE. An illustrative example is provided to demonstrate the tradeoffs that can be recognised and quantified using the MSE framework. The example explores the implications of a change in management strategy. This change not only has a direct impact on the targeted sectors, but also indirect impacts, including surprises.

Introduction

Environmental management is characterised by multiple and conflicting objectives, multiple stakeholders with divergent interests, and high levels of uncertainty about the dynamics of and interactions amongst the resources being managed. This conjunction of issues can result in high levels of contention and difficulties in the management process. MSE and risk assessment can assist in the resolution of some of these issues. MSE involves assessing the consequences of different management options and making clear the tradeoffs in their performance across a range of management objectives and uncertainties in resource dynamics (Butterworth et al. 1997; Cochrane et al. 1998; Butterworth and Punt 1999; Sainsbury et al. 2000).

The approach involves: close co-operation and collaboration with stakeholders and management agencies to map broad objectives into specific and quantifiable performance indicators; identifying and incorporating key uncertainties in determining effects on these performance indicators of proposed management measures and strategies; and communicating the results effectively to client groups and decision makers. At a technical level, the MSE framework facilitates dealing with multiple objectives and uncertainties in prediction (Butterworth and Punt 1999). At the implementation level, it fails if it cannot accommodate effective stakeholder participation and acceptance. MSE can be used to develop adaptive monitoring and management strategies, to develop effective management procedures, and to provide guidance on key gaps in scientific knowledge (as exemplified by Sainsbury 1991).

Methods

The core computational components of the MSE framework are: a model of the natural system; a model of each of the important sectors of human activity; and a model of how decisions are made, including associated monitoring activities. These models are used together to determine how the natural system might respond to both natural events and human activity. The computer program used for MSE traces the impact of a particular management strategy on the actions of sector firms or agencies, the effect that these actions have on the natural environment and the impact on the performance indicators and measures. In so doing, the program keeps track of details on sector response to management actions, sector performance, the way the natural system responds to sector-specific actions and important random or periodic events, and any strategy-mandated adjustments by managers as a result of sector and/or system response.

Central to the program's output is the choice of environmental and sector performance indicators. The set of indicators includes attributes of water quality, habitats and biodiversity, food-chain integrity, endangered species/habitats/communities, public amenity and ecosystem services.

The implications of a particular management strategy are recorded through many scenarios, each of which involves numerous model runs that evaluate specific cases of management action and specific uncertainty in resource and environmental dynamics. These simulations capture the response of performance indicators to major data, model, and process uncertainties so that it is possible to examine the performance measures that can be used to evaluate how well that management strategy meets management objectives. Multiple simulations with alternative management strategies provide the performance measures response profile with which one can compare strategies.

The flow diagram for the MSE framework is shown in Figure 13.1. The biophysical model emulates the physical and biological features of the natural marine ecosystem, including the bathymetry, currents, waves, type of sea-bottom, benthic flora and fauna, local animal populations and migratory animals. This model also includes a representation of the impact of natural forces and human activity. Outputs from this model include information about the state of the physical environment, about stocks of both renewable and non-renewable resources, and about other important features of the ecosystem. An assessment of variability is also included, thus providing an indication of the implicit uncertainty in models, observations and the bio-physical processes themselves.

Figure 13.1. The MSE Framework

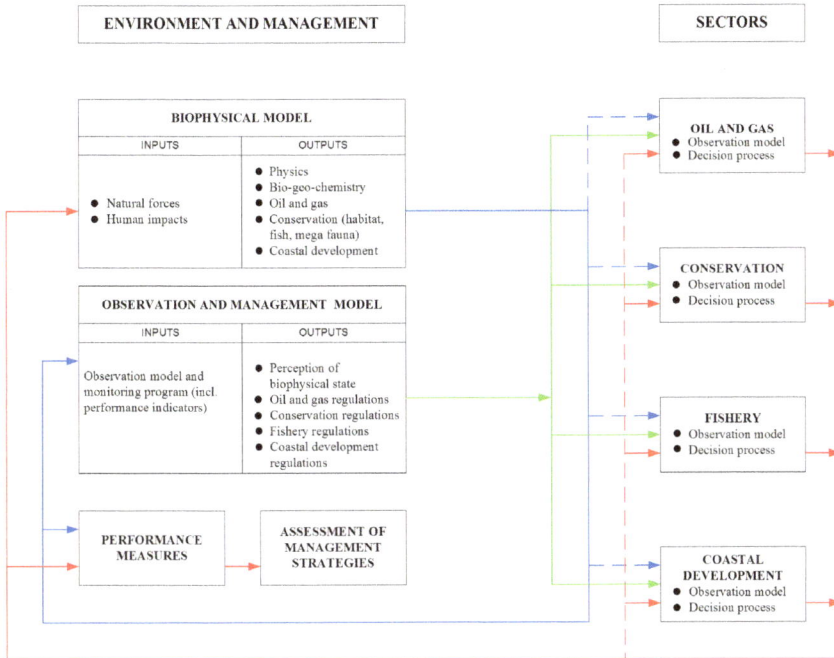

The sectors represent human activity in petroleum exploration and extraction, conservation, fisheries and coastal development (which were developed by drawing on work in Sainsbury 1988; McDonald 1991; Iledare and Pulsipher 1999). Key players in each of these sectors observe the natural system imperfectly and make decisions about levels and locations of their activities. Sector activities have an impact on the natural system and also on public management agencies that monitor and regulate the activities of the sectors.

The observation and management model simulates the actions of public management agencies and the response of performance indicators are noted via a simulated management monitoring strategy. In this model management objectives are given quantitative interpretation, as are the management strategies and decision rules. Each of the sectors is then faced with implementation of a management strategy, which constrains their activities and, therefore, their impact on the natural system.

InVitro

The computer software for the MSE framework comprises a set of linked dynamic models, a user interface, visualisation tools, and a data retrieval and storage capability. An example of a dynamic modelling software used for MSE is the agent-based model called InVitro. InVitro is a spatially explicit agent-based framework containing many embedded sub-models in continuous space with an individually variable time step.

Traditionally there have been two main types of ecological models: aggregate state models (Silvert 1981; Jørgensen 1994) and individual based models (Caswell and John 1992; DeAngelis and Gross 1992). Separation of these model types is not always simple. Sometimes, within the latter form of model, the *individuals* represent schools, patches of homogeneous ground cover, flocks, patches of reef, or some other portion of the whole population that could be treated as equivalent to an *entity*. From this it is clear that most of these aggregate state models can be seen as a proper subset of the set of all individual (or agent) state models. Consequently, we can treat these aggregate state models as agents within the system. This is the approach that has been taken in InVitro.

InVitro exhibits aspects of both of the traditional classes of ecological models, conceptually embedding them in a time-sharing universe. Agents are allowed to operate at time and space scales appropriate to the nature of the processes in question. Seasonal cycles, for example, are not forced to adhere to time steps more appropriate to larval migration or tides, and the spatial scale of interaction is set at the native resolution of the processes and their associated data sets. This treatment of a general model as an agent within a modelling framework has consequences: models may have inherently different spatial or temporal characteristics, and these must be made compatible. To achieve this, all the agents in InVitro act within the floating point representation of a continuous three dimensional spatial environment, and a temporal environment which arbitrarily has a fundamental step of one second. It would be inefficient to make all the agents run in a lock-step fashion with a temporal step of one second, so a scheduler was used as a means of executing larger blocks of time in a consistent way. Issues of concurrency also arise. What happens when agents are to interact? How do we prevent agents from getting less or more time than they are allowed? These questions also arise in the context of operating systems, and the InVitro scheduler does bear a marked resemblance to such an operating system. Like most operating systems and ecological models, it does not exhibit truly parallel execution, and careful attention has been given to the issues of temporal order of action and interaction to ensure that dynamic processes occur *when* and *where* they should be.

Within the broad MSE-InVitro model there are submodels that reflect the bio-physical and anthropogenic activity in a coastal ecosystem; namely biophysical interactions, fishing, shipping, industrial/coastal development and contaminants. The software details particular sectors of interest: oil and gas production, fisheries, conservation, and coastal development. These sectors are engaged with the issues of contaminant release, contact and uptake, sustainable harvest, ecosystem health, habitat quality and damage, economic growth and amenity. These are central to the assessment of various management strategies. Our explicit functional groups must, therefore, specify inter-group processes at appropriate space and time scales, a task made easier by the flexible handling of these factors by InVitro.

A practical part of capturing these critical scales is achieved in InVitro through the use of a hierarchical agent structure. The most significant division in this structure is the separation of *Monitors*, *Things* and *Environments*. A *Thing* is something that usually has a single location or sphere of influence. Typical *Things* are ships, schools of fish, turtles, and oil rigs. They may or may not be mobile, and they can usually be thought of either as an individual or some aggregate representation of individuals (for example, a fish school). An *Environment* is characterised by a fundamentally spatial nature: it covers an area, has some significant spatial structure, or its heterogeneous distribution may be of particular importance. Typical *Environments* include biogenic habitats (such as sponge beds and mangrove forests), bathymetry, plumes of contaminants, marine parks and road networks. *Monitors* are possibly periodic things that *happen* to either all the *non-Monitor* agents or some subset thereof. They are characteristically things like biomass assessors, water quality monitoring stations, models imposing recreational fishing mortality and management agents. *Monitors*, unlike other agents, are guaranteed that the agents they are interested in are temporally synchronous at the time the *Monitor* acts.

It is possible to broadly summarise the components of the NWS system using these 3 classes, with each agent represented by one of these classes. However, processes or entities within the framework may have more than one representation, which may come from any of the 3 classes or may be multiple representations from within one of the three principal classes. For example, the framework has several quite distinct potential submodels representing fish, especially juvenile fish. This is explained more fully below, but essentially the alternative representations of adult and juvenile fish can either lie closer to traditional population models or to individual-based models. The choice of representation used in any particular model simulation depends on what it is we wish to model, but each of these representations still needs to address the fundamental aspects of *being fish*. Each of the adult agents ought to be able to spawn to the chosen representation of juvenile fish, for example. This range of alternative representations that interlock with the other components of the model is exemplified by the fish, but not restricted to them. For instance, currents might be represented as a mathematical model of tides, or as a time series of snapshots of the current field; either way they must still be accessible by, and to, the other agents. This multi-faceted, modularised form not only allows for the adaptation of the modelling framework to answering a variety of management questions, but has also facilitated an exploration of alternative model assumptions and structures.

Environments

Environment agents are in some ways the most complex agents in the InVitro framework. These agents' representations are typically either vertex-based (lists of polygons, lines or points) or grid-based (grids or sets of coincident grids). Most of the archetypal environmental characteristics are modelled as *Environment* or classes derived from *Environment*, but a range of other data may also be modelled by these classes, for example, ocean currents, management control zones, primary production, contaminant plumes, and sampling regions. Anything represented using vertex-based agents may exhibit diffusion, advection, self-motility, or some combination of them all. They may also engender new polygons. Any of the *Environment* representations may also be reset at predetermined times to some given state.

It is easiest to convey some of the dynamics associated with *Environment* by example. One of the management options considered in the MSE-InVitro model is the use of Marine Protected Areas (MPAs). In MPAs, there are constraints on fishing and development, and the effects of external development on the habitat and populations within the MPA are monitored. These MPAs may consist of either a list of polygons or a set of time-tagged lists of polygons. In either case the polygons may be subject to independent management decisions (for example, 'The MPA is in very good shape, so we can allow recreational fishing'). The boundaries of the MPA may vary through time to accommodate changes to legislation or administrative policy using time-tagged source data. The presence of these data can prompt the underlying framework to reset the data to known, possibly quite different, conditions at predetermined times.

A particular subclass, *Tracers*, is used to represent both contaminant plumes and, via derived classes, most of the *Environment*-represented biological entities. By deriving the biological classes from the class used to track things like contaminants, the mechanics of the motion and oceanographic influences on the entities are made consistent. This is particularly important when trying to simulate organisms suspended in the water column and the outflow from some stationary source under the influence of tides (as shown by Condie et al. 1999; Walker 1999; Bruce et al. 2001).

Ecology

The ecology in MSE-InVitro is primarily represented via a combination of *Environment* and *Thing* classes. The main groups of interest include fish (various species with varying life history characteristics, from prawns through to large fish and elasmobranchs), turtles, benthic habitat (beds of sponges, coral, seagrass and macroalgae) and mangroves. The first 2 are usually represented by *Things*, while the latter two are *Environments*. Regardless of the class used, the modelled processes include reproduction, growth, predation (either implicitly in the probability of mortality or explicitly with predator/prey interactions), foraging and larger scale migrations (for example, for spawning).

As mentioned above, there are several representations for some members of the virtual ecosystem. This is most evident for fish, for which there are two distinct representations for adults, and three distinct representations for juveniles. Adult fish may be modelled either as a more-or-less traditional population model (*Population*), or as schools of varying sizes and locations (*Fish*). Juveniles are represented as separate from the adults rather than as simply the youngest age class as in traditional population models; this allows for ontogenetic changes in habitat, diet and behaviour. Juveniles can be modelled as new instances of *Fish*, or as *Larvae* (a mobile, polygonal list based representation), or *Blastula* (a region covering a fixed area of suitable habitat) with different possibilities for interaction with other agents.

The other major part of the ecological submodel in MSE-InVitro are the biogenic habitats (*Benthics*), such as mangroves and seafloor benthic communities. These are all currently modelled as sets of polygons with appropriate attributes. These polygons may be irregular and discontinuous, or they can be a fixed grid. If represented by a fixed grid, sub-grid scale processes are represented through a meta-population based size-age model which allows for changes in community structure and coverage within each grid cell. Thus both forms of *Benthic* deal with spatial coverage and the effects of competition, growth, predation and physical destruction.

Fisheries

There are 4 fisheries represented in MSE-InVitro at present: trawling for scale fish, trap fishing, prawn trawling, and recreational fishing. The first 3 are modelled using explicit vessels with specific process models of the physical act of fishing. The trawlers trawl through the fish or prawn schools catching a proportion of the school and the trappers set their traps to catch fish. The vessels used in the process-based fishery representations choose likely locations for trawling based on past experience, and learn as conditions change, and should their expectation fall too low, they will *prospect for fish*. The steps involved in fishing (the hauling and processing of fish) are also modelled, and there is room for the mechanisms for returning discarded fish to the water as carrion. Turtles, sharks and all other suitably sized animals may also be caught in the net, and benthic habitats suffer damage along a swath beneath the path of the boat and may be cleared completely. In addition, to allow for realistic changes in fisheries behaviour, vessels are parameterised with their real physical constraints (hold capacity, fuel capacity, steaming speed, net characteristics, etc.), so that a change in regulation or technology (say, mesh size, zonation, or turtle exclusion gear) will flow naturally through the model behaviour. Other modelled vessel behaviour includes the explicit use of navigation channels, since this has bearing on the location of marine parks, the potential anthropogenic impact on marine parks and traffic congestion with respect to heavy shipping.

In contrast to the process representation of these fisheries, the recreational fishery is a simple catch equation. The recreational fishery is implemented as a *Monitor* that runs once a week and takes from each school a tithe that is a function of the human population in the nearest towns, distance of the schools to access points (jetties and boat ramps) and the distance from these points to the towns via a road network.

Other human sectors

Heavy shipping is modelled as a manifestation of economic activity. Ships with deep draughts require deeper navigation channels and resultant spoil grounds, which have their own environmental impacts. Increasing congestion associated with increased production may prompt a range of management responses (controlled access to ports, additional transit channels, jetties/causeways, for example). Congestion is detected and reported to a Monitor which plays the role of the Department of Transport. It sits collecting information about port utilisation, shipping movements, near misses and collisions. Residence time and the size of a vessel in a port are related to the contaminant level within the port (for example, toxins associated with anti-fouling paint and fuel). Ships form the base class for the fishing vessels and share many of their properties. While the ships have largely predetermined paths once they leave the port's control, the courses of all vessels are subject to the control of the management strategy of the *fixture* (oil platforms, islands and shallow water, for example) they are approaching.

The coastal and industrial development sector is to some extent treated as one of the *forcing functions* of the system. One set of runs may proceed on the basis that there will be no increase in industrial activity (oil/gas, salt production, or coastal development), while another may have a 50 per cent increase in production over a given period. This is largely accomplished by increasing contaminants, increasing the economic worth of various activities, changing underlying environment (removal of habitat, causeways, roads, etc.), and increasing the human population.

The development sector is represented in the model by *Ports, Rigs, Roads, Jetties, Contaminant Plumes* (which are emitted by *Fixtures* or are present as a background part of the environment), *Transit Channels* (which may need dredging from time to time), the human population, and some notion of economic worth.

The modelling of contaminants in marine systems is a complex issue. Not only are there many mechanisms and rates of uptake for the diverse toxicants, but the combined effect of toxicants on mortality rates at a given level of contamination is quite complex. Contaminants are modelled as *Environment* agents that have a list of contaminants associated with them. These agents search for all other agents whose spatial extent impinges upon the contaminant agent and they then affect the agent appropriately. These contaminants are not necessarily all toxicants, some may be nutrients. Moreover, the issue of intoxication is fraught with difficulties. Not only are there quite different representations for the various agents (prawns are *Fish* schools with a fairly restricted radius, while *Benthic* sponge beds may cover quite large areas), but the contaminants with which an agent type is concerned may vary widely. Vessels may be quite concerned about contact with oil slicks, but relatively disinterested in contact with bitterns, while, to fish larvae, the bitterns are a poisoned chalice. The source of the contamination can also be important—the contaminant load from food may have a quite different rate of uptake and residence time than contamination through contact. Work is continuing on the implementation of mortality rates for given levels of contaminants across the set of contaminants deemed important for a given agent type. Issues such as dealing with the difference in toxicity from acute environmental contact and that from chronic tissue load are still unresolved. In part, these issues are a matter of extending the amount of data maintained in each individual's state vector and acting appropriately, but deeper problems remain in that much of the information we have about toxicity and environmental effects is available from specific laboratory studies of single toxicants.

There have been many prosaic issues for which solutions have required much effort and time. These include problems not encountered by the modelling group before—for example, getting the model to transparently accept either a single grid of data or a time-series of grids of data that update at the right time, or coalescing 3 different variants of essentially the same class into one class that *Does the Right Thing*. Ultimately however, the work has not only led to a greater understanding of the regional social-environmental system and the interaction of its defining processes, but has increased our understanding and appreciation of relationships among alternative model forms and the dynamics they can represent. It has given us a new perspective on the issue of scale, as well as insight and experience with truly complex systems and their emergent properties.

Management of the Australian NWS regional ecosystem

Management of the North-West Shelf regional ecosystem is conducted by government agencies at federal, state and local levels. Management agencies correspond to the activity sectors in the main, although all four sectors are regulated by more than one agency. For each of the management issues identified in the course of the NW Shelf Joint Environmental Management Study (JEMS), the following are specified for MSE:

- *management objectives* expressed in terms of their intended impact on the regional ecosystem and/or local environment;
- *management strategies* for achieving specified objectives (including identification of feasible control variables, monitoring programs and feedback mechanisms, as well as specification of decision rules); and
- *indicators and performance measures* from observation and monitoring for assessing how well management objectives have been achieved.

MSE has been carried out while considering biophysical uncertainty, although accounting explicitly for the response of managers to this uncertainty has not been done as yet. Currently the MSE handles management response as part of the matrix of scenarios which evaluate uncertainties in system state, forcing and response. By way of example, we consider ecosystem response under 2 model specifications, 2 management strategies and 2 development scenarios. Space constraints prevent a full description of these in the present chapter. Suffice to say, we examine optimistic versus pessimistic model specifications, status quo versus enhanced management strategies and present infrastructure versus expanded industrial development.

Results

Fisheries management

Under the current stock assessment protocol it is possible for the assessment model to become ill informed and diverge from the actual stock size (from the biophysical model) (Figure 13.2, status quo). With the introduction of a vessel survey, which provides additional information on stock status, the agreement between the stock assessment trajectory and the actual trajectory can be improved substantially (Figure 13.2, enhanced).

Figure 13.2. Estimated and actual stock sizes for the primary fishery target group (Lutjanids)

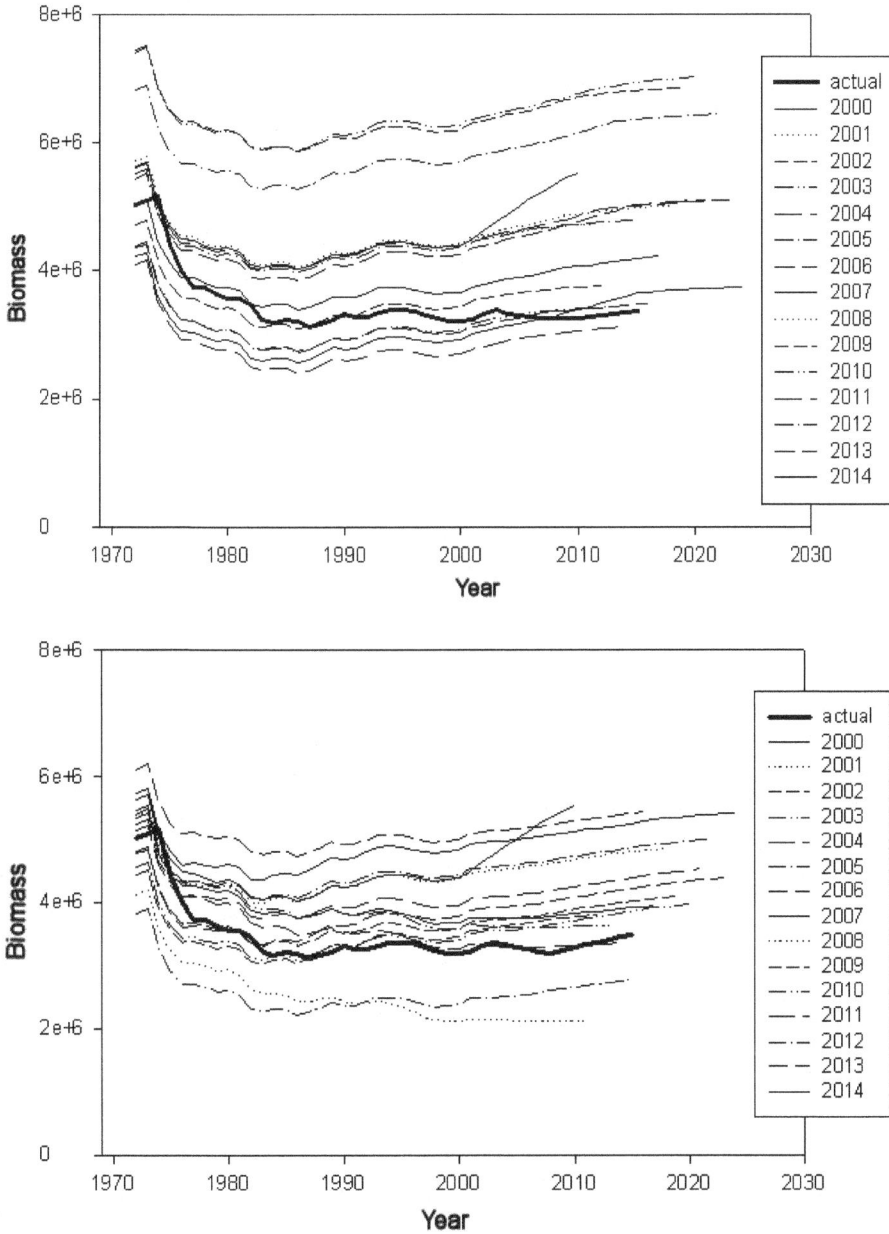

The estimated and actual stock sizes for the primary fishery target group (Lutjanids) under the (a) status quo and (b) enhanced management strategies.

Contamination

Status quo management of contamination means that the sector is effectively unlimited, which can lead to significant contact rates for animals in the vicinity of the outfalls (line with dots in Figure 13.3). The enhanced management strategy for this sector ties allowed outfall rates to the level of contamination found in sample animals. If an animal is found with contaminant levels above the regulated trigger point, then the outfall rate is reduced, which ultimately leads to much smaller contact rates.

Figure 13.3. Relative levels of contaminant in prawns within 20 km of an outfall under the status quo and enhanced management strategies

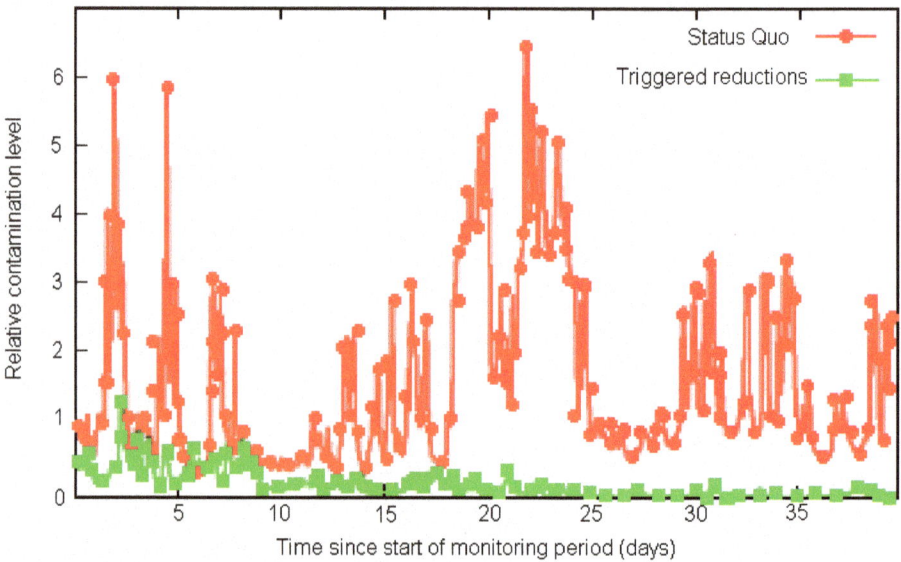

Coastal development—capacity and shipping traffic

While revenue suffers with increasing vessel traffic flow under status quo management, there are clear revenue benefits from increasing port capacity by 50–100 per cent (Figure 13.4).

Figure 13.4. Profit relative to profit at historical capacity for the port of Dampier under the range of port capacity levels allowed by the status quo and enhanced management strategies

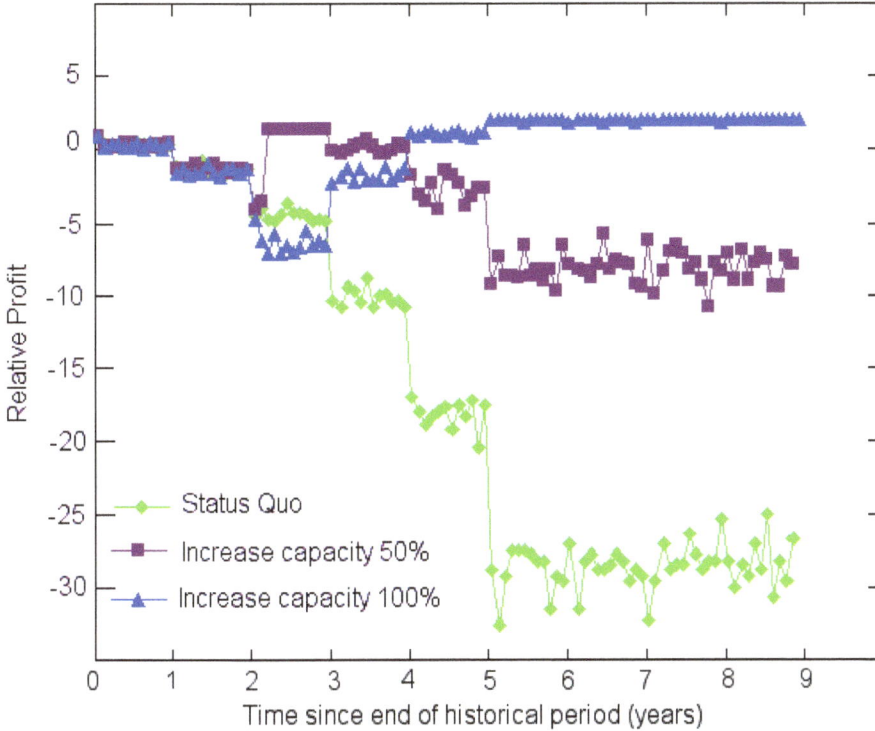

Conservation implications

The enhanced management strategies for fisheries and contaminant flows have positive conservation implications. A better knowledge of the state of the fished stock leads to more targeted management and less fishing effort, thus leading to fewer incidental effects on bycatch groups (such as turtles, sharks and the seabed habitat). Similarly, the enhanced contaminant control methods lead to less mortality due to toxic contamination, particularly for juvenile animals (which often live in shallower water than the adults) and the local seabed habitat groups.

The enhanced management strategy for the shipping traffic has mixed implications (due to the requirement for additional dredging). However, as the new channels lie in an area which is already poor habitat for benthic animals, then the negative effects of dredge spoil are more than compensated for by the benefits of vastly reduced potential collision (and major spill) rates.

Conclusions

The MSE framework implemented within the context of an integrated agent based modelling system provides a powerful tool for evaluating the response of systems under a range management strategies, model specifications and scenarios of system dynamics.

The InVitro modelling system provides a highly flexible and modular tool that can simulate the gamut of behaviours of agents ranging from traditional to rule-based individuals. Likewise the characterisation of the environment can be structured in a variety of ways with extensive facilities to monitor impacts.

Performance measures and indicators based on management needs and concerns allow stakeholders and managers to play a key role in defining explicit and objective evaluations that can improve their management practices. It also allows participants to appreciate the interactions amongst the different sectors.

References

Bruce, B.D., S.A. Condie and C.A. Sutton (2001) Larval distribution of blue grenadier (*Macruronus novaezelandiae* Hector) in south-eastern Australia: further evidence for a second spawning area. *Marine and Freshwater Research* 52: 603–10.

Butterworth, D.S., K.L. Cochrane and J.A.A. De Oliveria (1997) Management procedures: a better way to manage fisheries? The South African experience. In E.K. Pikitch, D.D. Huppert, and M.P. Sissenwine (eds), *Global Trends: Fisheries Management* , pp. 83–90. Bethesda, Maryland: American Fisheries Society Symposium 20.

Butterworth, D.S. and A.E. Punt (1999) Experiences in the evaluation and implementation of management procedures. *ICES Journal of Marine Science* 56: 985–98.

Caswell, H. and A.M. John (1992) From the individual to the population in demographic models. In D. DeAngelis and L. Gross (eds), *Individual-Based Models and Approaches in Ecology*, pp. 36–61. New York, NY: Chapman and Hall.

Cochrane, K.L., D.S. Butterworth, J.A.A. De Oliveria and B.A. Roel (1998) Management procedures in a fishery based on highly variable stocks and with conflicting objectives: experiences in the South African pelagic fishery. *Reviews in Fish Biology and Fisheries* 8: 177–214.

Condie, S.A., N.R. Loneragan and D.J. Die (1999) Modelling the recruitment of tiger prawns *Penaeus esculentus* and *P. semisulcatus* to nursery grounds in the Gulf of Carpentaria, northern Australia: implications for assessing stock-recruitment relationships. *Marine Ecology Progress Series* 178:55–68.

DeAngelis, D. and L. Gross (eds) (1992) *Individual-Based Models and Approaches in Ecology*. New York: Chapman and Hall.

Iledare, O.O. and A.G. Pulsipher (1999) Sources of change in petroleum drilling productivity in onshore Louisiana in the US, 1977–1994. *Energy Economics* 21: 261–71.

Jørgensen, S.E. (1994) *Fundamentals of Ecological Modelling* , 2nd ed., Developments in Environmental Modelling 19. Amsterdam, London, New York: Elsevier.

McDonald, A.D. (1991) A technique for estimating the discount rate in Pindyck's stochastic model of nonrenewable resource extraction. *Journal of Environment Economics and Management* 21: 154–68.

Sainsbury, K.J. (1988) The ecological basis of multispecies fisheries, and management of a demersal fishery in tropical Australia. In J.A. Gulland (ed.), *Fish Population Dynamics* (2nd ed.), pp. 349–82. London:Wiley.

Sainsbury, K.J. (1991) Application of an experimental approach to management of a tropical multispecies fishery with highly uncertain dynamics. *ICES Marine Science Symposium* 193: 301–20.

Sainsbury, K.J., A.E. Punt and A.D.M. Smith (2000) Design of operational management strategies for achieving fishery ecosystem objectives. *ICES Journal of Marine Science* 57: 731–41.

Silvert, W.L. (1981) Principles of ecosystem modelling. In A.R. Longhurst (ed.), *Analysis of Marine Ecosystems* . New York: Academic Press.

Smith, A.D.M., K.J. Sainsbury and R.A. Stevens (1999) Implementing effective fisheries management systems: management strategy evaluation and the Australian partnership approach. *ICES Journal of Marine Science* 56: 967–79.

Walker, S.J. (1999) Coupled hydrodynamic and transport models of Port Phillip Bay, a semi-enclosed bay in south-eastern Australia. *Marine and Freshwater Research* 50: 469–81.

Viewpoint from a Practitioner

Meat and Livestock Australia (MLA) works in consultation with producers to develop and implement research, development and extension aimed at creating an innovative and sustainable red meat industry in Australia. In order to achieve ongoing industry leadership, MLA promotes a *triple-bottom-line* approach to meat and livestock production, recognising that pastoral businesses need to be economically, environmentally and socially sustainable.

A large proportion of Australia's red meat and livestock are produced from the rangelands. Increasingly producers are faced with having to produce more meat at lower cost to remain economically viable. Increasing property size and rangeland consolidation is one response of producers to declining returns. However, pastoral businesses operate in a complex dynamic system, interacting with highly variable landscapes, climate, markets, prices, government policies and community concerns.

In this chapter the authors adopt a Complex Systems Science (CSS) approach to explore the pressures that influence land consolidation patterns in the rangelands. The utility of CSS in this context is that incorporating interactions between different rangeland sub-systems allows the emergence of global behaviour that may not be anticipated from the behaviour of the components in isolation. Using agent-based models, the authors show that land parcel sales were influenced strongly by seasonal variability and increased during periods of below average rainfall and drought when forage availability and profits were reduced.

Most importantly, model simulations for a range of initial property sizes demonstrated there was a non-linear relationship between the initial level of fragmentation and the degree of subsequent consolidation. In other words, in areas where initial property size was less than the threshold range for economic viability (< 20,000-25,000 ha), consolidation would proceed so a region would be dominated by a few very large enterprises. Similarities between consolidation patterns simulated from high, medium and low fragmentation scenarios were made with contrasting rangeland areas in southwest Queensland, the Dalrymple shire and the Victoria River District respectively. For policy makers these results demonstrate the need to understand the historical setting of land fragmentation in order to better understand rangeland consolidation responses.

Understanding rangeland consolidation process is important to the red meat and livestock industry because of the interactions with economic viability, and environmental and social sustainability. This knowledge may also assist agricultural policy formulation in areas of northern Australia where rangeland intensification and fragmentation is still proceeding. However, in addition to declining profits, there are complex demographic, social and psychological reasons many producers have for remaining on land that also need to be considered. Additionally, incorporating agricultural policy influences is also particularly important. Drought policy outcomes on one hand may distort economic signals promoting consolidation process, but also conflict with rural adjustment policy objectives on the other. Therefore an important challenge for future research is to incorporate non-economic and rural policy influences affecting rangeland consolidation patterns into CSS approaches.

Rodd Dyer B. Agric. Sc., M. Agric. Sc.
Project Manager
Northern Beef Program Meat and Livestock Australia

14. Rangeland Consolidation Patterns in Australia: An Agent-Based Modelling Approach

Ryan R. J. McAllister, John E. Gross and Chris J. Stokes

Abstract

In Australian rangelands, post-European pastoral land-use systems have been characterised by private-property regimes, which to varying degrees have created fragmented and disconnected landscapes. There are both environmental and economic risks associated with productive land fragments being too small. These risks necessitate an understanding of fragmentation's driving forces, made difficult because the problem involves social, economic and environmental factors, interacting over a range of temporal and spatial scales. We developed an agent-based model to help us explore the complex rangeland dynamics. In this paper we apply our model to the problem of landscape fragmentation and subsequent consolidation. Our simple model contains pastoralists, livestock, key ecological processes, and governance. Over the last century, declining enterprise sizes, increasing costs and declining returns have all contributed to increasing pressures on the viability of pastoral enterprises. We found strong regional variation is driven by localised historical and economic patterns which seed the consolidation process differently in different regions. Anecdotal evidence from South-West Queensland, Victoria River District, and Dalrymple Shire, show our model can explain regional variation. Our simulations represent a starting point for developing a more complete and insightful understanding of ongoing dynamics in rangelands, with the aim of informing policy.

Introduction

In most semi-arid and savanna ecosystems, extensive grazing dominates land use. European-style development of these ecosystems, with its privatised land tenure systems, has often profoundly altered the scale of land use through the subdivision of landscapes into discrete blocks of land that have been subdivided, traded and amalgamated to form enterprises of different sizes. In this chapter we take a complex systems approach to analysing the drivers of change in spatial scale of extensive rangeland enterprises, with the aim of informing policy. One working hypothesis is that development of rangelands is initially accompanied by fragmentation as land units are sub-divided, but later it is followed by a period of consolidation (Ash et al. 2004). We focus on the later process of land consolidation of cattle enterprises of northern Australia. While consolidation can occur at any stage during land ownership, in this later stage consolidation is the norm. This stage is characterised by external pressures that decrease profit margins, particularly for smaller enterprises. To remain viable, small or inefficient enterprises must increase profits by increasing the scale of operation, diversification and/or economies of scale (i.e., increasing in physical size). The point in the process where consolidation pressures begin to outweigh fragmentation pressures is important because at this point the environment is at most risk. An understanding of rangeland dynamics is necessary to guide development of wise policy, and it helps identify the causes of differences in regional patterns of consolidation. However, as we discuss below, different regions within Australian rangelands, despite sharing many aspects of governance and global economic trends, are presently experiencing polar differences in terms of consolidation/fragmentation pressure (Stokes et al. in press).

Since Europeans settled in Australia, land-use systems in rangelands have been gradually transformed from a semi-nomadic indigenous system to one characterised by a private-property regime. Private-property regimes are generally associated with the division of land for the exclusive use of individuals, often with boundaries clearly defined by fences. The expansive nature of production in rangelands, which is typically dominated by cattle and/or sheep production on enterprises over 20,000 hectares in area, has meant that the physical bounding of enterprises has, in some cases, taken over a century. The transition from semi-nomadic to the present day system has involved several stages (Stokes et al. 2004). These stages have generally seen rangelands initially pass through a period of fragmentation followed by a period of consolidation. The consolidation process is the focus of this paper, but the starting point of consolidation is a fragmented landscape, and we discuss the risks associated with fragmentation here. According to Stokes et al. (2004) there are two main risks of fragmentation: it allows the enterprise to fall below a minimum viable size; and it disconnects parts of the landscape limiting access to heterogeneity (Janssen et al. 2006). We discuss these risks in turn.

Rangelands are not suitable for intensive primary production and are generally managed as low-input systems. In such cases profits are generated from very small margins per area of land and, accordingly, very large enterprises are required to support a family. Policy has traditionally been an important driver of Australian rangeland fragmentation where land was allocated as parcels just large enough to support a family. The problem is that declining terms of trade (Macleod 1990) and increasing property prices have seen the minimum viable property size increase over time (Caltabiano et al. 1999). The threshold which defines a viable property size is dynamic, and if actual property fragments become too small because of a change in the viability threshold, then financial, environmental and social pressures seriously undermine the whole system's viability. Because of the inherent variability in the natural system, even highly fragmented landscapes can thrive in periods of above average rainfall, but when periods of unfavourable seasons coincide with downturns in markets, fragmented landscapes are at most risk (Stokes et al. 2004).

The second risk of fragmentation is that it disconnects large scale ecological processes. Most notably, rangelands are generally limited by having only patchy, irregular and infrequent rain events. When rangeland systems are not fragmented, herbivores can migrate around the landscape in pursuit of its patchy resources. However, fragmentation has bounded the movement of livestock, now the dominant herbivores in these landscapes. In rangelands throughout the world, institutions have developed that help human-managed herbivores move around the landscape, sometimes over massive distances, but such systems are imperfect (McAllister et al. 2006). The risk of fragmentation therefore, is that it bounds enterprises at a scale that is too small to buffer the systems high level of resource variability.

The negative effects that accompany over-fragmentation can be a strong motivator for change in rangelands (Landsberg et al. 1998). In practice, 2 adaptations to fragmentation are observed: production intensification and land consolidation.

The benefits of intensification have become more numerous since technological advances have made intensified production cost effective. Most innovations have sought a more uniform utilisation of pastures across the landscape. European breeds of cattle, which dominated Australian rangelands until the 1970s, did not venture far from water sources, creating heavy utilisation in some parts of the landscape and light utilisation in others. In this regard, production has been intensified by incorporating Indian cattle breeds, which can travel much farther from water, and by increasing the distribution of available water (Abbot and McAllister 2004). A poor distribution of water points itself can be a source of fragmentation, so water point development can reduce fragmentation

Land consolidation is the focus of this paper and there is evidence of increasing consolidation of pastoral properties in some regions in Australia's rangelands, but not all regions (Stokes et al. 2004). Consolidation is generally driven by the economic desire for greater returns (not necessarily economies of scale), but many pastoral enterprises are owner-operated and personal circumstances often seem to dictate when consolidation is required (for example, succession planning). Strategic-business decisions can also drive consolidation, particularly when an enterprise is seeking to hold complimentary properties (breeding and fattening, for example). Nevertheless, the primary driving force for consolidation is simply that an enterprise with given economic and environmental conditions, is not sufficient in area to generate enough cash flow to support its owners. In other words the enterprise fragment is too small. The process of consolidation is ongoing. Furthermore, the fragmentation-consolidation story is experienced differently across Australia.

What we discuss next is how we employ complex systems science methods to inform policy on the difficult and important issue of land consolidation. We used an agent-based model, which included pastoral enterprise agents with the ability to sell and purchase land parcels, to test how the system responded to various pressures. This model was designed to provide a platform for supporting policy in rangelands (Gross et al. in press), and could be used to explore various other issues. Here we limit our analysis to the important problem of land consolidation. The results from this application of the model helped us to develop a theoretical model which explains why the process of land consolidation emerges differently across contrasting Australian rangeland regions. We first describe the issue of land fragmentation in Australian rangelands, and then explore the issue using an agent-based model, and finally present the core results and findings.

Methods

Both the biophysical and human-dominated components of rangeland systems respond and adapt to system states, thus the processes of evolution and learning are critical in rangeland systems (Walker and Abel 2002). Such an adaptive system can be evaluated within the framework defined by the emerging science of complex systems, and the application of complex system science methods provides an alternative perspective on rangeland dynamics. In particular, agent-based models and computer-assisted reasoning show potential for gaining new insights into systems in which decisions are based on criteria with widely differing currencies, and where a multitude of individual decisions leads to emergent properties at a higher level (Levin 1998; Lempert 2002). We take a complex systems approach because these methodologies have already proved useful in exploring various aspects of rangeland dynamics (Janssen et al. 2000; Anderies et al. 2002; Walker and Janssen 2002; Janssen et al. 2004).

We developed a simple and quite general agent-based model of rangeland systems typical of northern Australia savanna. An earlier implementation of the model is described by Gross et al. (in press) and we embellished this model to more thoroughly explore land consolidation processes. In addition to the implementation described by Gross et al., our model considers: how costs and spatial heterogeneity drive consolidation; how different consolidation drivers affect land management practices; and how temporal heterogeneity affects the timing of consolidation. We summarise the important elements of our agent-based model here.

Pastoral enterprise submodel

Pastoral enterprise actions were represented by a strategy set that defined management actions and characteristics. This strategy set included criteria for setting a target stocking rate and making other financial decisions (sale of cattle, borrowing, destocking and restocking, etc.). Enterprises accumulated either debt or wealth depending on the overall success of their respective rules through time. The focus of this model was on the process of consolidation of Australian rangelands. To deal with this issue we needed our model enterprises to be equipped with rules for selling and buying sub-enterprise blocks of land. We did this using simple rules.

If an individual pastoral enterprise's debt position rose above 80 per cent (debt position is defined as debts or savings divided by the value of cattle and land assets) then that property was forced to sell one block of land. If that enterprise sold its last block of land, then it exited the system. If, in a given period, there were multiple enterprises with a debt position above 80 per cent, only the enterprise with the worst debt position sold land.

If a block was for sale, it was offered to the enterprise in the system with the highest estimated post-purchase debt position—i.e., what debt position an enterprise would have after purchasing the block. We assumed this enterprise bought the property with a probability of 50 per cent and if they did not purchase the property then it was not offered to other enterprises in the same month. If an enterprise had a positive debt position, then when the purchasing decision was made their debt position was weighted down by the total enterprise area (after purchase). This reflected a diminished desire for successful enterprises to change and grow.

The biophysical submodel was largely driven by precipitation, and each property received some random variation of the landscape-wide average reflecting spatial variability (details below). If a block was purchased, then the variation in precipitation applied to both the buying and selling enterprises was adjusted according the new total area of land owned.

The most important decisions that pastoralists have to make relate to how many cattle to stock. Stock numbers affect the biophysical condition of properties, which affects live weight gains. Also, seasonal depletion of forage affects live weight gains and/or supplement use. Live weight gain was the primary link between the biophysical agent (the property) and the pastoral enterprise agent. Live weight gains affected supplementary feed requirements, fecundity and mortality rates, and the weight of cattle sold.

We assumed that pastoralists made stocking decisions in May and that all cattle breeding (and mortality) occurred only in November. Simulated pastoralists had a target stocking rate that guided their cattle trading decisions. In the month when decisions were made, our model pastoralists sold and purchased cattle in order to maintain a target stock, but filtered their buying and selling rules by destocking during droughts (and subsequently restocking). Rates of destocking (or restocking) were estimated from forage utilisation (Littleboy and McKeon 1997) as:

$$\text{\% Change} = 100 - a1 * \ln(\text{Utilisation}) \tag{1}$$

where Utilisation is the one-year average of forage utilisation (proportion of biomass production consumed by livestock) on a property;

and a1 is a shape parameter that determines how reactive individual pastoralists' stocking rates are to heavy utilisation (associated with low forage production under drought).

The parameter a1 varied for individual pastoralists in the system, reflecting heterogeneous approaches to dealing with drought. In these simulations, a1 varied (uniformly) from −35 to −25.

Cattle were purchased or sold in response to rates of forage utilisation (equation 1), and differences between target and actual stocking rates. If the achieved stocking rate was less than the target and % Change was positive, cattle were purchased. In this case, the number of cattle purchased was the lesser of the number needed to reach the target stocking level, or the absolute value of % Change (i.e., the number of cattle proportional to % Change). Cattle were always sold when the rate of stocking exceeded the target and cattle were always sold when % Change was negative. The number of cattle sold was the greater of the number needed to achieve the target stocking rate or the proportion of the herd equal to the absolute value of % Change.

The mass of each cow that was purchased was assumed to be 400 kg. The weight of cattle sold was taken as the total monthly live weight gains on a property over the past 3.5 years. These live weight gains included the effects of supplementary feeding, which bound the minimum monthly gains at 4.167 kg. Supplementary feeding incurred costs, which were related to what the gains would be without supplementary feeding, and we applied rules used by Macleod et al. (2004) to determine the level of costs incurred.

In our simulations, mortality and branding rates followed Gillard and Monypenny (1988):

$$\text{\% Mortality} = 3.2 + 87.0*e^{-0.03*(LWG+50)}, \tag{2}$$

$$\text{\% Branding} = 0 \leq 15.6 + 0.488*LWG \leq 80.0, \tag{3}$$

where LWG is the one-year average of live weight gains on a property. All births and deaths occurred at the same time each year.

Biophysical submodel

The underlying conceptual structure of the plant and livestock submodels was to define optimal growth rates, and then reduce these rates by factors that represented the most important constraints and stressors. Simulated plant growth could be limited by precipitation or seasonal changes in temperature. Plant growth varied seasonally in a manner consistent with growth of C4 species found in tropical Australia. Multi-year changes in potential plant growth resulted from changes in grass basal area, which responded on an annual basis to aboveground productivity and to utilisation. Maximum plant production (kg ha^{-1} mo^{-1}) was a function of basal area, thus the change in green vegetation (G, kg ha^{-1}) was:

$$dG/dt = B_a\ B_{gmax}\ M_{precip}\ M_{seas} - S\ G \qquad (4)$$

where B_a is basal area (percentage cover), B_{gmax} is maximum growth per unit of B_a (kg/ha), M_{precip} is a precipitation multiplier (0-1), M_{seas} is a seasonal growth factor (0-1), and S is senescent rate (proportion).

A key process in rangelands is degradation and the gradual run-down of production. A dynamic basal area function, based on that in the grassland simulation model GRASP (Littleboy and McKeon 1997), was used to represent the effects of heavy grazing or sustained drought. High rates of utilisation or low production (typically the result of drought) led to a reduction in B_a. We placed no restrictions on how rapidly B_a can decline, but increases in B_a were limited by averaging production over the previous 2 years, and by maximum growth rate. Consequently, rates of decline in basal area were usually more rapid than rates of recovery, representing the hysteresis that typically accompanies ecosystem degradation.

Grass was partitioned into pools of green and dry material (kg/ha). These pools varied in nutritional quality and were important to adequately represent livestock dynamics. Parameters in the plant submodel were calibrated to observations of biomass dynamics near Charters Towers, and to outputs from GRASP (Littleboy and McKeon 1997; parameter set Burdy11). The plant submodel we implemented reproduced patterns and quantities of forage that roughly matched observations from the black speargrass paddocks in the Charters Towers region.

Spatial heterogeneity was represented in the model by modifying precipitation as a function of property size. As property size increased, variation in precipitation declined according to equation 5. Equation 5 varies precipitation to property size to account for the observation that larger properties buffer spatial variation in precipitation (Perevolotsky 1987). Precipitation for each property was determined by multiplying average precipitations for all properties by:

$$1 + r(e^{-Area/1000000}), \qquad (5)$$

where r is a random number drawn annually from a uniform distribution [-0.5,0.5] and Area is the total size of a property.

The livestock submodel partitioned cattle into 2 age classes. The model simulated live weight gain (LWG, kg) and it included simple diet selection and animal growth functions that interact with the plant submodel through selection and consumption of green and dry grass. Growth rates responded to forage availability and quality.

Diet selection is a key process in tropical pastures, and the quality of the diet consumed in our model was determined by the proportion of green forage in the diet. Diet selection was implemented following the basic conceptual model proposed by Blackburn and Kothmann (1991), but selection for green forage is driven only by the proportion of green in the sward and absolute biomass (Chacon and Stobbs 1976; Hendricksen et al. 1982). Diet quality was determined from the proportion of the diet composed of grass that grows in the current time step (highest quality green grass), green grass that grew in previous time steps (intermediate quality), and dry grass (lowest quality). Diet selection and diet quality are key drivers of animal growth and condition, which subsequently influences growth and survival of livestock in the absence of supplemental food. However, if average live weight gains fell below 4.167 kg for a given month, following MacLeod et al. (2004) we assumed that feed was supplemented at a level such that exactly 4.167 kg of weight was gained. The exact diet deficiency affected the amount of supplementary feed required (hence costs, see above) but not the minimum level of live weight gain.

Results

Reference conditions

The simulations presented in this paper were based on 113-year simulations with a model time step of one month. The biophysical model was calibrated to output from GRASP, a grassland simulation model (Littleboy and McKeon 1997), to reflect characteristics of the Dalrymple Shire in North-East Australia. Our simulations were designed to mimic a hypothetical region consisting of 2.5 million hectares, which is a region approximately one-third of the size of the Dalrymple Shire. Our reference conditions used monthly precipitation observations from Charters Towers, Queensland, Australia, recorded between 1890 and 2003. For this case we used fixed cattle prices and fixed supplementary feeding costs and other variable cost prices, and did not allow trading of properties. Results are presented for simulations where property size was held constant at 20,000 ha (Figure 14.1). We then relaxed our constant size rule by dividing each property into 4 blocks of land and permitting land sales and purchases. These simulations indicated that land sales were related to weather patterns (Figure 14.2) and, after a sustained period of simulation, the distribution of property sizes was right skewed (Figure 14.3).

Figure 14.1. Results of reference treatment in a heterogeneous rangeland system

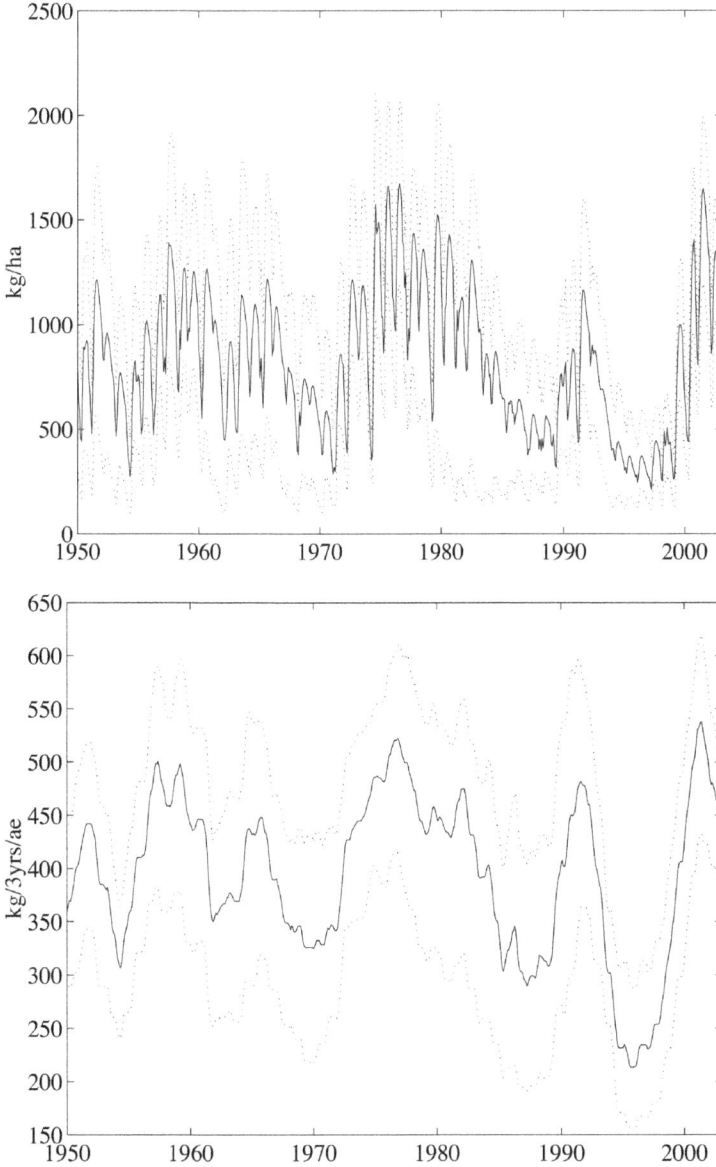

(a) standing dry matter; and (b) cumulative 3-year live weight gain per adult equivalent. The solid lines show the mean of all enterprises in the system, while the dotted-lines above and below represent 95 per cent confidence intervals for individual enterprises (percentile bootstrap method).

Figure 14.2. Number of block sales for the reference treatment (solid line)

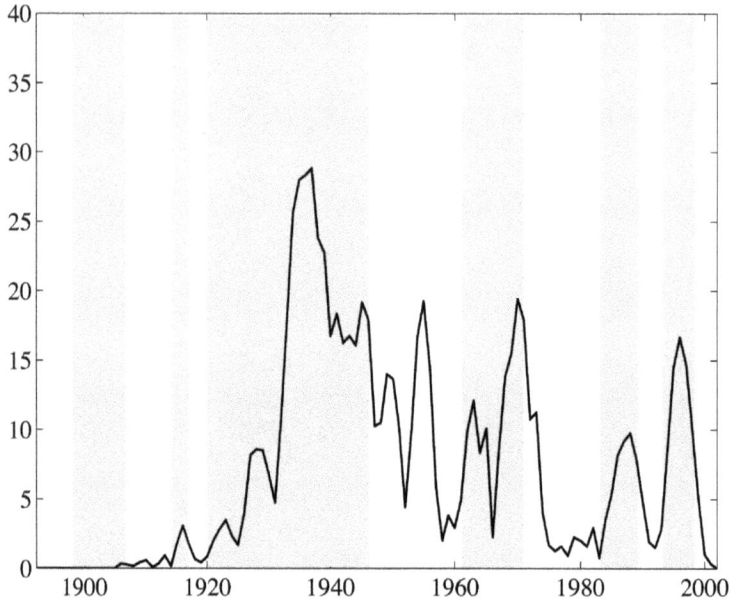

Simulations were driven by observed monthly Charters Towers precipitation records. Shaded areas indicate periods where both the 3-year and 1-year average precipitation was below average (periods with below average precipitation separated by < 2 years were joined).

Figure 14.3. Consolidated property size distribution at the end of the simulations using reference conditions

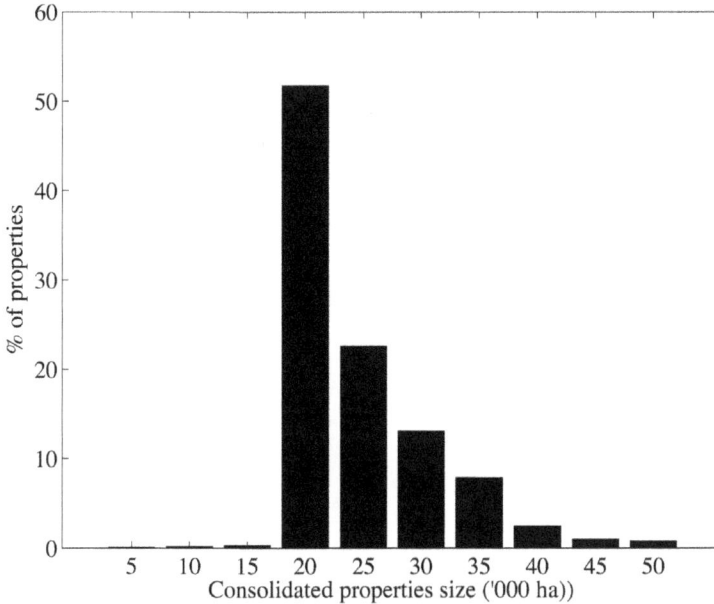

Sensitivity analysis

Financial success was the only direct driver of consolidation in our model, though financial success was a function of property size, precipitation, management strategy and biophysical condition, as well as interactions of these factors. To explore the direct effect of financial pressure, we simulated treatments with different variable and fixed costs. In these simulations, monthly variable costs (other than supplementary feeding costs) per adult equivalent were fixed at $1.00, $1.25, $1.50, $1.75, or $2.00 (Figure 14.4). In the same way, we also examined the effects of supplementary feeding costs and fixed costs. All changes in costs that we examined led to similar results: increased costs uniformly led to greater rates of consolidation. As an exemplar of these simulations, Figure 14.4 demonstrates how increased costs led to consolidation of properties.

Figure 14.4. Average consolidated property size response to the variable costs

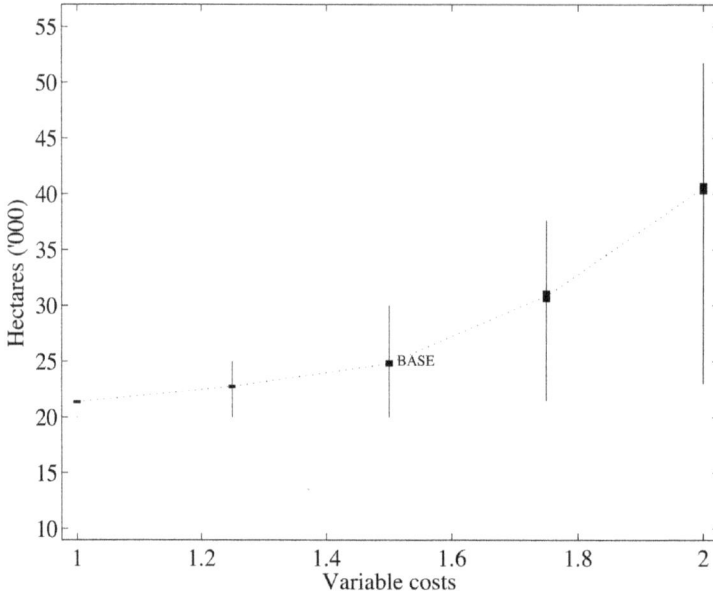

Thin vertical lines show percentile ranges (25 and 75 percentiles) for 100 repetitions, while the thicker bars show 95 per cent confidence intervals for individual enterprises (percentile bootstrap method).

We explored the effects of the initial distribution of property size on consolidation by maintaining 2.5 million hectares in the system, but dividing land evenly between various numbers of enterprises, hence varying the initial property size in the simulation. We compared 5 starting enterprise sizes: 10,000; 15,000; 20,000; 25,000 and 30,000 hectares (all divided into tradable units of 5,000 ha) and we repeated our simulations 100 times per treatment. Results from these simulations showed that the relationship between starting size and the mean size of enterprises at the end of simulation was convex (Figure 14.5). Also, the distribution of the resulting enterprise sizes for the treatment with the smallest assumed starting size was heavily skewed towards being small. This is shown by the percentile range; in particular, the upper percentile range fell below the mean result, indicating that consolidation activity was dominated by a relatively small number of enterprises. Clearly in a treatment dominated by sub-economic sized enterprises, any enterprise that took the initial step towards overcoming this was then in a relatively better position to succeed and continue to grow.

Figure 14.5. Mean consolidated enterprise size at the start and end of 113-year simulations

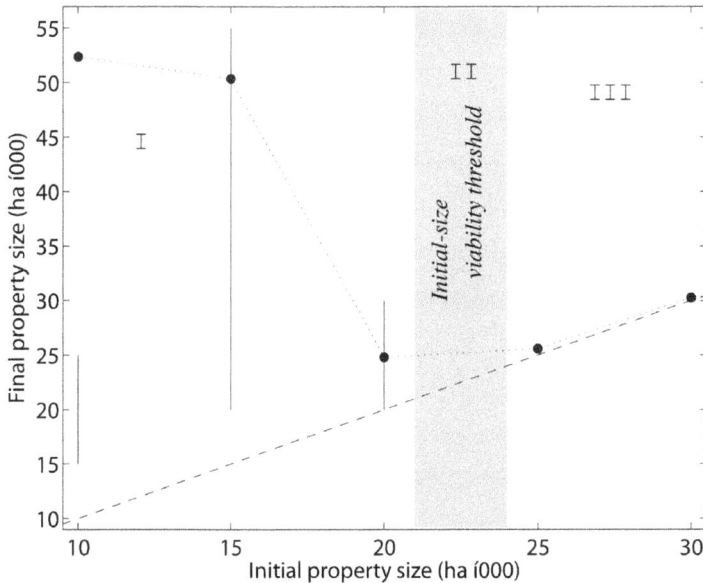

Thin vertical lines show percentile ranges (25 and 75 percentiles) for 100 repetitions. The dashed reference line marks points where the initial and final property size are equal. The shaded band indicates a threshold property size range for enterprise viability; property trading and consolidation only occurs when smaller, non-viable properties are forced to sell land.

Discussion

Consolidation is one reaction to decreasing returns from pastoralism (Figure 14.4). Consolidation requires both a demand for property by existing enterprises and a supply of land. In our simulations, supply was strongly related to weather conditions (Figure 14.2), which influenced forage availability and thus profits. Simulations reported here resulted in right-skewed distributions of property sizes after the consolidation process (i.e., the mean property size is larger than the median). This distribution is consistent with observations from Australian rangelands (Bortolussi et al. 2005).

There is a discernible threshold range of property size (20,000–25,000 ha: Figures 14.3 and 14.5) separating viable and non-viable enterprises. When simulations were initialised with enterprises too small to be economically viable (< 20,000 ha), the simulation usually ended with a few very large enterprises that skewed the final distribution of property sizes. When simulations started with very large enterprises (> threshold), these large enterprises were generally profitable enough to retain all their land and there was little opportunity for consolidation to occur. In this case, initial and final enterprise sizes were nearly the same.

There is some variation in the *initial-size viability threshold* between enterprises because properties are managed by heterogeneous agents, each with unique practices. This threshold would likely become less clearly defined if other sources of heterogeneity, such as differences in land resources between properties, were included in the model.

The model demonstrated that there is a non-linear relationship between the level of fragmentation in land-use systems, and the degree of their subsequent consolidation. We can compare our model results against historical observations from Australia. Three strongly contrasting scenarios of fragmentation in Australian rangelands, described by Stokes et al. (in press), serve as a useful basis for this comparison: the Dalrymple Shire (DS); the Victoria River District (VRD); and, south-west Queensland (SWQ) (Figure 14.6). From the late 1800s to the mid 1900s, land use in Australian rangelands was strongly influenced by policies of 'closer settlement'. Policies of this period were aimed at increasing rural populations, encouraging development of the pastoral industry and promoting equity in land distribution. This led to the subdivision of initially vast pioneering pastoral stations into progressively smaller enterprises. Over the last century, declining enterprise sizes, increasing costs and declining returns have all contributed to increasing pressures on the viability of pastoral enterprises. One of the consequences of these changes in recent decades has been pressure to consolidate pastoral enterprises. But there is strong variation between different regions of Australian rangelands in the extent to which enterprises became fragmented and in the subsequent consolidation of properties. South-west Queensland has had a long history of pastoral development, is relatively close to population centres, and has experienced several periods of strong profitability in the sheep industry. It was therefore very strongly influenced by policies of land subdivision. This corresponds to scenario I in Figure 14.5, where properties became subdivided to sizes well below the threshold for viability, and there has since been a growing trend towards enterprise consolidation. At the opposite extreme, the Victoria River District is remote from population centres, was amongst the last areas in Australia to be developed for pastoralism, and is only suitable for cattle, an industry that has not been as profitable as the sheep industry in the past. As a result this area has escaped the influence of policies of closer settlement and land fragmentation. This corresponds to scenario III in Figure 14.5. Initial property sizes were vast and have remained so, without being strongly influenced by pressures to consolidate. Dalrymple Shire represents an intermediate scenario, where rangelands only ever became moderately fragmented. Property sizes were probably viable at the time of the last subdivision, but were close to the viability threshold (Figure 14.5 II) and have subsequently become subject to consolidation pressures under the influence of rising costs of production (Figure 14.4).

Figure 14.6. Location of sites

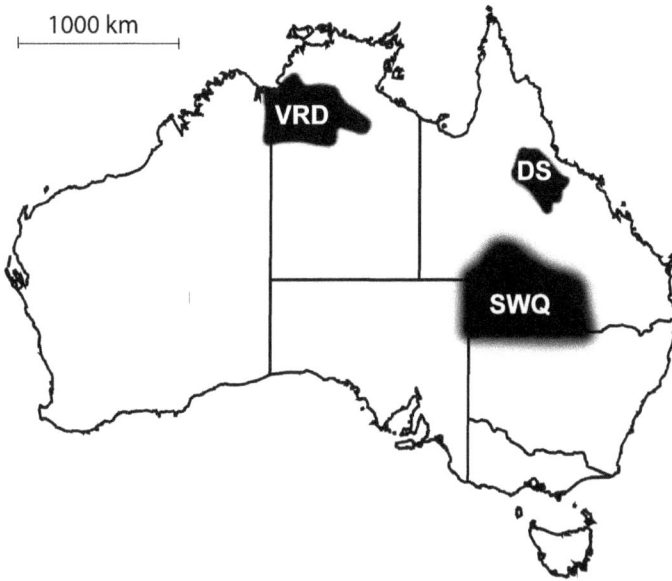

Conclusions

To some degree, different regions within Australian rangelands demonstrate our simulated pattern, but localised historical and economic patterns have seeded the consolidation process differently in different regions. This process is ongoing and has various implications for regional Australia, in particular for social capital. If we seek to guide policy in rangelands we need to understand this process in a strong context of its historical setting. Enterprise size buffers against climatic variability and financial risk, and may improve economies of scale. But the drive to increase the scale of operations need not be met by consolidation alone. Anecdotal evidence suggests that kin-based networks may be important. Agistment networks also allow pastoralists to overcome spatial resource variability (Janssen et al., 2006; McAllister et al. 2005, 2006), which is more prominent for small enterprises. The size-related viability of rangeland enterprises portends important consequences for the social and environmental health of Australia's rangeland communities. In the past, small enterprises were more likely than large enterprises to perform poorly (Hinton 1995), and land consolidation was an effective response to the constraints imposed by small property sizes.

Our simulations represent a starting point for developing a more complete and insightful understanding of the dynamics of rangelands. This understanding is the key to interpreting events in the past and to forecasting consequences of policy into the future. Our model results, and equally the thought process involved in constructing the model, contributed to this understanding.

References

Abbott, B.N., and R.R.J. McAllister (2004) Using GIS and satellite imagery to estimate the historical expansion of grazing country in the Dalrymple shire. In G. Bastin, D. Walsh and S. Nicolson (eds) *Proceeding of the 13th Biennial Australian Rangeland Society Conference*, pp. 405–6. Alice Springs: Australian Rangeland Society.

Anderies, J.M., M.A. Janssen and B.H. Walker (2002) Grazing management, resilience, and the dynamics of a fire-driven rangeland system. *Ecosystems* 5(1): 23–44.

Ash, A.J., J.E. Gross and M. Stafford Smith (2004) Scale, heterogeneity and secondary production in tropical rangelands. *African Journal of Range and Forage Science* 21(3): 137–45.

Blackburn, H.D. and M.M. Kothmann (1991) Modelling diet selection and intake for grazing herbivores. *Ecological Modelling* 57(1-2): 145–63.

Bortolussi, G., J.G. McIvor, J.J. Hodgkinson, S.G. Coffey and C.R. Holmes (2005) The northern Australian beef industry, A snapshot: 1. regional enterprise activity and structure. *Australian Journal of Experimental Agriculture* 45: 1057–73.

Caltabiano, T., Hardman, J.R.P. and R. Reynolds (1999) Living Area Standards. Queensland Department of Natural Resources, Coorparoo.

Chacon, E. and T.H. Stobbs (1976) Influence of progressive defoliation of a grass sward on the eating behaviour of cattle. *Australian Journal of Agricultural Research* 27(5): 709–27.

Gillard, P. and R. Monypenny (1988) Decision support approach for the beef cattle industry of tropical Australia. *Agricultural Systems* 26: 179–90.

Gross, J.E., R.R.J. McAllister, N. Abel, D.M. Stafford Smith, and Y. Maru. (2006) Australian rangelands as complex adaptive systems: A conceptual model and preliminary results. *Environmental Modelling and Software* 21(9): 1264–72.

Hendricksen, R., K.G. Rickert, A.J. Ash and G.M. McKeon (1982) Beef production model. *Animal Production in Australia* 14: 204–8.

Hinton, A.W. (1995) *Land condition and profit: a study of Dalrymple Shire properties*. Brisbane, Australia: Queensland Department of Primary Industries.

Janssen, M.A., B.H. Walker, J. Langridge and N. Abel (2000) An adaptive agent model for analysing co-evolution of management and policies in a complex rangeland System. *Ecological Modelling* 131(2-3): 249–68.

Janssen, M.A., J.M. Anderies and B.H. Walker (2004) Robust strategies for managing rangelands with multiple stable attractors. *Journal of Environmental Economics and Management* 47(1): 140–62.

Janssen, M.A., Ö. Bodin, J.M. Anderies, T. Enquist, H. Ernstson, R.R.J. McAllister, P. Ryan, and P. Olson (2006) Towards a network perspective of the study of resilience in social-ecological systems. *Ecology and Society*, 11(1): 15.

Landsberg, R.G., A.J. Ash, R.K. Shepherd and G.M. McKeon (1998) Learning from history to survive in the future: management evolution on Trafalgar Station, north-east Queensland. *The Rangeland Journal* 20(1): 104–18.

Lempert, R.J. (2002) A new decision sciences for complex systems. *Proceedings of the National Academy of Sciences of the USA* 99: 7309–13.

Levin, S.A. (1998) Ecosystems and the biosphere as complex adaptive systems. *Ecosystems* 1(5): 431–6.

Littleboy, M and G.M. McKeon (1997) Appendix 2. Subroutine GRASP: grass production model. In K.A. Day, G.M. McKeon and J. O. Carter (eds), *Evaluating the Risks of Pasture and Land Degradation in Native Pastures in Queensland*. Canberra: Rural Industries Research and Development Corporation.

Macleod, N.D. (1990) Issues of size and viability of pastoral holdings in the Western Division of New South Wales. *Australian Rangeland Journal* 12:(2): 67–78.

Macleod, N.D., A.J. Ash, and J.G. McIvor (2004) An economic assessment of the impact of grazing land condition on livestock performance in tropical woodlands. *The Rangeland Journal* 26(1): 49–71.

McAllister, R.R.J., I.J. Gordon, M.A. Janssen (2005) Trust and cooperation in natural resource management: The case of agistment in rangelands. In A Zerger and RM Argent (eds) *Proceedings of the International Congress on Modelling and Simulation*, pp. 2334–9. Modelling and Simulation Society of Australia and New Zealand.

McAllister R.R.J., I.J. Gordon, M.A. Janssen and N. Abel (2006) Pastoralists' responses to variation of rangeland resources in time and space. *Ecological Applications*, 16(2): 572-83.

Perevolotsky, A. (1987) Territoriality and resource sharing among the Bedouin of southern Sinai: a socio-ecological interpretation. *Journal of Arid Environments* 13: 153–61.

Stokes, C.J., R.R.J. McAllister, A.J. Ashand and J.E. Gross, Changing patterns of land use and tenure in the Dalrymple Shire, Australia. In K. A. Galvin, R. Reid, R. H. Behnke, and N. T. Hobbs (eds), Fragmentation in semi-arid and arid landscapes: consequences for human and natural systems. Kluwer, in press.

Stokes, C.J., A.J. Ash and R.R.J. McAllister (2004) Fragmentation of Australian rangelands: Risks and trade-offs for land management. In G. Bastin, D. Walsh, and S. Nicolson (eds). 13th Biennial Australian Rangeland Society Conference. 39–48. Alice Springs, Australian Rangelands Society.

Walker, B.H. and N. Abel (2002) Resilient rangelands–adaptation in complex systems. In L.H. Gunderson and C.S. Holling (eds) *Panarchy: Understanding Transformations in Human and Natural Systems,* pp. 293–313. Washington: Island Press.

Walker, B.H. and M.A. Janssen (2002) Rangelands, pastoralists and governments: Interlinked systems of people and nature. *Philosophical Transactions of the Royal Society of London B* 357: 719–25.

List of Contributors

Hussein Abbass obtained his PhD from the Queensland University of Technology, Australia. He is an Associate Professor at the School of Information Technology and Electrical Engineering, University of New South Wales (UNSW), Canberra. His current research interests are in complex adaptive systems, data mining, multi-agent systems, and multi-criteria decision making. He has authored/co-authored more than 120 fully refereed papers.

Gabriele Bammer is a Professor at the National Centre for Epidemiology and Population Health at The Australian National University (ANU) and a Research Fellow at the Hauser Center for Nonprofit Organisations at Harvard University. Her main interest is in effective ways of bringing different disciplinary and practice perspectives together to tackle major social issues. She is seeking to develop more formal processes for doing this by establishing a new specialisation: Integration and Implementation Sciences.

Michael Barlow (who prefers to be called Spike) is a senior lecturer in the School of Information Technology and Electrical Engineering, UNSW at the Australian Defence Force Academy (ADFA). He is also the Deputy Director of the UNSW Security and Defence Applications Research Centre, and Director of the Virtual Environments and Simulation Lab at ADFA. Spike's research areas include multi-agent systems, Commercial Off-The-Shelf games (COTS), scientific visualisation, virtual environments, and speech processing.

David Batten holds a PhD in Economics from Sweden and is an Adjunct Research Fellow with the Commonwealth Scientific and Industrial Research Organisation (CSIRO). Following 16 years at the CSIRO, in 1986 he moved to a Chair in Infrastructure Economics in Sweden. From 1991–95, he also held the position of Professorial Fellow at the Institute for Futures Studies, a scientific think tank in Stockholm. His work there focused on co-evolutionary learning across complex infrastructure systems—transport, energy, water and urban planning. In 2002, he was invited to return to the CSIRO and is currently involved in projects funded by CSIRO's Centre for Complex Systems Science and COSNet (the ARC's Complex Open Systems Network). Until recently, he led the Energy Flagship's NEMSIM Project, a multi-agent-based simulation model that represents Australia's National Electricity Market (NEM) as a co-evolving system of complex interactions between NEM participants and their effects on technical infrastructures and the natural environment. He also coordinates CSIRO's Agent-Based Modelling Working Group and COSNet Theme 5: Cellular Automata, Agent-Based Modelling and Simulation.

Roger Bradbury is a Professor in the Resource Management in Asia-Pacific Program at the ANU. A zoologist by training, he has worked extensively on the analysis of coral reefs and other complex systems, ranging from natural resources to infrastructure to national security systems.

Neville Curtis obtained his PhD in Chemistry from the University of Alberta in 1981. He joined the Defence Science and Technology Organisation in 1984 and has worked in the areas of Energetic Materials, Weapons Systems and Military Operations Research. His current research interests are in concept exploration and development studies, and problem structuring methods ('soft OR'). He has authored over 100 publications in a wide variety of areas.

Katherine Daniell is currently completing a PhD in water resources management and sustainable development under a cotutelle program between the Centre for Resource and Environmental Studies, ANU, and Ecole du Génie Rural des Eaux et Forêts (ENGREF) in France, in collaboration with Centre d'etude du Machinisme Agricole du Génie Rural des Eaux et Forêts (CEMAGREF) and the CSIRO. Her research interests include sustainability assessment, whole systems thinking and knowledge integration, as well as developing methods of collective decision aiding for communities, policy makers, managers, and technical experts.

Patrick D'Aquino is Director of Research in Social Geography at CIRAD. For 10 years he focused on the conception of supports (not only tools) to accompany (not supply solutions) decentralised processes within land use planning, especially about common resources (pastures, lands, water). His concern is to reconcile social geography and the development of technical tools adapted to local stakes and constraints. Focused on social processes and local knowledge for the management of common resources, he studied and assisted local organizations that were implementing and regulating land use rights in the Sahelian zone of Africa. Now he is spreading his approach in Pacific areas. He works on geographic and other participatory modelling tools (maps, Global Imaging Systems, Agent-Based Models) and on novel planning practices to support local organisations and social dynamics.

Anne Dray is an agronomist working as junior scientist for CIRAD and a Visiting Fellow at the ANU. Her work aims at developing a negotiation support tool based on Agent-Based Modelling and Role-Playing Games joint use to facilitate equitable water allocation in Tarawa (Republic of Kiribati). She is also involved in training courses for Australian students on the Common-pool Resources and Multi-Agent Systems (CORMAS) platform.

Bernadette Foley is a PhD Research Scholar in the School of Civil and Environmental Engineering at the University of Adelaide. Her research focuses on using alternative thinking techniques and tools to advance the concept of sustainability and its assessment, with a view to better incorporating these concepts and techniques into decision-making processes for projects, companies and other applications.

Elizabeth Fulton joined the CSIRO in 2001 and has played a leading role in the development of marine ecological models, including agent-based models. She has made a significant contribution to the international peer-reviewed Marine Science and Ecological Modelling literature. She received the PhD Award of the Royal Society of Tasmania for her thesis entitled 'The effects of model structure and complexity on the behaviour and performance of marine ecosystem models'.

Randall Gray joined the CSIRO in 1990 and soon after introduced the team to the use of agent-based modelling for simulating the interaction between humans and marine populations. He has led the development of the In Vitro software being used by the CSIRO to assess strategies for managing multiple uses of coastal and marine ecosystems. Randall has been active in presenting his agent-based work at international conferences.

John Gross is a systems ecologist, previously with the CSIRO and now with the US National Park Service. His work has focused on modelling, from individual animals to ecosystem scales, with an emphasis on plant-animal interactions and population biology.

George Grozev is a research scientist with the CSIRO's Manufacturing and Infrastructure Technology (CMIT) in Melbourne. His background is in operations research, particularly network and graph theory. His recent research has focused on sustainable development issues within electricity markets, including approaches for modelling and simulation of future developments in Australia's National Electricity Market. He has participated in the development of several simulation, optimization and mapping tools for cellular mobile networks.

L. Richard Little joined the CSIRO in 1999 and has played a leading role in the development of agent-based models, particularly those related to harvesting of wild fish stocks and networking among fishers. He has published in the international peer-reviewed Marine Science and Ecological Modelling literature and received the Early Career Research Excellence Award of the Modelling and Simulation Society of Australia and New Zealand in 2005.

Ashley Kingsborough is an Environmental Consulting Engineer for KBR, Adelaide, in the Natural Resources Management Group. Particularly interested in environmentally responsible development projects for communities and ecosystems, he holds positions on the board of Engineers Without Borders South Australia, the Australian Society for Sustainability and Environmental Engineering, and the College of Environmental Engineers Australia.

Christophe Le Page is an agronomist working for CIRAD, in Montpellier France. He is a member of the Green Research Unit, which is promoting a companion modelling approach for natural resources management. With a background in fish population dynamics, he has progressively specialised in building agent-based models to simulate the interplay between ecological and social dynamics in ecosystems holding renewable resources used or managed by different categories of stakeholders. He is participating in the development of the CORMAS platform, with a special interest in spatial aspects. He is also involved in several training courses introducing the usefulness of Agent-Based Modelling for the simulation of agro-ecosystems.

Vincent Lyne joined the CSIRO in 1985 and has played a leading role in the development of methods for use of remote-sensing data in biological oceanography, particularly with respect to fish migration patterns and fishery dynamics. He has led the development of Australia's marine habitat classification and bioregionalisation schemes. Vincent has published in the internationally peer-reviewed literature on Marine Ecology and Oceanography.

Ryan McAllister is a research scientist with the CSIRO. His research focus has been on applying complex system methodologies to Australian rangelands, in particular, exploring how pastoralists adapt to the high degree of resource variation inherent in rangelands.

A. David McDonald joined the CSIRO in 1994. He has published in international peer-reviewed journals in Economics, Statistics, Marine Science and Environmental Modelling. He is a Fellow of the Modelling and Simulation Society of Australia and New Zealand and is a member of the Editorial Board of Environmental Modelling and Software.

Holger Maier is an Associate Professor in the School of Civil and Environmental Engineering at the University of Adelaide. His research interests fall under the umbrella of sustainable infrastructure and water resources management, and include prediction and forecasting using artificial neural networks, optimisation using evolutionary algorithms, agent-based modelling, risk, uncertainty and sensitivity analysis and multi-criteria decision-analysis.

David Malovka recently graduated with a double degree in Civil and Environmental Engineering and Economics at the University of Adelaide and will soon take up a position as an Environmental Consulting Engineer for KBR Halliburton, Adelaide, in the Natural Resources Management group. His domains of interest include system design and economic analysis techniques for water engineering projects and optimisation applications such as irrigation networks.

David Newth, is a research scientist working for the CSIRO. He is currently a member of the CSIRO Centre for Complex Systems Science where he is the co-ordinator of the Network Theory Working Group. A computer scientist by training, his most recent work focuses on the analysis and modelling of complex systems, including ecosystems, social systems, gene regulatory networks and large scale infrastructure.

Pascal Perez is currently seconded by CIRAD to the ANU. He is the convenor of the HEMA international network. An agronomist by training, his most recent work focuses on human ecosystems modelling. He has developed projects in northern Thailand, Indonesia and Micronesia. He is currently teaching agent based modelling techniques at ANU.

Keith Sainsbury joined the CSIRO in 1977 as a marine ecologist and mathematical modeller. He developed one of the first applications of actively adaptive management to large-scale trawl fishing on Australia's North West Shelf in the 1980s. He has made influential contributions to the international peer-reviewed Marine Science and Ecological literature and has played a leading role in development and application of methods of assessing cumulative impacts and multiple uses of marine ecosystems. Keith was awarded the 2004 Japan Prize for ecological and sustainability research.

Ruhul Sarker obtained his PhD from DalTech, Dalhousie University, Canada. He is currently a senior lecturer at the School of Information Technology and Electrical Engineering, University of New South Wales, Canberra. Dr Sarker's research interests are in applied operations research, evolutionary optimization and multi-agent systems.

Heath Sommerville is a Water and Environmental Consulting Engineer for Sinclair Knight Merz in Melbourne. His current work revolves around finding sustainable and environmentally sound solutions to water related problems across Victoria. He guards a particular interest in developing models to represent these complex problems that can be used to explore and develop improved management processes.

Chris Stokes is a systems ecologist with the CSIRO. He is interested in the way that landscape scale heterogeneity and interactions affect ecosystem processes and behaviours, and in dealing with the challenges that this complexity presents for extrapolating detailed mechanistic understanding to broader, real-world scales.

Ian White is Professor of Hydrology at the ANU. He is a distinguished member of several international organisations, including the UNESCO International Hydrological Program. His major research theme is the prediction and measurement of the downstream impacts of landuse, including the acidification of coastal streams, modelling the groundwater dynamics of salinity and waste disposal schemes, prediction of water use by trees, sustainability of water extraction from shallow groundwater systems in coastal areas, and quantitative techniques for soil water measurement in the root and vadose zones at appropriate scales.

Ang Yang holds a Bachelors and two Masters degrees and a Postgraduate Diploma. He is currently pursuing his PhD degree in Computer Science at UNSW. His current research interests include complex adaptive systems, multi-agent systems, modelling and simulation, evolutionary computation, network theory and web-based intelligent systems.

Index